GOD AND REASON IN THE MIDDLE AGES

Between 1100 and 1600, the emphasis on reason in the learning and intellectual life of Western Europe became more pervasive and widespread than ever before in the history of human civilization. This dramatic state of affairs followed the long, difficult period of the barbarian invasions, which ended around A.D. 1000 when a new and vibrant Europe emerged. Of crucial significance was the invention of the university around 1200, within which reason was institutionalized and where it became a deeply embedded, permanent feature of Western thought and culture. It is therefore appropriate to speak of an Age of Reason in the Middle Ages, and to view it as a forerunner and herald of the Age of Reason that was to come in the seventeenth century.

The object of this book is twofold: to describe how reason was manifested in the curriculum of medieval universities, especially in the subjects of logic, natural philosophy, and theology; and to explain how the Middle Ages acquired an undeserved reputation as an age of superstition, barbarism, and unreason.

Edward Grant is Distinguished Professor Emeritus of History and Philosophy of Science and Professor Emeritus of History at Indiana University. He is the author of *The Foundations of Modern Science in the Middle Ages: Their Religious, Institutional, and Intellectual Contexts* (Cambridge University Press, 1996); *Planets, Stars, & Orbs: The Medieval Cosmos* (Cambridge University Press, 1994); *Mathematics and Its Applications to Science and Natural Philosophy in the Middle Ages* (Cambridge University Press, 1987), and numerous other books and articles. He was awarded the George Sarton Medal of the History of Science Society in 1992.

GOD AND REASON IN THE MIDDLE AGES

EDWARD GRANT

Indiana University

CAMBRIDGE UNIVERSITY PRESS
Cambridge, New York, Melbourne, Madrid, Cape Town, Singapore,
São Paulo, Delhi, Dubai, Tokyo

Cambridge University Press
The Edinburgh Building, Cambridge CB2 8RU, UK

Published in the United States of America by Cambridge University Press, New York

www.cambridge.org
Information on this title: www.cambridge.org/9780521003377

First published 2001

A catalogue record for this publication is available from the British Library

Library of Congress Cataloguing in Publication data

Grant, Edward, 1926–
God and reason in the Middle Ages/Edward Grant.
p. cm.
Includes bibliographical references and index.
ISBN 0-521-80279-2 – ISBN 0-521-00337-7 (pb.)
1. Reason – History. 2. Faith and reason – Christianity – History of doctrines.
3. Learning and scholarship – History – Medieval, 500–1500.
4. Universities and colleges – Europe – History. I. Title.

B738. R42 G73 2001
189–dc21 00-065116

ISBN 978-0-521-80279-6 Hardback
ISBN 978-0-521-00337-7 Paperback

Transferred to digital printing 2009

To Sydelle,
once again

CONTENTS

ACKNOWLEDGMENTS

In writing any book, an author assumes obligations that he or she is often unaware of and that only careful reflection brings to mind. My reflections tell me that I owe thanks in varying degrees to a number of individuals and to one institution. The latter is the Österreichische Nationalbibliothek for its kind permission to reproduce the image of God designing the universe, which occurs in Latin MSS, MS 2554, fol. 1r, and appears on the cover of the paperback version. As always, I am indebted to the librarians of Indiana University, who have helped immeasurably in locating and acquiring all manner of research materials. I am especially grateful to the three anonymous readers whose perceptive insights and criticisms proved enormously helpful. The numerous publications of Professor John E. Murdoch (Harvard University), a friend and professional colleague of long standing, were especially relevant and helpful. University colleagues who made their expertise available to me are Professors Paul Vincent Spade (Department of Philosophy), Michael Berkvam (Department of French & Italian), and Leah Shopkow (Department of History). I am indebted to my departmental colleagues, Professors Michael Friedman and Michael Dickson, and to a former colleague and friend of many years, Professor John Winnie, Sr., for generous help on a number of technical, philosophical problems. For the pleasure and privilege of participating in stimulating and fruitful discussions over the years, I wish to express my gratitude to Professors Roger C. Buck, Frederick Churchill, H. Scott Gordon, Noretta Koertge, and Jack Moore, longtime friends, colleagues, and weekly luncheon companions.

And, finally, for patiently reading certain sections of my book, and responding to innumerable queries about it, I owe my greatest debt to Syd (Sydelle), my wife of 50 years, to whom this book is gratefully and lovingly dedicated.

INTRODUCTION

MOST WHO STUDY THE POLITICAL, SOCIAL, INSTITUTIONAL, AND intellectual developments in Western Europe during the Middle Ages find it easy to believe that "Western civilization was created in medieval Europe." George Holmes, the author of that sweeping statement, argues further that

> [t]he forms of thought and action which we take for granted in modern Europe and America, which we have exported to other substantial portions of the globe, and from which indeed, we cannot escape, were implanted in the mentalities of our ancestors in the struggles of the medieval centuries.[1]

Just what was implanted in the peoples of the Middle Ages between approximately 1050 to 1500? Nothing less than a capacity for establishing the foundations of the nation state, parliaments, democracy, commerce, banking, higher education, and various literary forms, such as novels and history.[2] By the late Middle Ages, Europe had also produced numerous laborsaving technological innovations. The profound problems involved in reconciling church and state, and natural philosophy and Scripture were first seriously encountered in this same period. Indeed, it was during the Middle Ages that canon and civil law were reorganized and revitalized. Not only did these newly fashioned disciplines lay the foundations of Western legal systems, but from the canon law also came the concept of a corporation, which enabled various institutions in the West – commercial, educational, and religious – to organize and govern themselves in a manner that had never been done before.

1. George Holmes, ed., *The Oxford History of Medieval Europe* (Oxford: Oxford University Press, 1992), v. These are the opening words of the book in the "Editor's Foreword."
2. See ibid., v–vi.

REASON AND SOCIETY

Why did Western Europe emerge in the tenth and eleventh centuries to begin the development of all these institutions and activities? We may never really know, but one factor that undoubtedly played a significant role was a new self-conscious emphasis on reason that is already apparent in the educational activities of the eleventh century and in the emerging theology that began at approximately the same time. The new emphasis on reason affected all the subject areas that formed the curriculum of the universities that came into being around 1200.

Concurrent with these developments was the application of reason to societal activities. In his splendid book, *Reason and Society in the Middle Ages,* Alexander Murray takes a very broad approach to reason and shows it operating in various aspects of society. In the first part, he treats of reason in economics, devoting separate chapters to money, avarice, and ambition, following with a significant chapter titled "Reason and Power." In this chapter, Murray seeks to show that from the late eleventh century onward, there developed "the concept that the mind, quite apart from any pleasure or edification its exercise may afford, is an efficacious weapon in man's battle with his environment."[3] Technology, magic, and astrology were all used to do battle with the natural environment. By using one or more of these three tools, one could exercise power over nature. Murray also regards the study of history as an illustration of the use of reason because "history helped you avoid mistakes."[4] The study and use of arithmetic in commerce and government was another powerful illustration of the application of reason.[5] In all this, and in subsequent treatment of the intellectual elite, the universities, and the nobility, Murray emphasizes that the use of reason was viewed as a means to power and upward mobility. He explicitly avoids academic discussions about faith and reason, explaining that "academic disputes were relatively esoteric; and our business is with reason on the broadest-possible social stage."[6]

The broad manner in which Murray uses the term *reason* does not distinguish the ways in which the West used reason differently than it had ever been used before. After all, mathematics, especially arithmetic and algebra, was used extensively by the ancient Mesopotamians. The peoples of ancient Mesopotamia also kept extensive economic records on clay tablets, thus

3. Alexander Murray, *Reason and Society in the Middle Ages* (Oxford: Oxford University Press, 1978), 111.
4. Ibid., 131.
5. See ibid., Part II, Chapters 6–8.
6. Ibid., 6–7.

recording commercial dealings. And yet there is no evidence that the Mesopotamian peoples self-consciously emphasized reason, although they were clearly using it. In late antiquity and in Islam, magic and astrology played significant roles. If these activities constitute an exercise of reason, then we might well conclude that these other societies also emphasized reason, perhaps as much as did the West. Murray has interpreted the use of reason so broadly that we find little to distinguish the medieval Latin West from Islam, the Byzantine Empire, and any other society in which magic, astrology, and mathematics were practiced and used, and where upward mobility may have been a factor. There is nothing distinctive about the use of reason in the societal activities that Murray distinguishes.

In this volume, I shall largely confine my study of reason to medieval intellectual life as it developed within the universities. In emphasizing the curriculum of the medieval universities, I shall focus on the disciplines of natural philosophy, logic, and theology and their interrelations, which inevitably involved faith and reason. Murray omitted discussions of these subjects because they were too esoteric and would draw attention away from the use of reason on "the broadest-possible social stage." And yet, I shall attempt to show that it was in the esoteric domain of university scholasticism that reason was most highly developed and perhaps ultimately most influential. Indeed, it was permanently institutionalized in the universities of Europe. Reason was interwoven with the very fabric of a European-wide medieval curriculum and thus played its most significant role in preparing the way for the establishment of a deep-rooted scientific temperament[7] that was an indispensable prerequisite for the emergence of early modern science. Reason in the university context was not intended for the acquistion of power over others, or to improve the material well-being of the general populace. Its primary purpose was to elucidate the natural and supernatural worlds. In all the history of human civilization, reason had never been accorded such a central role, one that involved so many people over such a wide area for such an extended period. To explicate how reason functioned in the university environment and how it was related to revelation and faith, and, to a much lesser extent, how it was related to observation and sense perception, is the major objective of my study.

PERCEPTIONS OF HISTORICAL EPOCHS

The urge to cut history into tidy, manageable segments and to characterize each segment by a memorable catch-phrase has been with us for some time,

7. Because I find the phrase "scientific temperament" descriptive of an important aspect of the medieval approach to the world, I shall use it a number of times in this study.

and will probably remain with us into the foreseeable future. Two widely used phrases that purport to capture the essence of two historical epochs are "The Age of Faith" for the Middle Ages and "The Age of Reason" for the seventeenth and eighteenth centuries. If these are apt descriptive phrases, we may properly infer that in moving from the Middle Ages to the seventeenth and eighteenth centuries, we have shifted from faith to reason, that we have somehow emerged from an age of uncritical belief, and even ignorance, to one of knowledge based on the use of reason.

There is an element of truth in these pithy descriptive phrases. The Middle Ages did stress faith, and the seventeenth and eighteenth centuries did lay emphasis on reason and reasoned discourse. But a single feature, however prominent, cannot characterize an historical epoch. In every period of history, many things develop and evolve concurrently. Although faith was a powerful force in the Middle Ages, so was reason. In this study, my aim is to describe and interpret the role that reason played in the medieval effort to understand the physical and spiritual worlds.

We are many things simultaneously. Indeed, the dominance of science and technology in our own age might tempt one to infer that ours is a preeminently rational age. Closer inspection reveals how rash such an inference would be. Think of all the irrationalities that pervade our society, many of them masquerading under the very science that epitomizes rationality. New Age religions abound and alternative medical treatments promise to accomplish what traditional medicine fails to achieve and cannot promise. Indeed, the ultimate health claim is immortality, a state of existence that is promised to all who join People Forever (headquartered in Scottsdale, Arizona). The claim of this aptly named organization is to have discovered the secret of immortality. According to one of its spokespersons, the human species "has the ability to perpetually renew itself" by "tapping into the intelligence of the cells themselves." People Forever claims that it has members in 16 countries with a mailing list of 10,000 and a monthly magazine. Although three of its members had died when a reporter wrote about the group, and one of the group confessed that they could not guarantee immortality, another member insisted that "the minute you decide you want to live forever, everything else falls into place."[8] Indeed, in this year of 2000, six years after the article about the group appeared in 1994, the group is still around. From their website, where the last dated entry I found is 1998, we can see that their leader, or one of their leaders, Mr.

8. From an Associated Press article in the *Herald-Times* of Bloomington, Indiana, on July 24, 1994.

Charles Paul Brown, published a book titled *Together Forever: An Invitation to Physical Immortality.*[9] Moreover, "every Wednesday night, Charles and a growing body of cellularly-connected individuals gather in Scottsdale, Arizona to share the adventure of infinite immortal life."[10] Apparently, audiotapes of "Charles' expressions at the weekly gatherings are available for $10 per set."

Whatever else they may be, we can safely assume that the organizers of this movement are not irrational – the dream of immortality has perhaps made them comfortable, if not downright rich. But what about the deluded individuals who join People Forever hoping, and expecting, to achieve physical immortality? Do they mark an advance over medieval gullibility and superstition? It does not appear so. In behalf of the denizens of the Middle Ages, we might mention that they too were seized with a great desire for immortality. But they achieved it the old-fashioned way: by first dying, a method not susceptible to counterinstances.

Also noteworthy are past efforts by the now-defunct Soviet Union and the United States to use psychic power to achieve state and military objectives. Indeed, such powers are not reserved for governments alone. For a few dollars, you can dial your favorite psychic and learn all about your future, or, if you prefer, read your horoscope in a daily newspaper. The indubitable fact that what we know today about the world and its operations dwarfs what was known about it during the Middle Ages, might lead us to believe that this enormous disparity in knowledge would also produce an analogous disparity in the use of, and reliance on, reason. If we confine our comparison to the literate in both periods, we moderns of the twenty-first century ought to be eminently more rational than our counterparts in the Middle Ages. But this is hardly obvious, and is very likely untrue. Sheer magnitude or quantity of information cannot in itself guarantee a more rational society. The Age of Information that has engulfed us is, alas, not synonymous with knowledge and wisdom. While science itself requires a rational methodology, the success of science is no guarantee that those who live in a society in which science is dominant and pervasive will usually act rationally. Untold mischief has been done, and will continue to be done, under the good name of *science.*

9. The URL address in which excerpts from Mr. Brown's book appear is: *http://www3.pair.com/genesis/charlespaulbrown/excerpts/index.html.* A check of library catalogs shows that the book does exist under this title. It was published in Scottsdale, Arizona, in 1990 by the Eternal Flame Foundation. On the Web site, the title of the book is erroneously given as *Together Forever: An Invitation to Be Physically Immortal.*

10. Cited from a different, though no longer existing, hyperlink that ended with */gatherings/index.html.*

THE DARK SIDE OF REASON

Many aspects of human behavior carried out in the name of reason and science are irrational, although not devastating. Nevertheless, we must recognize that reason has an ominous, dark side. During the Middle Ages, much else went on that was less lofty and noble than reason, sometimes even masquerading as reason. Superstition, religious persecution, brutality, and ignorance were reason's constant companions. Many, if not most, of the medieval authors who will be cited here for their emphasis on reason in one context or another may have been far from rationalistic in many other aspects of their lives. It is rare that one attribute – reason, or superstition, or brutality, or whatever – dominates our behavior to the exclusion of all others.

If nothing else persuades us that those who lived in the Middle Ages were no less rational than we moderns, and were perhaps even more rational, we should recall the grotesque atrocities carried out in the twentieth century by the likes of Adolf Hitler, Josef Stalin, Pol Pot, and a host of lesser murderers. During the Middle Ages, heretics were considered dangerous to the faith and, therefore, in the absence of a tradition of tolerance, often persecuted. Few in the Middle Ages would have judged the torture and execution of heretics and witches as unreasonable. The twentieth century was no stranger to such behavior. All too often, it witnessed the coexistence of reason and irrational persecution. Even as they tortured and murdered millions and millions, Adolf Hitler and Josef Stalin ruled over governments that relied heavily on science, and, therefore, on the reason that made it all possible. The dark side of reason will, unfortunately, always be with us. The misapplication of reason to gain knowledge, to resolve problems, or to control our lives better seems an unavoidable aspect of human society. We cannot forget the perverse medical experiments that were performed on innocent victims by Nazi and Japanese doctors during World War II, nor indeed the infamous Tuskegee syphilis experiments in the United States carried out on African Americans from 1932 to 1972.

The dark side of reason, which often draws upon ignorance, fear, prejudice, and hatred, is an all-too-common feature of the human condition. Because of this perennial dark side, there is no effective means of measuring and comparing the rationality of one age against the rationality of another. But it is worth mentioning that witchcraft persecutions intensified in the seventeenth century and magic played a greater role in the sixteenth century than it ever did in the late Middle Ages, during the thirteenth and fourteenth centuries. The Middle Ages had its brutal atrocities and egregious stupidities. It also had the Inquisition. By comparison to their modern counterparts, however, the murderous irrationalities of the Middle Ages seem less flagrant. Those who lived during the Middle Ages simply lacked the capacity to kill and destroy on the scale of modern societies.

If this were a study of human society as a whole, the dark side of reason would have to play a significant role. But this is a book about the positive side of reason, rather than its ominous aspects, and I shall, therefore, say no more about the societal impact of the darker recesses of the human intellect.

THE APPROACH TO REASON IN THIS STUDY

Without the rigorous use of reason to interpret the natural phenomena of our physical world, Western society could not have developed science to its present level. Indeed, our society cannot survive without science and the reasoning that makes it possible. Even the problems science causes can only be remedied by science itself. But *when, how,* and *why* did Western civilization place reason at the center of intellectual life and thereby make possible the development of modern science? The answer to the "when" part of this query is straightforward: the late Middle Ages, from around 1100 to 1500.

In a book that is well known to scholars of eighteenth-century intellectual history, Carl Becker showed rare insight into the nature of medieval thought when he characterized it as highly rationalistic. "I know," he explained, "it is the custom to call the thirteenth century an age of faith, and to contrast it with the eighteenth century, which is thought to be preeminently the age of reason."[11] Becker explains that "since eighteenth-century writers employed reason to discredit Christian dogma, a 'rationalist' in common parlance came to mean an 'unbeliever,' one who denied the truth of Christianity. In this sense Voltaire was a rationalist, St. Thomas a man of faith." But Becker explains that Voltaire and Thomas did share something rather important, namely, "the profound conviction that their beliefs could be reasonably demonstrated." Because of this shared conviction, "in a very real sense, it may be said of the eighteenth century that it was an age of faith as well as of reason, and of the thirteenth century that it was an age of reason as well as of faith."[12] Much of this study is an effort to provide evidential support for Becker's perceptive insights by demonstrating that medieval university scholars and teachers, spread over four centuries or

11. Carl L. Becker, *The Heavenly City of the Eighteenth-Century Philosophers* (New Haven: Yale University Press, 1952), 8.
12. Ibid.

more, placed a heavy reliance on reason. Moreover, in the history of civilization, they were the first to do so self-consciously on a grand scale.

In the modern incarnation of Western civilization, a new attitude emerged toward reason and rationality. By the "modern incarnation of Western civilization" I mean the new society that emerged from the transformation of the Roman Empire in Western Europe during the turbulent centuries of the barbarian invasions – from approximately the sixth to tenth centuries.[13] By the late eleventh century an energetic new society and civilization had come into existence and the momentous events that will be mentioned and discussed in this study were under way. A major feature of the new European society was an extraordinary emphasis on the use of reason to understand the world and to solve problems, both practical and theoretical. Although the scope of reason would be greater in the seventeenth and eighteenth centuries, the period traditionally described as the Age of Reason, I shall argue that that age began in the late Middle Ages, which deserves to be regarded as the unqualified starting point for what would become a growing and evolving emphasis on reason as the arbiter of disputes and disagreements. The differences that seem to distinguish the use of reason in the seventeenth and eighteenth centuries from its use in the late Middle Ages derive largely from major changes in European history – the Protestant Reformation and the Scientific Revolution,[14] to name two of the most significant. The range of uses to which reason could be applied undoubtedly expanded in the later period, but it could do so only because the ground had been solidly prepared in the preceding centuries. Reason was not a newly emphasized activity that burst forth in the so-called Age of Reason in contrast to its relative absence in the late Middle Ages. I shall argue that the Age of Reason is hardly imaginable without the central role that reason played in the late Middle Ages. If revolutionary rational

13. Historians of the Roman Empire no longer speak of a "decline and fall" of that empire in the West, to allude to Edward Gibbon's famous title, but about a transformation into something else. The Roman Empire in the East, the Byzantine Empire, was remarkably strong and resilient, continuing on until 1453, when it was captured by the Turks and became part of the Ottoman Empire. See Glen W. Bowersock, "The Vanishing Paradigm of the Fall of Rome," in *Bulletin of the American Academy of Arts and Sciences* 49 (May 1996), 29–43.

14. In recent years, the concept of a Scientific Revolution in the seventeenth century has been challenged. But whatever term, or terms, we may use to embrace the dramatic changes that occurred in science in that century, there could be no doubt that something significant occurred. For lack of a better term to describe those changes, I have retained the expression Scientific Revolution.

thoughts were expressed in the Age of Reason, they were made possible only because of the long medieval tradition that established the use of reason as one of the most important of human activities.

Reason, however, is not a medieval invention. Indeed, it is an activity that is manifested in every civilization and in every culture. Humans could not survive without it. What differentiates Western civilization from other societies and cultures that used reason is the self-consciousness with which it was used, and the scope, intensity, and duration of its application.

The achievements of Western society were made possible because of the intellectual gifts it received from the pagan Greeks, the Byzantine Christian Greeks, and the civilization of Islam. Although reason was valued in these civilizations, it was consciously esteemed by a relatively small number of scholars who were never sufficiently influential to give reason the intellectual standing that it would receive in the medieval West. The West did what no other society had previously done: It institutionalized reason in its universities, which were themselves an invention of the West.

But what is reason? How should it be understood for the purpose of this inquiry? One cannot approach the use of reason in the Middle Ages without simultaneously thinking of its opposite activity, revelation. Strictly speaking, revelation, that is, the articles of faith, is not subject to reason. Revelation is true because it embraces truths that are believed to come directly from God, or from His revealed word in Holy Scripture. Such truths – the Trinity, Incarnation, Redemption, and Eucharist – are beyond the comprehension of human reason. Where reason applies logical analysis to problems about the physical world and to aspects of the spiritual world, the same kinds of analyses are of no avail when applied to articles of faith. Reason, Christians argued, could neither prove nor disprove such revealed truths. Nevertheless, as we shall see, Christian scholars, usually theologians or theologian-natural philosophers, often tried to present reasoned analyses of revealed truths. They did so ostensibly better to understand, or to demonstrate, what they already believed on faith. We shall see that the use of reason in medieval theology and natural philosophy was pervasive and wide-ranging. Indeed, medieval scholars often seem besotted with reason. But there was one boundary line that reason could not cross. Medieval intellectuals, whether logicians, theologians, or natural philosophers, could not arrive at conclusions that were contrary to revealed truth – that was heresy. Not until the seventeenth century, and then far more pervasively in the eighteenth century, was reason applied to revelation without restriction or qualification.

So far as my study is concerned, that is the major difference in the way scholars used reason in the Middle Ages as compared to the way they used it in the Age of Reason. While it is a significant difference, it should not obscure the more fundamental truth that reason, and reasoned argumentation, lay at the heart of medieval intellectual life. Reason was the weapon of choice at medieval universities. Its systematic application to all the disciplines taught at the university gives us a sound basis for claiming that the academic use of reason on a broad, even vast, scale was a medieval invention.

With the exception of revealed truth, reason in the Middle Ages could be used to analyze virtually anything without fear of repression. By relating reason to revelation, however, we only learn about the bounds within which reason had to operate. From that, unfortunately, we do not learn what reason is. Since my objective is to describe how reason was viewed by medieval scholars, the role they assigned to it, and how they actually used it, it will be useful to characterize briefly the medieval attitude toward reason. During the late Middle Ages, reason in its traditional sense was regarded as "a faculty or capacity whose province is theoretical knowledge or inquiry; more broadly, the faculty concerned with ascertaining truth of any kind."[15] The medieval understanding of theoretical knowledge was derived from Aristotle, and it embraced metaphysics, or theology as it was also called, natural philosophy, or physics, and mathematics. Overarching all these disciplines was logic, which was regarded by Aristotle and his medieval followers as the indispensable instrument for demonstrating theoretical knowledge.[16] During the Middle Ages, reason was "contrasted sometimes with experience, sometimes with emotion and desire, sometimes with faith."[17] In this study, I shall contrast reason with experience and faith, but ignore emotion and desire.

Although logic, reason's most precise expression, was the supreme tool for the application of reason to theoretical knowledge, reason was regarded as much broader than formal logic. Reason in the Middle Ages was not tied to any particular theory of knowledge. Nominalists, realists, empiricists, and partisans of other theories of knowledge in the history of philosophy have regarded themselves as consciously applying reason to the resolution of philosophical problems. A modern philosopher has presented a good sense of what the broader aspects of reason and rationality imply for all philosophers, including those of the Middle Ages. "Rational inquiry," he has declared,

15. From A. R. M. (Alfred R. Mele), "theoretical reason," in *The Cambridge Dictionary of Philosophy*, 796.

16. In the course of this study, I shall deal with all of these except mathematics.

17. A. R. M., "theoretical reason," 796. Although A. R. M. does not mention the Middle Ages, what he says serves to illuminate the role of reason in that period.

is to be viewed as an impersonal search for truth. It is impersonal in a number of respects. First, there is some method of inquiry that can be used by anyone. Second, the method yields evidence that would convince any rational person of the truth or falsity of a particular theory. Finally, the product of applying this method is a true theory that describes things adequately for any rational being and that, by virtue of discounting the influence of any particular being's contingent perspective, furnishes a picture of the universe from a cosmic or "God's eye" point of view.[18]

The importance of rationality in Western thought cannot be overestimated. For philosophers, it has been a "special tool for discovering truth,"[19] and for modern scientists, it has been the key to the transformation of society. Modern science is the outcome of a rigorous and successful application of reason to myriad problems that have confronted the human race over the centuries.

REASON FROM ANTIQUITY TO THE THIRTEENTH CENTURY

The self-conscious, explicit use of reason and the emphasis on rationality go back to the ancient Greeks.[20] But the path of reason was never smooth and incremental. Already in antiquity, Sophists and Skeptics were critical of rationality and the claims that had been made for it by Platonists and Aristotelians.[21] Nor did the ship of reason sail on smooth seas during the late Middle Ages. But it did sail and survive the storms that battered it from time to time.

With perhaps a few exceptions, philosophers, scientists, and natural philosophers in the ancient and medieval periods believed unequivocally in the existence of a unique, objective world that, with the exception of miracles, was regarded as intelligible, lawful, and essentially knowable. Thus, the powers of reason could be applied to a real, external world that had changeable and unchangeable characteristics. The parts of that world were not regarded as of equal value and virtue. To the contrary, almost all medieval scholars, following Aristotle, believed in a hierarchical universe where, at the very least, the celestial region was regarded as incomparably superior to the terrestrial region – the

18. From Stephen Nathanson, *The Ideal of Rationality* (Atlantic Highlands, NJ: Humanities Press International, 1985), 10.
19. Robert Nozick, *The Nature of Rationality* (Princeton: Princeton University Press, 1993), xii.
20. If earlier peoples had similar attitudes toward reason, they have not been preserved in any literary traditions of which I am aware.
21. See Calvin O. Schrag, *The Resources of Rationality: A Response to the Postmodern Challenge* (Bloomington: Indiana University Press, 1992), 1.

part of the world that lies below the moon, or between the moon and the center of the earth.[22] While reason was not considered the only means of understanding this hierarchical cosmos, it was viewed favorably because it seemed the most powerful tool available for attaining knowledge about the regular day-to-day workings of the real, natural world.

The instruments that reason used for understanding the hierarchical, external world of the Middle Ages were logic, natural philosophy, metaphysics, and the exact sciences. At first, the ecclesiastical authorities in Paris, the most important intellectual center of medieval Europe by virtue of being the location of the University of Paris, viewed with alarm the secular learning that had begun entering the West in the latter half of the twelfth century. As evidence of this, Aristotle's natural philosophy was banned at the University of Paris during most of the first half of the thirteenth century. By 1245, however, natural philosophy and the sciences were fully embraced, and they became the pillars on which the university arts curriculum was built. But theology and natural philosophy, or theologians and natural philosophers, found themselves in conflict again in the 1270s. Conservative theologians were alarmed at the tone and content of discussions in natural philosophy. They were concerned that Aristotle's natural philosophy was circumscribing God's absolute power to do anything He pleased, short of a logical contradiction. They feared that natural philosophers in the faculty of arts at the University of Paris were too captivated by Aristotle's opinions and would adopt his ideas at the expense of Church tenets and revealed truth. There was also an ongoing interdisciplinary struggle at the University of Paris between the two faculties of arts and theology. The arts masters regarded themselves as the guardians of reason as embodied in the natural philosophy of Aristotle. They were professional natural philosophers, some of whom devoted their lives to that discipline. The theologians in the faculty of theology, most of whom had also studied natural philosophy, were responsible for interpreting revelation as embodied in the Bible and Church doctrine and law. In the 1260s, conservative theologians became alarmed at some interpretations of Aristotle that were either written or transmitted orally.

In 1270, acting on the appeals of some of his theologians, the bishop of Paris, Etienne Tempier, condemned 13 articles that had been drawn from the writings of Aristotle and his Islamic commentator, Averroës. In 1272, the faculty of arts, trying to circumvent more drastic action, instituted an

22. On the celestial and terrestrial regions and their relations, see Edward Grant, *Planets, Stars, & Orbs: The Medieval Cosmos, 1200–1687* (Cambridge: Cambridge University Press, 1994).

oath that required all masters of arts to avoid theological questions in their teachings and writings. If perchance theological ideas and concepts were unavoidable, it was made incumbent on the oath taker to resolve the dispute in favor of the faith.[23] The tensions were apparently unresolved during the next few years, because in March, 1277, the bishop of Paris condemned 219 propositions drawn largely from the works of Aristotle, as well as from current ideas and distortions of those ideas, that were circulating in Paris. The Condemnation of 1277 not only set theologians against arts masters but also exacerbated rivalries among the theologians themselves, pitting neoconservative Augustinian theologians, perhaps influenced by St. Bonaventure, who had died in 1274, against the Dominican followers of St. Thomas Aquinas, who also died in 1274. So bitter was the controversy that some of St. Thomas's opinions were among those condemned in 1277.

Despite these difficulties, natural philosophy was welcomed within Western Christendom and became a powerful tool for both natural philosophers and theologians. It was valued precisely because it represented a rational approach to the world and was, therefore, viewed as supplementing revelation, and occasionally even explaining it. No better tribute was paid to the utility and importance of natural philosophy than its adoption as the basic subject of study in the curriculum of all arts faculties of medieval universities.

REASON AND REVELATION

Although, as we have seen, Carl Becker believed that in the Middle Ages reason played a significant role in addition to faith, he explained how that reason was employed. "Intelligence," by which Becker means reason,

> was essential, since God had endowed men with it. But the function of intelligence was strictly limited. Useless to inquire curiously into the origin or final state of existence, since both had been divinely determined and sufficiently revealed. Useless, even impious, to inquire into its ultimate meaning, since God alone could fully understand it. The function of intelligence was therefore to demonstrate the truth of revealed knowledge, to reconcile diverse and pragmatic experience with the rational pattern of the world as given in faith.[24]

23. The statute was translated by Lynn Thorndike, *University Records and Life in the Middle Ages* (New York: Columbia University Press, 1944), 85–86. Thorndike's translation is reprinted in Edward Grant, *A Source Book in Medieval Science* (Cambridge, MA: Harvard University Press, 1974), 44–45.
24. Becker, *The Heavenly City*, 7.

Becker's insistence that "intelligence," or reason, was solely confined to "demonstrate the truth of revealed knowledge" is untenable. The entire span between the "origin" and "final state" of existence is equivalent to the period that the created cosmos had endured. It was not theology's role to investigate the workings of that cosmos. That task fell to natural philosophy and science, which relied most heavily on reason, and to a lesser extent on experience and observation (see Chapter 5), to carry out their mission. Indeed, it was natural philosophy (and logic) that provided the theologians with the reasoned arguments they needed – or thought they needed – to investigate revealed knowledge. Why natural philosophy was so vital to the dissemination of reason will be made apparent later (see Chapter 3).

Much of what can be construed as reason, or reasonable, in the late Middle Ages is similar to our own ideas about reason and rationality. Nevertheless, there are significant differences that derive from radically different views about the relationship between reason and revelation and reason and experience. As we shall see in the following chapters, the earliest emphasis on reason in the intellectual life of the Middle Ages was a by-product of the turmoil that afflicted Europe from the sixth to tenth centuries. The problems that eventually served to project reason into the forefront were associated with the disarray and disorganization of knowledge in crucial areas of human activity, most notably in theology and law. Even more important was the age-old Christian problem between faith and reason. That relationship is of momentous significance in understanding how reason was used and the scope it had. But reason fared as well as it did because the same Christian society that had to cope with the relationship between reason and revelation eagerly, though often with some trepidation and anxiety, embraced the most monumental collection of rationalistic works assembled anywhere in the world prior to the seventeenth and eighteenth centuries, namely, the works of Aristotle, the greatest of Greek philosophers. Aristotle's ideas and attitudes transformed the way the West thought about the world and its operations.

During the late Middle Ages, those who applied reason to the solution of problems in theology knew that, in the final analysis, reason was subordinate to faith, the Christian faith based on the revelation of fundamental truths that were assumed to be beyond the ken of reason. Not until the eighteenth century could one suggest with impunity, though not without some hostile reaction, that unhindered reason was the only appropriate means of investigating all phenomena, including revealed religion.

Reason was important in the Middle Ages because the domain of thought was divided between truths presented by revelation and truths made available by reason. But if revelation was truth beyond compare, the rock on which Christian society was built, reason became the means to understand that revelation and its associated spiritual matters. In explicating the mysteries of revelation, reason was clearly subordinate, although its role would be significant and it often took on a life of its own. But reason was much more than a mere handmaiden for the explication of revelation. It was the essential tool for explaining the operation of the entire physical cosmos. Indeed, that was its primary role, a role that was given to it by Aristotle in his natural books. Because Christians wisely avoided Christianizing, or theologizing, natural philosophy, natural philosophers pursued knowledge about the universe in a remarkably secular and rationalistic manner with little interference from the Church and its theologians, who were themselves often engaged in the same activity: trying to understand the workings of the physical world. Much of this study will be devoted to explaining and illustrating these activities.

Does reason's subordination to religion during the Middle Ages signify that the Age of Reason could not have occurred in that period? It does, if by the Age of Reason we mean that everything, including revelation, is subject to analysis by reason. "In the field of religion," during the Age of Reason, "reason was considered capable of finding in itself and by itself the essential truths touching the nature of God and the duties of man; as a guiding principle it was sufficient in itself."[25] This book, however, is not about the Age of Reason, but about its beginnings. Without the momentous events that unfolded in the Middle Ages, during the period from approximately 1100 to 1500, the seventeenth-century version of the Age of Reason could not have occurred. In this study, I shall attempt to demonstrate this profound truth, and also to show how the Middle Ages came to be seen through a distorted historical lens. Rather than the proper perception of the late Middle Ages as one in which reason was regarded as the most powerful investigative resource available to the human intellect, the Middle Ages came to be viewed as an age in which reason was largely absent, an age in which superstition, ignorance, and empty rhetoric flourished in place of reason.

This study will be divided into seven chapters and a conclusion. The first describes the low point of European civilization in the early Middle Ages and the vibrant, new society that emerged in the eleventh and twelfth centuries. In the second chapter, I shall describe the emergence of reason as a

25. Ernest Campbell Mossner, *Bishop Butler and the Age of Reason: A Study in the History of Thought* (New York: The Macmillan Company, 1936), 14.

potent factor in the early Middle Ages to the end of the twelfth century; and in the third chapter, I shall briefly describe the new elements – the translations, especially the works of Aristotle; Aristotle himself and why he was so important; and the universities – that allowed European society to institutionalize reason and to perpetuate its impact and influence. In the following three chapters, I describe the way medieval scholars used logic (Ch. 4), natural philosophy (Ch. 5), and theology (Ch. 6), indicating the manner in which these disciplines employed reason and the extent to which these three disciplines relied on, or used, each other.[26]

Despite the great emphasis on reason that was characteristic of medieval thought, and that will be described in this study, the Middle Ages is hardly known as an age of reason. Indeed, it is more often thought of as an age of unreason. One need only mention the word "Inquisition" to arouse in a modern audience ideas of ignorance, superstition, cruelty, and fear. Indeed, one need only utter the word "medieval" to cause the same feelings. How did the Middle Ages come to be viewed as a period that was antithetical to reason? How did such a patently false idea about the Middle Ages take root? Answers to these questions will be attempted in the seventh, and concluding, chapter.

I have reserved the Conclusion for examining perhaps the most important consequence of a widespread and intensive use of reason: the culture and spirit of "poking around."

In these chapters, my aim is to focus on the positive use of reason as it shaped the intellectual life of medieval Europe. I shall emphasize those aspects of reason that exhibit the "scientific temperament," and pay little attention to the manner in which reason was used to organize and disseminate the knowledge that reason itself had produced. When all is done, I hope that I shall have successfully balanced accounts by showing that just as the Middle Ages laid the foundation for the irrational witchcraft persecutions of the seventeenth century, so also did it lay the foundations for the Age of Reason, which, in its most positive and laudable aspects, was associated with the new science of the seventeenth century and the Enlightenment of the eighteenth century.

26. For the two other disciplines, law and medicine, see Chapters 2 and 3.

I

THE EMERGENCE OF A TRANSFORMED EUROPE IN THE TWELFTH CENTURY

By the eleventh and twelfth centuries, a new civilization had emerged in Western Europe. That new civilization was largely a product of the peoples of northern Europe, who had been at the fringes of Roman civilization for many centuries. In the course of a lengthy period of upheaval and transformation, from around 400 to 1000, a new Europe was formed in the West, the product of a fusion of the new, largely Germanic, peoples with the inhabitants of the older Roman civilization.[1]

CENTURIES OF DISSOLUTION: EUROPE AT ITS NADIR

The birth of the new Europe was a lengthy process because Germanic tribes – Ostrogoths, Visigoths, Burgundians, Lombards, Franks, and others – from the fourth to the seventh centuries were constantly at war in the northern part of continental Europe, or in process of migration, as imperial Rome weakened and gradually dissolved in Western Europe. Just when it seemed that the Franks under Charlemagne would bring a much greater degree of stability and peace than had hitherto been known in Europe, the death of Charlemagne in 814 brought further disintegration. The tendency toward central government ended, and the trend toward feudal states accelerated as noble families sought to retain whatever power and land they possessed, and to add whatever they could by fair means or foul.

Superimposed on these intrafrankish struggles was an even greater danger: the scourge of the Norsemen, who began invading various parts of Europe in the late eighth century and increased their raids and conquests during the

1. For a summary account of this early formative period, see Edward James, "The Northern World in the Dark Ages 400–900," in George Holmes, ed., *The Oxford History of Medieval Europe* (Oxford: Oxford University Press, 1992), 59–108.

ninth and tenth centuries. They struck into France, where they laid siege to Paris in 885–886, the Low Countries, Britain, Ireland, and Spain. As Norsemen transformed into Normans, they invaded Sicily and southern Italy in the eleventh century; and as warriors known as Rus, they invaded Russia in the East and gave it their name. But "after the thieving and the killing and the land-taking, they [the Norsemen] farmed and gradually became Englishmen, Irishmen, Scotsmen, Frenchmen, and Slavs."[2] So thorough was their absorption into the surrounding native populations that the Vikings left hardly any trace of their native culture and language. As they settled down in various locales, however, they showed an unusual aptitude for governing, as was especially evident by the Norsemen who became the Normans of northern France and carried their talents into England and Sicily. As if the Germanic tribes and the Norse were not enough of an affliction, Europe was also assaulted by Magyars in the East and by Muslim pirates in the Mediterranean.

Because of the barbarian invasions that endured almost continuously from the fifth to the tenth centuries, Roman civilization in the West suffered grievously as the peoples of that region sought to preserve what they could of an imperial Roman legacy that had become little more than a vague historical memory. With meaningful centralized government virtually nonexistent, feudal kingdoms largely governed Europe. But somehow, by the eleventh and twelfth centuries, Europe was transformed into a new and vibrant society. How is this phenomenon to be explained? How was it possible for a Europe that had been wracked by invasions and chaotic conditions for centuries to emerge as a new and vigorous society? Whatever the explanation, the firm foundations of a new society had been laid by the eleventh and twelfth centuries. The barbarian invasions were over and the integration of peoples – Romans and barbarians and barbarians with barbarians – was completed. A new society had come into being with characteristics that were radically different from the Roman Empire, and from any other society that had ever existed. What made it exceptional is the emphasis it would place on reason and rationality. In a real sense, Western Europe became a society obsessed with reason, which it consciously employed in many, if not most, of its activities. Nothing like it had ever been seen.

THE GRADUAL EVOLUTION TOWARD A NEW EUROPE

With the cessation of destructive invasions from outside, Europe entered a period of relative peace and equilibrium. Political conditions improved

2. See F. Donald Logan, "Vikings," in Joseph R. Strayer, ed., *Dictionary of the Middle Ages*, vol. 12 (New York: Charles Scribner's Sons, 1989), 436. The brackets are mine.

considerably. Much of France was reasonably governed because many French feudal lords provided a stable political environment that brought order from disorder. Many French knights and adventurers – Normans and others – left France to seek their fortunes elsewhere. They brought French-style feudal government to Anglo-Saxon England, to Italy, Sicily, Spain, and Portugal. They seem to have had a flair for governing and thus helped stabilize Europe. As an indication of the new penchant for government, one need only call to mind the famous Domesday Inquest instituted in 1086 in England by William the Conqueror, the Norman ruler of that land. Only a strong government could have carried out such a detailed inventory. Europe's renewal is also evidenced by the reconquest of Spain from the Muslims, which was well under way by the end of the eleventh century and would continue through the twelfth and thirteenth centuries, until relatively little was left of the Muslim conquest of Spain.

The Revival of the Economy: Agriculture and the Cities

With the establishment of greater security, Europe's economy revived and the standard of living rose. Significant advances were made in agriculture, so that European farmers were able to produce far more food than they required for their own needs. They could do this because of agricultural improvements in the early Middle Ages. Northern Europe saw the advent of the heavy plough, which was made feasible for turning over heavy soils by new ways of using horsepower effectively.[3] Two advances made the horse useful for the new agriculture: the nailed horseshoe, available in the West by the end of the ninth century; and the replacement of the yoke-harness with the collar-harness. The horse collar was padded and rigid and rested on the shoulders of the horse, thus enabling it to pull a load without choking, as often happened with the yoke-harness, one strap of which was wrapped around the horse's neck. As the horse pulled, the neck strap tightened, thus restricting the flow of air and often causing it to suffocate. By contrast, a horse harnessed around the shoulders with a padded, rigid collar not only avoided this cruel fate but was also capable of pulling a load four or five times heavier than it could pull with a yoke-harness. With the adaptation of the horse for the plough, agriculture was ready for a great leap for-

3. See Lynn White, Jr., *Medieval Technology and Social Change* (Oxford: Oxford University Press, 1962), 57; also see Lynn White, Jr., "Agriculture and Nutrition," in Joseph R. Strayer, ed., *Dictionary of the Middle Ages*, vol. 1 (1982), 89–92, and David Nicholas, *The Evolution of the Medieval World* (New York: Longman Publishing, 1992), 287.

ward. Since the horse is quicker and has more endurance than an ox, it was a far more efficient source of power for farmers, who could now plough more land more quickly than ever before. "By the end of the eleventh century," Lynn White informs us, "the plough horse must have been a common sight on Europe's northern plains."[4]

To these advances, we must add the replacement of the two-field system of crop rotation with that of the three-field system. In the two-field system, half of the arable land was ploughed and half left at rest, whereas in the three-field system, two-thirds of the land was ploughed and cultivated in any given year, with only one-third left uncultivated. On a 600-acre farm, for example, use of the two-field rotation system meant that 300 acres were cultivated in any given year and 300 left fallow. In the three-field system, 400 acres would be cultivated with 200 left fallow. By planting the extra 100 acres, farmers increased their productivity by one-third. With horseshoes and collar-harnesses, horses could assume the extra burden. By means of these cumulative advances, European agricultural productivity was enormously increased.

With more food available on a regular basis, the general population grew, especially in the cities and towns. In time, it became necessary to build hundreds of new towns. Europeans began to colonize previously unpopulated, or lightly populated, lands, or they drove eastward against the Slavs, as the Germans did in their movement beyond the Elbe River. As the quest for land intensified, inhabitants in the Low Countries began to reclaim land from the sea. Europeans were on the move and significant migrations occurred. Free men populated many of the new towns:

> By 1100 Europe had a surplus of agricultural products, an increasing population, and a surplus of labor. The surplus of food made it possible to support large numbers of men who were not directly engaged in agriculture; the surplus of labor encouraged manufacturing and trade. As the towns grew they stimulated agriculture by affording markets which could absorb all the food produced for miles around. So more land was cleared, and more new villages were founded, and this in turn made possible a new increase in urban economic activities.[5]

This process continued on through the twelfth and thirteenth centuries. By the end of the twelfth century, the level of commerce and manufacturing in Europe was probably greater than it had been at the height of the

4. White, *Medieval Technology*, 63. For more on horses, see White, "Agriculture and Nutrition," in *Dictionary of the Middle Ages*, vol. 1 (1982), 92–93.

5. Joseph R. Strayer and Dana Carleton Munro, *The Middle Ages 395–1500* (New York: Appleton-Century-Crofts, 1942), 191.

Roman Empire. There can be no doubt that between the ninth and thirteenth centuries, Europe had been transformed. A money economy had come into being.

Significant changes in government also occurred. Two major conflicts developed: one between towns (or cities) and neighboring feudal rulers, the other between secular and ecclesiastical rulers. The increasingly free, urban populations sought as much self-government as they could get and struggled to free themselves from taxes imposed by hereditary nobles. The concept of a commune was developed, with attendant rights of citizenship. To increase their power and protect their rights, the cities of Europe opportunistically aligned themselves with whomever could advance their objectives: popes, kings, emperors, or independent princes. The rise of towns was momentous because

> [i]t upset the balance of power in every European country. The towns were wealthy, they had great strategic value, and some of them had important military or naval forces. A ruler who could use the new wealth and new power of the towns could break any internal opposition; a ruler who could not control his towns had little authority. This is one of the most important keys to an understanding of the political events of the twelfth and thirteenth centuries. The kings of France and England, who gained control of their towns, were able to create powerful unified monarchies. In the Low Countries, control of the towns went to the feudal lords, who became practically independent princes. In Italy the towns aided by the pope, gained complete freedom and in doing so destroyed the power of the emperor. The political future of every European country was determined by the relations between its king and the towns.[6]

Thus it was that cities became a powerful force in the economic, political, religious, and cultural life of the European continent. When combined with other significant changes, we may appropriately speak of a new Europe that had emerged by the eleventh and twelfth centuries, a Europe that had developed striking, even momentous, features, most of which proved permanent. Some of these had roots in ancient Greek thought, in the early centuries of Christianity, and also in Greco-Islamic (or Arabic) culture; others only emerged in the eleventh and twelfth centuries. To understand and appreciate the extraordinary role of reason and rationality in Western Europe, it is essential to identify and describe those features that were conducive and, in some instances, even essential, to the emergence and subsequent preservation of reason and rationality.

6. Ibid., 198–199.

The Separation of Church and State

One feature of Western civilization that facilitated the emergence of reason as a significant component in intellectual and social life was the separation of church and state embodied in the momentous words of Jesus to the Pharisees: "Render therefore unto Caesar the things which are Caesar's; and unto God the things that are God's" (Matt. 22.21). The papacy acknowledged the two separate jurisdictions, as when Pope Gelasius (492–496) declared that "there are ... two by whom principally this world is ruled: the sacred authority of the pontiffs, and the royal power."[7] Here Pope Gelasius gave voice to the concept of "two swords," one secular, the other religious.[8] Although each recognized the independence of the other, one almost inevitably tried to dominate. Some secular and religious leaders sought to control both realms by imposing a theocracy. Among secular rulers, "Charles the Great [Charlemagne]," for example, "conceived of his empire as a theocracy, in which the emperor was God's representative, and in which the Church, whose one concern was religion, was one of the instruments of the state."[9] From the sixth to eleventh centuries, the state tended to impose its will on the Church. Secular rulers, kings and powerful nobles, often appointed higher clergy – abbots and bishops – who, not infrequently, were relatives.

But if the Church was not always able to enforce its will on the secular authorities, its claims of supremacy were more outrageous than those that had been proclaimed by the state. It envisioned itself as heir to the Roman Empire's global dominion, wherein the Church replaced the state and popes would rule over the vast domain once governed by emperors. This was nothing less than a Christian theocracy. In the fifth century, Pope Gelasius, whom we have already encountered, unqualifiedly declared the greater importance of priests over secular rulers.[10] But for many centuries, secular rulers ignored papal claims to supremacy and intervened in, and often controlled, the elections of bishops by the investiture process. In theory, when a bishop was elected to his office, he was supposed to receive a

7. Quoted from Williston Walker, *A History of the Christian Church* (New York: Scribner's, 1949), 135.
8. For the biblical origins of the "two swords" concept, see John L. LaMonte, *The World of the Middle Ages: A Reorientation of Medieval History* (New York: Appleton-Century-Crofts, Inc., 1949), 254–255.
9. James Westfall Thompson and Edgar Nathaniel Johnson, *An Introduction to Medieval Europe 300–1500* (New York: W. W. Norton, 1937), 379.
10. Walker, *A History of the Christian Church*, 7.

double investiture, a spiritual investiture from his ecclesiastical superior, and a temporal investiture from the secular ruler, initially the Roman emperor, later kings, dukes, and other feudal nobility. From the days of the Roman Empire, however, secular rulers, usually emperors, had insisted on investing a bishop with both the spiritual and temporal authority of his office. How effectively this was carried out depended on the relative strengths of the secular and spiritual powers. From the early ninth to eleventh centuries, the secular authorities usually enjoyed the power to enforce a double investiture and thus control the election of bishops.

The Papal Revolution

With the papacy of Gregory VII (1073–1085), the Church entered the famous Investiture Struggle (1075–1122), a protracted, and largely successful, conflict with the secular authorities for control of Church offices. Gregory VII employed his formidable energy to assert the supremacy of papal over secular power. In a work he wrote for himself in 1075, called the *Dictatus Papae (The Pope's Dictate),* Gregory VII proclaimed that "the Roman church has never erred, nor will it err to all eternity"; that the pope "himself may be judged by no one," and that "a sentence passed by him may be retracted by no one." Gregory claimed moreover that a pope "may absolve subjects from their fealty to wicked men" and that "of the pope alone all princes shall kiss the feet"; that he "may be permitted to ... depose emperors" and that "he alone may use the imperial insignia."[11]

Gregory VII began the process that culminated in 1122 in the Concordat of Worms (during the reign of the French pope, Calixtus II [1119–1124]), whereby the Holy Roman Emperor agreed to give up spiritual investiture and allow free ecclesiastical elections. The process manifested by the Investiture Struggle has been appropriately called the *Papal Revolution.*[12] Its most immediate consequence was that it freed the clergy from domination by secular authorities: emperors, kings, and feudal nobility. With control over its own clergy, the papacy became an awesome, centralized, bureaucratic powerhouse, an institution in which literacy, a formidable tool in the Middle Ages, was concentrated.

11. All of these brief quotations are from Thompson and Johnson, *An Introduction to Medieval Europe,* 378–379.

12. This expression is employed by Berman in Chapter 2, titled "The Origin of the Western Legal Tradition in the Papal Revolution," in his *Law and Revolution: The Formation of the Western Legal Tradition* (Cambridge, MA: Harvard University Press, 1983), 85–119.

The Papal Revolution had major political, economic, social, and cultural consequences. With regard to the cultural and intellectual consequences, it "may be viewed as a motive force in the creation of the first European universities, in the emergence of theology and jurisprudence and philosophy as systematic disciplines, in the creation of new literary and artistic styles, and in the development of a new consciousness."[13] European universities, and the disciplines of theology, law, and philosophy that took root in those universities, were destined to play a vital role in the development of reason and rationality in European civilization, a story that will be told in Chapter 3.

As a result of the Papal Revolution, the papacy grew stronger and more formidable. It reached the pinnacle of its power more than a century later in the pontificate of Innocent III (1198–1216), perhaps the most powerful of all medieval popes, who unabashedly proclaimed that "The Lord Jesus Christ has set up one ruler over all things as his universal vicar, and, as all things in heaven, earth and hell bow the knee to Christ, so should all obey Christ's vicar, that there be one flock and one shepherd." In a more specific reference to the secular power, Innocent used a popular simile involving sun and moon:

> As God, the creator of the universe, set two great lights in the firmament of heaven, the greater light to rule the day and the lesser light to rule the night [Gen. 1.15, 16], so He set two great dignities in the firmament of the universal church, ... the greater to rule the day, that is, souls, and the lesser to rule the night, that is, bodies. These dignities are the papal authority and the royal power. And just as the moon gets her light from the sun, and is inferior to the sun in quality, quantity, position and effect, so the royal power gets the splendor of its dignity from the papal authority.[14]

While the Church was abolishing lay investiture and advancing its own claims with stunning success, the power of the secular states of Europe was also growing. Despite many claims and counterclaims that were made between church and state in Western Christendom, and the fact that one or the other tried to dominate when an opportunity arose, the separation endured, largely because the Papal Revolution had vaulted the Church into virtual parity with the secular authorities.

13. Berman, *Law and Revolution,* 100. For Berman's brief summaries of the political and socioeconomic aspects of the Papal Revolution, see 100–103.
14. Cited from Thompson and Johnson, *An Introduction to Medieval Europe,* 645. The authors also observe that "Frederick II [Holy Roman Emperor and bitter foe of Innocent III] used the same simile, but insisted that the sun should keep out of the moon's orbit." Ibid. I have added the bracketed description.

Although the Papal Revolution was an equalizing process in power and authority between church and state, the latter had much to learn from the former. Of the two entities, the Church was far more centralized and effective. In achieving its revolution, and developing an efficient and far-flung bureaucracy, the Church inadvertently laid the foundation for the modern state, the model for which, ironically, was the Church itself. For the Papal Revolution, as Berman describes it, laid "the foundation for the subsequent emergence of the modern secular state by withdrawing from emperors and kings the spiritual competence which they had previously exercised." As a consequence,

> [t]he Church had the paradoxical character of a church-state, a *Kirchenstaat:* it was a spiritual community which also exercised temporal functions, and whose constitution was in the form of a modern state. The secular state on the other hand, had the paradoxical character of a state without ecclesiastical functions, a secular polity, all of whose subjects also constituted a spiritual community living under a separate spiritual authority.[15]

The Papal Revolution allowed the Church "to create an autonomous legal order. It asserted a right to jurisdiction, a right to hear all cases within its domain, a right to legislate new laws, and a commitment to conduct its affairs according to law."[16] As a consequence, the Church "established the model by which secular states could organize their affairs, establish courts, elect officials, and enact their own laws, in order to govern their political, economic, and social domains."[17]

But the Papal Revolution achieved more than that. By insuring that secular authorities were excluded from ecclesiastical involvement, the Church inadvertently helped create a more positive environment for secular affairs. It enabled Western civilization to avoid the pitfalls of Caesaropapism, which had bedeviled the Byzantine Empire. It had helped create secular governing entities within which reasoned discourse, without revelation, could be carried on. In time, these secular governments would assume responsibility for most of the universities of Europe and assume many functions that had previously been conducted by the Church.

Education and Learning: The Cathedral Schools

From the time of Charlemagne, who was crowned emperor of the Romans in A.D. 800, the tradition of learning was gradually extended and intensi-

15. Berman, *Law and Revolution,* 115.
16. Toby E. Huff, *The Rise of Early Modern Science: Islam, China, and the West* (Cambridge: Cambridge University Press, 1993), 126.
17. Ibid.

fied. In 789, Charlemagne issued a capitulary in which he called for the establishment of schools in monasteries and cathedrals. The monasteries were the first to respond, and famous schools appeared at Fulda, St. Gallen, and Corbey.[18] By the eleventh and twelfth centuries, however, the locus of educational activity had moved from monasteries to schools connected with cathedrals. Since cathedrals were in the major cities of dioceses, the emergence of cathedral schools marks a significant shift of education from the countryside to the cities of Western Europe. Among cathedral schools, those at Paris, Liège, Rheims, Orleans, Laon, and Chartres achieved great fame in their day. Established initially to educate secular clergy, the cathedral schools soon attracted laymen who wished to learn Latin and other subjects useful for professional purposes in medicine, law, or civil or ecclesiastical administration. Many famous teachers were associated with cathedral schools, from Gerbert of Aurillac, founder of the school at Rheims in the late tenth century, to Peter Abelard of the cathedral school of Paris in the first half of the twelfth century, who was perhaps the most famous teacher in the Middle Ages. Sandwiched between these two was Fulbert of Chartres, who "is the patriarch among the masters of the great cathedral schools" and who was "the first to form a school with a distinctive tradition which persisted long after his death."[19]

Almost from their beginnings, the cathedral schools fulfilled a valuable need. Students flocked to them from all over Europe. Famous masters attracted large numbers of students. It was customary for students to move around and study with different masters, sometimes reinforcing and extending their knowledge of the same subject, but often studying different subjects. A significant aspect of cathedral schools was the diversity of subjects taught. In a striking paragraph, R. R. Bolgar has brilliantly captured the essential role played by these unusual and important schools:

> The teachers who thus collected round the great cathedrals and the more famous collegiate churches like St Geneviève in Paris gave instruction not so much in the elements of Latin as in advanced grammar which involved the reading of authors; and they also treated rhetoric, the subjects of the Quadrivium which served as a preparation for medicine, elementary law, and above all dialectic. Rheims under Gerbert and later the schools of Chartres were famous for their science. Bologna acquired a great reputation first for literary studies and later for Roman law, while

18. See R. R. Bolgar, *The Classical Heritage and Its Beneficiaries* (Cambridge: Cambridge University Press, 1954), 194; on the cathedral schools, see 193–201.
19. R. W. Southern, *The Making of the Middle Ages* (New Haven: Yale University Press, 1953), 197.

Paris became the centre for theology, dialectic, and philosophical learning in general. They all served professional rather than strictly religious aims. Even the education they offered to those whose careers were to lie within the Church was primarily technical in character. For theology, especially the philosophical theology of the twelfth century, Canon Law, and the niceties of ecclesiastical administration must in the last analysis be regarded as professional interests. And in addition they seem to have drawn into their classrooms an appreciable number of those who intended to spend their lives in definitely lay pursuits, in legal work, medicine, or municipal and feudal business. Their fundamentally non-religious character was long masked, however, by the fact that their students were all supposed to be clerics. The convenient clerical status conferred by the possession of minor orders enabled them to welcome not only the type of student who even at an earlier date might anyway have attended an ecclesiastical school, but also those categories who formerly would have been educated in a private and unorganised fashion by lay teachers.[20]

The original purpose of the cathedral schools had been to teach the reading and writing of Latin in the form of grammar and rhetoric. Students were also exposed to pagan Latin literature, which was subsumed under grammar and rhetoric. In time, however, and as available expertise permitted, the schools added courses in logic and natural philosophy, perhaps even some rudimentary science. To these possibilities, some schools added professional training in civil and canon law, theology, and medicine. Although it may be somewhat of an exaggeration, one might argue that the cathedral schools taught the seven liberal arts – that is, the trivium of grammar, rhetoric, and logic (or, dialectic); and the quadrivium of science subjects: astronomy, geometry, arithmetic, and music – along with law and medicine. The fare of the quadrivial sciences would have been modest indeed, since there was virtually no Euclidean geometry worthy of the name before the mid-twelfth century, and the astronomical texts that were available, although sometimes quantitative, did not apply geometry, the language of technical astronomy since Greek antiquity, to astronomical problems.[21] The teaching of arithmetic and music would have depended solely on Boethius's elementary texts bearing the titles *On Arithmetic* and *On Music*.

Cathedral schools specialized in the expertise that was available. At one cathedral school, a master trained in civil law could teach that subject to interested advanced students; at another, where such expertise was lacking,

20. Bolgar, *The Classical Heritage*, 194–195.

21. For an excellent account of astronomy in the early Middle Ages, see Stephen C. McCluskey, *Astronomies and Cultures in Early Medieval Europe* (Cambridge: Cambridge University Press, 1998). See especially pp. 140, 163–164, 206–208.

there might reside a master who was trained in medicine. Because of fortuitous circumstances, the schools varied greatly. Students flocked to schools that offered subjects they thought were relevant or interesting, or which were of potential professional use. They also sought out masters who had acquired great teaching reputations. During the period 1050 to 1150, when the cathedral schools reached their zenith, many students were on the move throughout Europe seeking the right schools and the best teachers.

The teachers who taught in a cathedral school were eventually licensed, often by the chancellor of the school, who was appointed by the bishop. By the twelfth century, the quality of the teachers was probably higher than it had been a century or two earlier.

The legacy of the cathedral schools was enormous, though paradoxical. They had evolved within the framework of the old Latin learning. Except for the reintroduction of the Roman law in the form of the Justinian code in the twelfth century, the intellectual fare of the cathedral schools was derived from a few Roman authors and a group of encyclopedic authors, often referred to collectively as the Latin Encyclopedists, who lived during the fourth to eighth centuries. Until the twelfth century, students and teachers at the cathedral schools subsisted on the *Timaeus* of Plato, embedded in the commentary of Chalcidius (fourth or fifth century A.D.); the commentary on the *Dream of Scipio,* which was the sixth book of Cicero's *Republic,* by Macrobius (fl. early fifth century A.D.); the *Natural Questions* of Seneca (ca. 4 B.C.–A.D. 65); the translations of Aristotle's elementary logical works by Boethius (ca. 480–525) (for more on this, see Chapter 2), along with Boethius's own elementary works on arithmetic and music; the *Institutions* of Cassiodorus (ca. 480–ca. 575), the second book of which was an encylopedic manual intended to enlighten monks about what they should know of the seven liberal arts; *The Marriage of Mercury and Philology,* on the seven liberal arts by Martianus Capella (fl. ca. 365–440); the *Etymologies* and *On the Nature of Things* by Isidore of Seville (ca. 560–636); and *On Times (De ratione temporum)* by the Venerable Bede (672/673–735).[22] There were other lesser treatises, but all were tied to the old Latin learning.

Missing, of course, were the intellectual treasures of the Greek world. With a few minor exceptions, its science, natural philosophy, literature, and history were absent from Western civilization. The Romans had never been sufficiently motivated to translate the great works of Greek thought into

22. Brief sketches of the life and works of all of these authors can be found in Charles Gillispie, ed., *Dictionary of Scientific Biography.* See under each author's name in volume 16 (Index).

Latin for the overwhelming mass of Romans who knew no Greek. When the western and eastern parts of the Roman Empire split more or less permanently in 395 with the death of the emperor, Theodosius I, knowledge of Greek became rare in the West and, aside from a few translations of Hippocratic medical works and a bit of Aristotle's elementary logic, chances of acquiring Greek learning by translation into Latin were virtually nil by the end of the sixth century.

Despite the absence of Greek learning and the rudimentary level of scholarship in the West, "the period of the Roman past glittered in their traditions as a golden age which having existed once could no doubt be restored."[23] And how mightily they labored to acquire an intellectual heritage that had never before existed in Rome, or in the West. What was needed was a coming together of all, or nearly all, the disciplines that had been taught sporadically in cathedral schools for approximately 150 years. The cathedral school was an evolutionary step on the path to the formation of the university, which was a wholly new institution that not only transformed the curriculum but also the faculty and its relationship to state and church.

REFLECTIONS ON THE ROLE OF REASON IN THE NEW EUROPE

It is an irony of medieval history that reason and rationality had, for better or worse, virtually everything to do with religion, theology, and the Church, and relatively little to do with the state. This was true in the early Middle Ages prior to the emergence of universities around 1200, but became even more pronounced after their formation. To understand this phenomenon, we must recognize that reason was very frequently contrasted with revelation, and that the latter was the basis of the Christian faith. Relations between faith and reason in Western Europe go back to St. Augustine and Boethius (on these two, see Chapter 2). But the role that reason would play in understanding the faith, and also in understanding the world, became a matter of conscious concern beginning in the eleventh century, and it emerged as a major problem in the twelfth century. The subjects that seemed to bear the stamp of reason were logic and natural philosophy, both ultimately derived from the world of ancient Greece, from the works and thoughts of Plato and Aristotle. How these disciplines were applied to theology and became intertwined with the Christian faith and religion is the most essential part of the formal history of reason in the

23. Bolgar, *The Classical Heritage,* 91.

Middle Ages.[24] It is a story in which many would ask whether the use of reason is an appropriate instrument for understanding the faith. But reason had a powerful existence in the late Middle Ages quite independently of the faith. It was embedded in the scholastic method that developed first in theology and law and then in natural philosophy. What made the Middle Ages the legitimate initiator of the Age of Reason is that the scholastic method, with its rationalistic character, and the subjects to which it was applied – natural philosophy, theology, law, and medicine – was institutionalized in the medieval universities. The history of these developments extends from the early Middle Ages to the end of the Middle Ages and will be described in the following chapters.

24. The relations between reason and empiricism, or observation, were also important but not as vital. For the relations between reason and empiricism, see the section devoted to this theme in Chapter 5.

2

REASON ASSERTS ITSELF

The Challenge to Authority in the Early Middle Ages to 1200

KNOWLEDGE OF SCIENCE AND NATURAL PHILOSOPHY IN THE early Roman Empire was largely an inheritance from the ancient Greeks. Christian beliefs and ideas about the world that emerged during the first few centuries after Christ interacted with the dominant pagan view and helped shape a new outlook and a new worldview. If the wisdom of the world had previously been embedded in pagan learning, the triumph of Christianity in the late fourth century changed all that. The new wisdom emanated from Sacred Scripture, the Bible, and in the fundamental belief that an omniscient and omnipotent God had created our world from nothing, a conception that would have been utterly incomprehensible and unintelligible to traditional Greek philosophers. An important part of our story concerns the interrelationship between pagan and Christian learning. The eventual explicit and self-conscious use of reason as a force in medieval intellectual life emerged from this interrelationship, with results that were profound for the late Middle Ages and for the future of Western society.

CHRISTIANITY AND LATE ANTIQUITY

To understand what happened in the twelfth century, it is necessary to begin with the early Middle Ages, with its roots in the late Roman Empire. During that period in Western Europe, intellectual life was shaped by concerns about the Christian religion and its theology, and by a modest amount of secular learning that was largely an inheritance from pagan Greek sources, such as Plato, Aristotle, Cicero, and the Stoics. Most of what was known of these classical authors was not derived directly from their works, which were largely unavailable, but was filtered through Latin authors of the period – St. Augustine, Martianus Capella, Macrobius, Boethius, Cassiodorus, Isidore of Seville, Venerable Bede, and others.

The Handmaiden Tradition: Christianity Coming to Grips with Pagan Thought

These scholars preserved some learning through the difficult centuries of the early Middle Ages. They did so by means of a tradition within Christianity that inadvertently emphasized the use of reason and thus began a process that would grow increasingly more independent by the late Middle Ages. I refer here to the "handmaiden" tradition, which had its roots in Philo Judaeus but was absorbed into Christianity through Clement of Alexandria, who, in his *Miscellanies (Stromata),* shaped it into the form it would take in the West: "philosophy is the handmaid of theology," a concept that became commonplace among the Greek and Latin Church fathers.[1]

Christianity emerged and developed within a pagan culture that was centuries old when the Roman Empire was established. Two major and conflicting Christian attitudes toward that empire developed. One is represented by those who, believing that they were protecting Christianity, sought to disengage it from the intellectual traditions of the pagan society in which it was born. The classic expression of those who advocated a total, or near total, separation was proclaimed by Tertullian (ca. 150–ca. 225) when he wrote:

> What indeed has Athens to do with Jerusalem? What concord is there between the academy and the Church? What between heretics and Christians? ... Away with all attempts to produce a mottled Christianity of Stoic, Platonic, and dialectic composition! We want no curious disputation after possessing Christ Jesus, no inquisition after enjoying the gospel! With our faith we desire no further belief.[2]

The same attitude is reflected in a biography of Saint Cyprian (ca. 200–258), a convert to Christianity, whose biographer, Deacon Pontius, discounted Cyprian's pre-Christian career as a rhetorician by declaring that

1. For a brief account of the handmaiden approach, see Edward Grant, *The Foundations of Modern Science in the Middle Ages: Their Religious, Institutional, and Intellectual Contexts* (Cambridge: Cambridge University Press, 1996), 3–5; also Etienne Gilson, *History of Christian Philosophy in the Middle Ages* (London: Sheed and Ward, 1955), 32–33; and David C. Lindberg, *The Beginnings of Western Science: The European Scientific Tradition in Philosophical, Religious, and Institutional Context, 600 B.C. to A.D. 1450* (Chicago: University of Chicago Press, 1992), 150-151, 223–234.
2. *On Prescription Against Heretics,* Chapter 7, translated by Peter Holmes in *The Ante-Nicene Fathers,* ed. Alexander Roberts and James Donaldson, 10 vols. (New York: Charles Scribner's Sons, 1896–1903), vol. 3, 246.

[t]he acts of a man of God must not be counted until the day when he is born in God. Whatever have been his studies, whatever influence the liberal arts have had upon him personally, I will omit all that, for it will serve no purpose except that of the world.[3]

The idea reflected in these passages will turn up at different times in the history of Christianity. Some churchmen and scholastic authors adopted a similar attitude by regarding Sacred Scripture as self-sufficient, requiring few, if any, external aids to interpret its meaning and significance.[4]

Although representatives of this view would never completely vanish during the Middle Ages, this negative attitude toward Greek philosophy was already overshadowed in the early centuries of Christianity. The sentiment that pagan philosophy could not be rejected arose from an early belief that pagan thought foreshadowed Christianity and that the latter might therefore receive guidance and insight from the secular knowledge and learning of pagan authors. The idea emerged that Christians might take what is of value in pagan thought and use it for their own benefit, just as in Exodus (3.22, 11.2, and 12.35), the Lord instructed Moses to plunder the wealth of the Egyptians. Another incentive for studying the philosophy and science of the pagans was to use their own words and ideas against them, just as David slew Goliath with the latter's own sword (1 Samuel 17.51). From such motivations, Christians adopted the fundamental idea of using philosophy and science as "handmaids to theology." The idea of using Greek studies in this manner is traceable to Philo Judaeus (ca. 25 B.C.–A.D. 50), a Hellenized Jew who lived in Alexandria. Philo firmly believed that a general education (consisting of what came later to be called the seven liberal arts)[5] was essential to the study of philosophy, which, in turn, was necessary for comprehending revealed theology.[6] Philo formulated the idea that Greek philosophy and science should be used to elucidate Scripture.

3. Yves M.-J. Congar, O. P., *A History of Theology,* translated and edited by Hunter Guthrie, S. J. (Garden City, NY: Doubleday, 1968), 37. Congar cites Deacon Pontius's *Life of Cyprian* (*Vita Cypriani,* 2).
4. Congar (ibid., 38) sees this attitude "in the writings of the Augustinians of the thirteenth century as well as those of Roger Bacon, Richard Fishacre, and St. Bonaventure."
5. Grammar, rhetoric, dialectic or logic, which came to be called the *trivium;* and the four scientific disciplines: geometry, arithmetic, music, and astronomy, which were later designated the *quadrivium.*
6. See Henry Chadwick, "Philo and the Beginnings of Christian Thought," in A. H. Armstrong, ed., *The Cambridge History of Later Greek and Early Medieval Philosophy* (Cambridge: Cambridge University Press, 1970), Part II, Chapters 8–11, 140.

That idea was adopted by such Greek Fathers as Clement of Alexandria, Origen, Gregory Nazianzen, Basil of Caesarea, and John Damascene.

Within this group, Clement of Alexandria (ca. 150–ca. 215), one of the earliest Church Fathers, is of great importance. Clement enthusiastically advocated Philo's idea of using philosophy for revealed theology. Since the Creator conferred a capacity for reason upon humans, it follows that philosophy, which is the embodiment of reason, is valuable for the study of theology. Clement informed his fellow Christians: "We shall not err in alleging that all things necessary and profitable for life came to us from God, and that philosophy more especially was given to the Greeks, as a covenant peculiar to them, being, as it were, a stepping-stone to the philosophy which is according to Christ."[7] But Clement went beyond by arguing that logic was an essential tool for theologians.[8] With seemingly less enthusiasm, Origen, sometime around 235, sent a letter to Gregory Thaumaturgus, soon to be bishop of NeoCaesarea. In his letter, Origen urges Gregory to "direct the whole force of your intelligence to Christianity as your end." But he also directs Gregory to take with him

> on the one hand those parts of the philosophy of the Greeks which are fit, as it were, to serve as general or preparatory studies for Christianity, and on the other hand so much of Geometry and Astronomy as may be helpful for the interpretation of the Holy Scriptures. The children of the philosophers speak of geometry and music and grammar and rhetoric and astronomy as being ancillary to philosophy; and in the same way we might speak of philosophy itself as being ancillary to Christianity.[9]

The last of the Greek Church Fathers, John of Damascus (John Damascene), about whom little is known except that he entered the monastery of St. Sabbas in Jerusalem around 730, used Greek philosophy extensively in an influential work, *The Fount of Knowledge (Fons scien-*

7. From Miscellanies, VI, 8 as translated in *The Ante-Nicene Fathers: Translations of the Writings of the Fathers Down to* A.D. 325. Vol. 2: *Fathers of the Second Century: Hermas, Tatian, Athenagoras, Theophilus, and Clement of Alexandria (Entire).* American edition, chronologically arranged, with notes, prefaces, and elucidations by A. Cleveland Coxe, D. D. (Grand Rapids, MI: Wm. B. Eerdmans Publishing Co., 1983), 495, col. 2; also quoted by Gilson, *History of Christian Philosophy,* 567, n. 8.
8. See Chadwick, "Philo and the Beginnings of Christian Thought," Chapter 10: "Clement of Alexandria," 169–170.
9. "Letter of Origen to Gregory," in Allan Menzies, D. D., ed., *The Ante-Nicene Fathers: Translations of the Writings of the Fathers Down to A.D. 325.* Original supplement to the American edition, fifth edition (Grand Rapids, MI: Wm. B. Eerdmans Publishing Co., 1980), vol. 10, 295, col. 1.

tiae), which is comprised of three parts: *The Philosophical Chapters, On Heresies,* and *On the Orthodox Faith.*[10] In the preface to his work, John reveals acceptance of the handmaiden concept when he declares that "I shall set forth the best contributions of the philosophers of the Greeks, because whatever there is of good has been given to men from above by God, since 'every best gift and every perfect gift is from above, coming down from the Father of lights.'"[11] In a by-now-standard theme, John justifies the taking of knowledge from Greek philosophers because whatever is good in the philosophy of the Greeks is good because it is a gift of God. Elaborating on his intentions a few lines below, John explains: "In imitation of the method of the bee, I shall make my composition from those things which are conformable with the truth and from our enemies themselves gather the fruit of salvation."[12] Here again lurks the theme of despoiling the Egyptians: Take all the good you can from your enemies and use it for the faith. "I shall add nothing of my own," John declares near the end of his preface, "but shall gather together into one those things which have been worked out by the most eminent of teachers and make a compendium of them."[13]

A glance at the 68 chapter titles of *The Philosophical Chapters* reveals a heavy emphasis on logic. Although he may have derived much of the content of his work from pagans, John pays tribute to reason and rationality in the very opening lines of the first chapter of the first treatise:

> Nothing is more estimable than knowledge, for knowledge is the light of the rational soul. The opposite, which is ignorance, is darkness. Just as the absence of light is darkness, so is the absence of knowledge a darkness of the reason. Now, ignorance is proper to irrational beings, while knowledge is proper to those who are rational.[14]

Saint Augustine and Boethius

The ideas of Clement and other Church Fathers were embraced in the West by two of the greatest luminaries of late antiquity: St. Augustine

10. The entire work is translated by Frederic H. Chase, Jr., *Saint John of Damascus, Writings,* in *The Fathers of the Church, A New Translation,* vol. 37 (New York: Fathers of the Church, Inc., 1958). See Chase's Introduction for a discussion of the life and works of John of Damascus.
11. Ibid., 5. The biblical quotation is from James 1.17.
12. Ibid.
13. Ibid., 6.
14. Ibid., 7.

(354–430) and Boethius (ca. 480–524/525), who together would dramatically shape medieval thought, the former exercising his formidable influence throughout the whole of the Middle Ages, the latter playing his most significant role in the early Middle Ages until the mid–twelfth century.

If it can be said that Augustine had a more significant impact in the domain of theology, this was more than counterbalanced, at least until the mid–twelfth century, by the contributions in logic associated with the name of Boethius. The role that these two scholars assigned to reason and rationality significantly influenced the way reason was viewed and used in the Middle Ages.

Although Augustine's mother, Monica, was a Christian, and his father, Patricius, abandoned paganism for Catholicism shortly before his death in 370, Augustine himself did not convert to Christianity until the summer of 386, and was not baptized until 387, by St. Ambrose in Milan.[15] Augustine was a teacher of rhetoric and Latin and obviously educated. Before accepting Christianity, he had absorbed philosophical and theological knowledge from the movements to which he had devoted himself, especially Manichaeanism and Neoplatonism.

St. Augustine was a prolific author, who, in the course of a long life, left to posterity a large body of literature. It is important to recognize at the outset that for Augustine, "the key constituents of Christian belief are credal statements concerning historical occurrences and, as such, lie outside the realm of the abstract, general truths accessible to philosophical reflection."[16] But credal statements form only a small part of the body of Scripture and of Christian literature and thought generally. How should a Christian approach the rest of it? Should one use the philosophical and scientific literature of the pagans, with its potentially subversive conclusions? Was such literature even relevant to Christianity? Like the Greek Fathers and Philo Judaeus before him, Augustine sought to address this monumental problem and, like many of them, adopted the "handmaiden" solution to secular learning. By so doing, he made a major contribution to the religious and intellectual life of the Middle Ages.

15. For a brief biographical section, including Augustine's works, see Frederick Copleston, *A History of Philosophy,* Vol. 2: *Medieval Philosophy: Augustine to Scotus* (Westminster, MD: The Newman Press, 1957), 40–50. For a full biography, see Peter Brown, *Augustine of Hippo, A Biography* (London: Faber, 1957).

16. Cited from R. A. Markus, "Part V: Marius Victorinus and Augustine," in A. H. Armstrong, ed., *The Cambridge History of Later Greek and Early Medieval Philosophy* (Cambridge: Cambridge University Press, 1970), 345. With regard to the resurrection, Markus cites Augustine's *On the Trinity,* bk. 4, Ch. 16.

In advising young people to avoid works suspected of error, or works associated with demons, Augustine urges that "they should also distance themselves from the study of the superfluous and dissolute arts which are of human institution," but "they should not neglect those humanly instituted arts and sciences which are of value for a proper social life,"[17] among which he includes history, natural history, logic, rhetoric, and mathematics. A few paragraphs later, Augustine admits "truthful" philosophy to the approved list in a significant passage that expresses the handmaiden theory in its most traditional form:

> If those ... who are called philosophers happen to have said anything that is true, and agreeable to our faith, the Platonists above all, not only should we not be afraid of them, but we should even claim back for our own use what they have said, as from its unjust possessors. It is like the Egyptians, who not only had idols and heavy burdens, which the people of Israel abominated and fled from, but also vessels and ornaments of gold and silver, and fine raiment, which the people secretly appropriated for their own, and indeed better, use as they went forth from Egypt; and this not on their own initiative, but on God's instructions, with the Egyptians unwittingly lending them things they were not themselves making good use of.[18]

Augustine's message is constant: Secular knowledge should not be sought for its own sake, or as an end in itself. For "all the knowledge derived from the books of the heathen, which is indeed useful, becomes little enough if it is compared with the knowledge of the divine scriptures."[19] A Christian should take only what is useful and ignore the rest. The quest for truth lay in a search for Christian wisdom. Only studies that furthered this goal ought to be pursued.

Since the handmaiden tradition clearly subordinated secular learning to the needs of the faith, reason for Augustine, and for many Christians, was not of the kind that seeks truth by following an argument wherever it may lead, at least not in matters of faith. In matters of faith, and in theology generally, the function of reason was to elucidate the faith by revealing its truth. But such a

17. From *On Christian Doctrine (De doctrina Christiana)* in *The Works of Saint Augustine, A Translation for the 21st Century*, ed. John E. Rotelle, O. S. A.: Vol. 11: *Teaching Christianity (De Doctrina Christiana)*, introduction, translation, and notes by Edmund Hill, O. P. (Hyde Park, NY: New City Press, 1996), bk. 2, 158–159, sec. 58. Hill, the translator, has changed the customary title *On Christian Doctrine* to *Teaching Christianity*. My references to this work will be in the form: *On Christian Doctrine (Teaching Christianity)*.
18. Augustine, *On Christian Doctrine (Teaching Christianity)*, 159–160, sec. 60.
19. Ibid., 62, sec. 63.

process presupposed that one accepted the faith before attempting to understand it. Faith must precede understanding. Only then could one successfully use reason and logic to understand it. In the eleventh century, Anselm of Canterbury (ca. 1033–1109), for whom Augustine was "the main source for the principles and content of his speculation,"[20] upheld this approach. As one scholar explains, when Anselm "says that men should use their reason to help them understand their faith he intends them to do so only in order to understand what they ought already to believe, and not to seek out new items of faith so as to extend the range of their beliefs."[21] It was a doctrine, however, that failed to circumscribe the theologians of the twelfth century.[22]

Augustine held logic and mathematics in high regard as belonging to the category of "rational discourse."[23] For him, "the discipline of rational discourse ... is of the greatest value in penetrating and solving all kinds of problems which crop up in the holy literature."[24] Logic was worthy because there are "many forms of argument called sophisms, false conclusions from reason which frequently look so like true ones that they can deceive not only the slow-witted but even sharp minds, when they are not paying careful attention."[25] To illustrate his point, Augustine offers this amusing example, in which one person says to another:

> 'You are not what I am.' And the other agreed; it was true after all, at least to this extent that one of them was crafty, the other simple. Then the first one added, 'But I am human.' When the other allowed this too, he concluded, 'Therefore you are not human.'[26]

For Augustine, the utility of logic for Scripture lay in its emphasis on rigorous inference, though he warned Christians that not only was it necessary to

20. A. H. Armstrong, ed., *The Cambridge History of Later Greek and Early Medieval Philosophy*, 614. The chapter on Anselm is by H. Liebeschütz.
21. G. R. Evans, *Anselm and a New Generation* (Oxford: Clarendon Press, 1980), 9.
22. Ibid., 10.
23. Bolgar, *The Classical Heritage and Its Beneficiaries* (Cambridge: Cambridge University Press, 1954), 36, explains that Augustine (and Martianus Capella, and later Cassiodorus) included logic as one of the seven liberal arts: "To include logic, which had previously been studied only in the schools of philosophy, as a basic subject for general education was to take a great step forward. It was to recognise the fact that clear thinking in private, rather than eloquence in public, was the activity which the problems of the age required."
24. Augustine, *On Christian Doctrine (Teaching Christianity)*, 153–154, sec. 48.
25. Ibid., 154, sec. 48.
26. Ibid.

draw the correct inferences, but that one also had to ascertain that the propositions are true. Augustine's own example reveals to us the way he thought logic could prove helpful to the study of Scripture. If someone assumes as an antecedent that "there is no resurrection of the dead," then it follows as a consequent "that neither has Christ risen again." Augustine agrees that

> the necessary consequent upon that antecedent statement, that there is no resurrection of the dead, is, *Neither has Christ risen again;* but this consequent is false, because Christ has risen again. Therefore the antecedent is also false, that there is no resurrection of the dead; accordingly there is a resurrection of the dead. This can all be put very briefly as follows: If there is no resurrection of the dead, then neither has Christ risen again; but Christ has risen again; therefore there is a resurrection of the dead.[27]

Augustine believed that if Christians made certain of the truth of the premises of an argument involving some Scriptural or doctrinal point, logic was a valuable tool that would enable them to infer the correct conclusions from the initial premises. So impressed was he with the "valid rules of logic" that he could not believe they were formulated by human beings. "They are," he boldly proclaimed "inscribed in the permanent and divinely instituted rationality of the universe."[28] In this Platonic interpretation, Augustine insisted that humans do not invent the valid rules of logic; they only discover them in the fabric of our rational universe. With this attitude toward logic and reason, Augustine was not reluctant to use analytic tools – especially Aristotle's categories – in his analysis of doctrinal truths, as he did in one of his greatest works, the fifteen books of *On the Trinity (De Trinitate).*[29]

If St. Augustine was the most significant influence in the process that would eventually lead to the rationalization of medieval theology, Boethius, who has been called "Last of the Romans, first of the scholastics,"[30] not only was influential in theology, where he reinforced Augustine's approach, with which he was familiar, but was responsible for composing and compiling the most important contribution to the history of reason and rationality in the early Middle Ages: the body of literature known as the old logic *(logica vetus).*

27. Ibid., 155, sec. 50.
28. Ibid., 154, sec. 50. Augustine seems also to have regarded arithmetic as part of the "divinely instituted rationality of the universe" (ibid., 158, sec. 56).
29. See Margaret Gibson, "The *Opuscula Sacra* in the Middle Ages," in Margaret Gibson, ed., *Boethius: His Life, Thought and Influence* (Oxford: Basil Blackwell, 1981), 214.
30. According to Henry Chadwick, *Boethius: The Consolations of Music, Logic, Theology, and Philosophy* (Oxford: Clarendon Press, 1981), xi, "the tag echoes a famous judgment by the humanist Lorenzo Valla in the fifteenth century."

Anicius Manlius Severinus Boethius (ca. 480–524/525) was descended from an old aristocratic Roman family.[31] In the course of his life, he was an honorary consul and served as the "master of offices" *(magisterium officiorum).* In 522, he was charged with treason by the Roman emperor, Theodoric, who imprisoned and then executed him in 524 or 525. It was during his incarceration that Boethius wrote his most famous work, *The Consolation of Philosophy.*

Boethius's importance for the early Middle Ages was immense. With a reasonable knowledge of Greek, he translated a number of Greek works into Latin and was thereby instrumental in preserving and making available numerous works that would otherwise have been unknown in the West. His reputation as a translator was emphasized in florid prose by Emperor Theodoric, who, prior to Boethius's fall from favor, invited him to supervise the conveyance of a mechanical water clock to the king of the Burgundians. In his invitation, Theodoric declared:

> From far away you entered the schools of Athens; you introduced a Roman toga into the throng of Greek cloaks; and in your hands Greek teachings have become Roman doctrine. For you have shown with what profundity speculative philosophy and its parts are studied, and with what rationality practical philosophy and its branches are investigated; and whatever wonders the sons of Cecrops bestowed upon the world, you have conveyed to the senators of Romulus. Thanks to your translations, Pythagoras the musician and Ptolemy the astronomer may be read as Italians; Nicomachus the arithmetician and Euclid the geometer speak as Ausonians; Plato the theologian and Aristotle the logician dispute in the language of the Qurinal; Archimedes the physicist you have restored to the Sicilians as a Latin: it is by your sole exertions that Rome may now cultivate in her mother tongue all those arts and skills which the fertile minds of Greece discovered."[32]

Although his goal was to translate all the works of Plato and Aristotle, a goal he never achieved, Boethius provided a substantial part of the philosophical and scientific fare for the early Middle Ages prior to the influx of

31. For a good, brief biography of Boethius, see Lorenzo Minio-Paluello, "Boethius," in Charles C. Gillispie, ed., *Dictionary of Scientific Biography,* 16 vols. (New York: Charles Scribner's Sons, 1970–1980), vol. 2, 228–236. See also John Matthews, "Anicius Manlius Severinus Boethius," in Margaret Gibson, ed., *Boethius: His Life, Thought and Influence* (Oxford: Basil Blackwell, 1981), 15–43.
32. Jonathan Barnes, "Boethius and the Study of Logic," in Margaret Gibson, ed., *Boethius: His Life, Thought and Influence* (Oxford: Basil Blackwell, 1981), 73. The letter was preserved by Cassiodorus (*Variae* I.45), and is published in *Corpus Christianorum Series Latina* (Turnhout, 1954), vol. 96.

Greco-Islamic science in the twelfth and thirteenth centuries. His most significant and influential achievement was in logic where he made a number of translations (from Greek into Latin), produced a series of commentaries on those translations, and composed five independent treatises on logic.[33] Included within this collection were at least five translations of Aristotle's logical works (*Categories, On Interpretation [De interpretatione], Sophistical Refutations, Prior Analytics,* and *Topics*).[34] Boethius also translated Porphyry's (ca. 234–304) *Introduction to Aristotle's Categories,* or *Isagoge,* as it was known. To these, we must add four commentaries on four different works: Porphyry's *Isagoge;* Aristotle's *Categories;* Aristotle's *On Interpretation;* and Cicero's *Topics.* As if this were insufficient, Boethius also wrote five independent treatises on logic, including two titled *On Categorical Syllogism* and *On Hypothetical Syllogisms.*[35] With the exception of his translations of Aristotle's *Sophistical Refutations, Prior Analytics,* and *Topics,* which did not really circulate until the twelfth century, the works of Boethius cited here, along with some works relevant to logic that were written prior to Boethius (for example, the *Topics* of Cicero [106–43 B.C.] and the *De definitionibus [On Definitions]* of Marius Victorinus [fl. 350–60]), were collectively known as the *logica vetus,* or the "old logic."

By his monumental achievement, Boethius guaranteed that logic, the most visible symbol of reason and rationality, remained alive at the lowest ebb of European civilization, between the fifth and tenth centuries. When, in the course of the eleventh century, the new Europe was emerging and European scholars, for reasons we may never confidently know, were aroused to an interest in logic and reason, the legacy of Boethius's "old logic" was at hand to make the revival possible, and was perhaps even instrumental in generating it. As Jonathan Barnes has expressed it, "Boethius' labours gave logic half a millenium of life: what logician could say as much as that for his work? what logician could desire to say more?"[36]

Boethius was also the author of five tractates on theology, and, even more than Augustine, applied reason and logic to that discipline. In *On the Trinity (De Trinitate),* Boethius used reason and logic to elucidate the Christian doctrine of the Trinity, a doctrine that lay at the heart of Western Christianity

33. For lists of these different categories of logical works, see Barnes, "Boethius and the Study of Logic," 85–86; Margaret Gibson, "Latin Commentaries on Logic before 1200," *Bulletin de philosophie médiévale* 24 (1982), 58–60; and Minio-Paluello, "Boethius," vol. 2, 230.
34. Here I follow Lorenzo Minio-Paluello, "Boethius," vol. 2, 230.
35. Minio-Paluello, ibid., 230; Gibson, "Latin Commentaries on Logic," 60.
36. Barnes, "Boethius and the Study of Logic," 85.

and was also the source of numerous heresies. Boethius believed that an inquiry ought to be pursued "only so far as the insight of man's reason is allowed to climb the height of heavenly knowledge."[37] One should apply reason to the ultimate limit that is possible, even if one fails in the attempt. After all, "Medicine ... does not always bring health to the sick, though the doctor will not be to blame if he has left nothing undone which should have been done."[38] In his explication of the Trinity, Boethius used the logical categories of difference *(differentia)*, genus *(genus)*, and species *(species)* to express the concepts of number and diversity.[39] He also employed the doctrine of predication, mentioning the nine categories that are predicable of a substance[40] and then explaining why these are not predicable of God. He concludes that in God there is no plurality because there are no differences; but "where there is no plurality there is unity."[41] Although he applied reason to faith, Boethius explains that he has done so "to an article which stands quite firmly by itself on the foundation of Faith."[42]

Substance and predication with respect to God and the Trinity play an important role in Boethius's theological tractates. In the second tractate, "Whether Father, Son, and Holy Spirit are substantially predicated of the divinity" *(Utrum pater et filius et spiritus sanctus ...)*, Boethius argues: "Everything, therefore, that is predicated of the divine substance must be common to the Three."[43] Thus, God can be predicated of the Father, the Son, and the Holy Spirit. But "what may be predicated of each single one but cannot be said of all is not predicated substantially, but in some other way.... For he who is Father does not transmit this name to the Son nor to the Holy Spirit. Hence it follows that this name is not attached to him as something substantial; for if it were substantial, as God, truth, justice, or substance itself, it would be affirmed of the other Persons."[44]

Boethius was ever ready to apply aspects of logic to theology, even in the form of mathematics. In the third tractate ("How substances are good in

37. *Boethius, The Theological Tractates,* with an English translation by H. F. Stewart, E. K. Rand, and S. J. Tester; *the Consolation of Philosophy,* with an English translation by S. J. Tester (London: William Heinemann Ltd.; Cambridge MA: Harvard University Press, 1973), Introduction, 5.
38. Boethius, *De Trinitate* I in *The Theological Tractates.*
39. *De Trinitate* I, *The Theological Tractates,* 7.
40. *De Trinitate* IV, ibid., 17.
41. *De Trinitate* V, ibid., 29.
42. *De Trinitate* VI, ibid., 31.
43. Boethius, *Utrum pater* in *The Theological Tractates,* 33.
44. *Utrum pater,* ibid., 35.

virtue of their existence without being substantial goods") *(Quomodo substantiae...)*, also known as *De hebdomadibus*,[45] Boethius proclaims that he has "followed the example of the mathematical and cognate sciences and laid down bounds and rules according to which I shall develop all that follows."[46] He promptly presents nine axioms and definitions from which he derives his conclusion. The first of his axioms is of great interest. Here, Boethius describes what he calls a "common conception of the mind" *(communis animi conceptio)*, which is "a statement which anyone accepts as soon as he hears it."[47] He distinguishes two kinds. The first kind is possessed by everyone. An example of this type is: "If you take equals from two equals, the remainders are equal,"[48] which is very similar to the axioms, or common notions, in Euclid's *Elements*.[49] Since Cassiodorus attributes a translation of Euclid's *Elements* to Boethius,[50] the use of this particular axiom may derive from Boethius's familiarity with the *Elements*. Indeed, the expression *communis animi conceptio* appears in Campanus of Novara's popular thirteenth-century version of Euclid's *Elements*, where it represents Euclid's axioms.[51]

45. Chadwick, *Boethius, The Consolations of Music, Logic, Theology, and Philosophy*, 203, explains: "In the middle ages 'De Hebdomadibus' was mistakenly taken to be the third tractate itself, and commentators offered fanciful explanations of the mysterious word." It is often cited as *De hebdomadibus*, but will be cited here as *Quomodo substantiae*.
46. Boethius, *Quomodo substantiae*, in *The Theological Tractates*, 41.
47. Boethius, *Quomodo substantiae*, ibid.
48. Boethius, *Quomodo substantiae*, ibid.
49. Among the five axioms, or "common notions," cited in Sir Thomas L. Heath's translation, Euclid has (1) "Things which are equal to the same thing are also equal to one another"; (2) "If equals be added to equals, the wholes are equal"; and (3) "If equals be subtracted from equals, the remainders are equal." See *The Thirteen Books of Euclid's "Elements,"* translated by Sir Thomas L. Heath, 3 vols., 2nd ed., revised with additions (New York: Dover Publications, 1956), vol. 1, 155. Boethius's axiom is not among Euclid's, though it is a mere variation on the same theme. See also Marshall Clagett, *Greek Science in Antiquity* (London: Abelard-Schuman Ltd., 1957), 152–153.
50. See John E. Murdoch, "Euclid: Transmission of the Elements," in *Dictionary of Scientific Biography*, vol. 4 (1971), 443. Despite the implication that a complete translation was made, the Euclidean geometry attributed to Boethius is fragmentary and incomplete (ibid).
51. See *Euclidis Megarensis mathematici clarissimi Elementorum geometricorum libri XV. Cum expositione Theonis in priores XIII a Bartholomaeo Veneto Latinitate donata. Campani in omnes, et Hypsicles Alexandrini in duos postremos* (Basel, 1546), 3. Boethius's axiom does not appear among Campanus's nine axioms, although it is similar to them. On Campanus's version of Euclid's *Elements*, see Murdoch, "Euclid," 4: 446–447, and G. J. Toomer, "Campanus of Novara," in *Dictionary of Scientific Biography*, 3, 24–25.

In addition to the "common conception of the mind" that anyone can understand, there is a second kind that "is intelligible only to the learned," as, for example, "Things which are incorporeal are not in space." Such conceptions, Boethius insists, "are approved as obvious to the learned but not to the common herd."[52] After enunciating his nine axioms, Boethius chooses not to invoke them, apparently convinced that "the intelligent interpreter of the discussion will supply the arguments appropriate to each point."[53] If Boethius really thought that his readers could invoke each axiom at the appropriate point in the treatise, he must have held them in high regard.

Boethius reinforced St. Augustine's emphasis on the application of reason to theology. Logic and mathematics were his models, and he used the former extensively, especially ideas about predication. It may be no exaggeration to claim, as does Henry Chadwick, that Boethius's third (theological) tractate

> taught the Latin West, above all else, the method of axiomatization, that is, of analysing an argument and making explicit the fundamental presuppositions and definitions on which its cogency rests. He taught his successors how to try to state truths in terms of first principles and then to trace how particular conclusions follow therefrom. The West learnt from him demonstrative method.[54]

The evidence from Boethius's logical and theological treatises leaves little doubt that he placed great value on reason and reasoned argument. It is fitting that his crowning work, the work for which he is most famous, *The Consolation of Philosophy,* written as he was imprisoned prior to his execution in 524 or 525, exhibits the same emphasis on reason that characterized all his works. The work is extraordinary because, although Boethius was a Christian, he gives no indication of it in a treatise written under great duress and in which it would have seemed quite natural to include direct and indirect references to his faith. But in treating the traditionally vexing dilemmas of faith – free will, providence, evil, God's foreknowledge of events, and others – Boethius played the role of the traditional philosopher and sought to resolve these problems by reason alone.

Boethius's influence on the early Middle Ages was immense, as is readily apparent in Henry Chadwick's pertinent assessment:

52. Boethius, *Quodmodo substantiae, The Theological Tractates,* 41.
53. Boethius, *Quodmodo substantiae,* ibid., 43.
54. Henry Chadwick, *Boethius,* 210.

In the twelfth century schools his influence reached its peak. His works became central to the syllabus of instruction, and strongly stimulated that thoroughgoing study of logic for its own sake which becomes so prominent a hallmark of the mediaeval schools. The *opuscula sacra* taught the theologians that they did not necessarily need to fear the application of rigorous logic to the traditional language of the Church. He made his readers hungry for even more Aristotle, and prepared the welcome given to the new twelfth century translations of the *Analytics* and the *Topics,* although his own versions were scarcely known at all. From the first of the *opuscula sacra* mediaeval philosophers learnt to draw up a hierarchy of the sciences and to see the different departments of knowledge, now being pursued together in community as the newly founded universities set themselves to their common task, as an organized and coherent scheme in which the various parts could be seen to be rationally related to each other.[55]

REASON AND LOGIC IN THE TWELFTH CENTURY

No one better illustrates the spirit of the twelfth century, with its emphasis on learning and its special fascination with logic, than does Peter Abelard (1079–1142). Peter's father was a soldier who had acquired some learning before entering upon a military career. He therefore wished his sons to have some learning before they too became soldiers. In his famous autobiography, *The Story of My Misfortunes (Historia Calamitatum),* Peter Abelard explains:

> As he loved me the more, being his first-born, so he saw to it that I was carefully instructed. The further I went in my studies and the more easily I made progress, the more I became attached to them and came to possess such a love of them that, giving up in favor of my brothers the pomp of military glory along with my right of inheritance and the other prerogatives of primogeniture, I renounced the field of Mars to be brought up at the knee of Minerva. Since I preferred the armor of logic to all the teaching of philosophy, I exchanged all other arms for it and chose the contests of disputation above the trophies of warfare. And so, practising logic I wandered about the various provinces wherever I heard the pursuit of this art was vigorous and became thereby like the peripatetics.[56]

55. Ibid., 252–253. For a similar endorsement of Boethius's importance for the application of logic to theology, see John Marenbon, *From the Circle of Alcuin to the School of Auxerre: Logic, Theology and Philosophy in the Early Middle Ages* (Cambridge: Cambridge University Press, 1981), 18–19.

56. J. T. Muckle, trans., *The Story of Abelard's Adversities, A Translation with Notes of the "Historia Calamitatum,"* with a preface by Etienne Gilson (Toronto: Pontifical Institute of Mediaeval Studies, 1964), 11–12.

Early Stirrings in the Ninth to Eleventh Centuries

Thus was Abelard filled with zeal for learning, especially logic. As we shall see, he was only one of many who shared this enthusiasm. Traveling around from school to school, seeking the best teachers of logic, was a common feature of twelfth-century intellectual life. But it did not spring full-blown. During the eighth to tenth centuries, there had been stirrings in Europe. According to John Marenbon, "At the courts of Charlemagne and Charles the Bald, and in the monasteries of Corbie and Auxerre, men of the early Middle Ages made their first attempts to grapple with abstract problems by the exercise of reason."[57] John Scotus Eriugena (ca. 810; d. after 877), the Irish philosopher and theologian who went to France to serve the court of King Charles the Bald, is one of the most significant early European thinkers emphasizing reason. In his great work, *On the Division of Nature (De divisione naturae),* Eriugena declares:

> For authority proceeds from true reason, but reason certainly does not proceed from authority. For every authority which is not upheld by true reason is seen to be weak, whereas true reason is kept firm and immutable by her own powers and does not require to be confirmed by the assent of any authority. For it seems to me that true authority is nothing else but the truth that has been discovered by the power of reason and set down in writing by the Holy Fathers for the use of posterity.[58]

Few utterances about reason in the later Middle Ages would equal in power this declaration by John Scotus Eriugena.

Reason would receive another boost from the teaching of logic. In the late tenth century, Gerbert of Aurillac (ca. 946–1003), who would become Pope Sylvester II (999–1003), was the most famous teacher in Europe. For ten years (972–982), he taught the seven liberal arts at the cathedral school of Rheims. Among these subjects, Gerbert regarded logic with special favor. He may have been the first to teach the works of the old logic, including Boethius's commentaries and original treatises.[59] These works

57. Marenbon, *From the Circle of Alcuin,* 139. Marenbon's book shows what these very early beginnings were like.

58. *Periphyseon (De divisione naturae) liber primus,* ed. and trans. I. P. Sheldon-Williams, with the collaboration of Ludwig Bieler in *Scriptores latini Hiberniae,* 7 (Dublin, 1968), 199. Cited from Peter Abelard, *A Dialogue of a Philosopher with a Jew, and a Christian,* translated by Pierre J. Payer (Toronto: Pontifical Institute of Mediaeval Studies, 1979), 82–83, n. 136.

59. See Southern, *The Making of the Middle Ages* (New Haven: Yale University Press, 1953), 175. Also on Gerbert, see Henry Osborn Taylor, *The Mediaeval Mind,* 2 vols., 4th ed. (Cambridge, MA: Harvard University Press, 1951), vol. 1, 282–294.

had been lying around Europe for centuries, but until Gerbert focused attention on them, they had been little used. Although Gerbert valued logic and emphasized it, he did not elevate it above the other liberal arts, as would happen in the twelfth century. Indeed, he made it subordinate to rhetoric. Gerbert was an intellectual star whose light shone throughout the eleventh century as his students – among them Adalberon of Laon, John of Auxerre, and Fulbert of Chartres, the most famous of them – disseminated his love of learning and teaching methods throughout northern Europe.[60]

Why did logic, of all subjects, emerge in such a forceful and dominant manner by the twelfth century? We might well ask, as R. W. Southern did,

> what it was that the study of logic, which the influence of Boethius did so much to foster, contributed to the intellectual formation of the Middle Ages, and why it was that, from the time of Gerbert, this study assumed an importance which it had never previously attained in the Latin world. The works of Boethius are immensely difficult to understand and repellent to read. Why should the subject have taken such a hold on the imaginations of scholars, so that they pursued it with unflagging zeal through all the obscurities of translation, heedless of the advice of many cautious men of learning?[61]

Perhaps this extraordinary phenomenon is partially explicable by the fact that prior to the translations of Greco-Islamic science and natural philosophy, in the mid–twelfth century, there was a paucity of treatises available in natural philosophy, theology, medicine, and law. The old logic, which had been handed down from Boethius, may have helped fill an intellectual void and provide some fare for hungry intellects. In the chaotic political and economic world of the tenth and eleventh centuries, perhaps logic "opened a window on to an orderly and systematic view of the world and of man's mind."[62] In its rigor and organization, logic stood in sharp contrast to the disarray of subjects like theology and law, which over the centuries had become filled with contradictions and inconsistencies. Logic was a model for the simplification and more rigorous organization of these vital subject areas, as will be seen later in this chapter. But it was indispensable for the study of philosophy, as Hugh of St. Victor explained when he declared that

60. See Southern, *The Making of the Middle Ages*, 177.
61. Ibid., 179.
62. Ibid., 180.

logic came last in time, but is first in order. It is logic which ought to be read first by those beginning the study of philosophy, for it teaches the nature of words and concepts, without both of which no treatise of philosophy can be explained rationally.[63]

Few would challenge the claim that the greatest representative of the new approach in the twelfth century was Peter Abelard, undoubtedly the most exciting and significant thinker of that century. Not only did he leave a profound impact on logic and theology, and in the application of logic to theology, but his famous love affair with Héloise, and the correspondence that emerged from it, along with his autobiography, *Historia Calamitatum,* or *The History of My Misfortunes,* have made him a romantic figure of near legendary dimensions.[64] In this study, I shall focus attention on Abelard's attitude toward the role of reason in human activity.

Abelard left works on logic and theology, the subjects for which he is famous. As a logician, he contributed significantly to the great ancient and medieval debate concerning universals, a problem about the significance of general terms, such as animal and horse – that is, terms that signify a genus or species. Are such terms mere names or do they represent something real? Do they exist or not?[65] As a logician, Abelard could hardly avoid the use of reason. His writings on logic, as well as those of his twelfth-century colleagues, are a testimony to the central role that reason, embodied in logic and dialectics, came to play.

John of Salisbury (ca. 1115–1180) on the Role of Logic

No scholar of the twelfth century has captured the power of logic in the educational system of the twelfth century better than John of Salisbury. John was an eminent figure in his own right. He had studied under the great masters of his day, among them Peter Abelard, of whom he spoke in

63. *Didascalicon,* bk. 1, ch. 11, in *The "Didascalicon" of Hugh of St. Victor: A Medieval Guide to the Arts,* translated from the Latin with an Introduction and Notes by Jerome Taylor (New York: Columbia University Press, 1961), 59. The *Didascalicon* was written in the 1120s.
64. In 1988, a movie of Abelard's life was made titled *Stealing Heaven,* starring Derek de Lint as Abelard. Although the movie is focused on the passionate romance between Abelard and Héloise, there is some discussion of theology in scenes involving Abelard as a teaching master.
65. On the problem of universals, see Julius Weinberg, *A Short History of Medieval Philosophy* (Princeton: Princeton University Press, 1964), Chapter 5: "Abelard and the Problem of Universals," 72–91.

admiring terms,[66] and was thoroughly knowledgeable about the seven liberal arts, especially logic. He also studied theology and was ordained a priest. He became secretary to the archbishop of Canterbury, Theobald, who upon dying in 1161 was succeeded by Thomas Becket, whose murder John witnessed on December 29, 1170. In 1176, John was made bishop of Chartres, where he died on October 25, 1180. The most influential of his many significant works was the *Metalogicon,* a work on educational theory. It is in this famous work, completed in 1159, that John informs us about the role of logic in medieval education.

The *Metalogicon* is directed against an individual whom John calls "Cornificius," who represents a group he calls the "Cornificians," those who opposed the study of the trivium (grammar, rhetoric, and logic). The Cornificians "would do away with logic," because they regard it as "the fallacious profession of the verbose, which dissipates the natural talents of many persons, blocks the gateway to philosophical studies, and excludes both sense and success from all undertakings."[67]

It is against the Cornificians and their attitude that John declares, in the prologue to the first book of the *Metalogicon,* that his purposes is "to defend logic."[68] At the outset, John sings the praises of "nature, the most loving mother and wise arranger of all that exists," who has "elevated man by privilege of reason, and distinguished him by the faculty of speech." Thus privileged, human beings are able to rise "on wings of reason and speech ... to outstrip all other beings, and to attain the crown of true happiness."[69] On a more mundane level, John defends logic against its detractors by appeal to St. Augustine, who "praised logic so highly that only the foolhardy and presumptuous would dare to rail against it." Not mincing words, John declares that "since logic has such tremendous power, anyone who charges that it is foolish to study this [art], thereby shows himself to be a fool of fools."[70]

66. See *The "Metalogicon" of John of Salisbury: A Twelfth-Century Defense of the Verbal and Logical Arts of the Trivium,* translated with an Introduction and Notes by Daniel D. McGarry (Berkeley/Los Angeles: University of California Press, 1955), bk. 2, ch. 10, 95, where John declares that as a youth, he betook himself "to the Peripatetic of Pallet [i.e., Abelard], who was then teaching at Mont Ste. Geneviève. The latter was a famed and learned master, admired by all. At his feet I learned the elementary principles of this art, drinking in, with consuming avidity, and to the full extent of my limited talents, every word that fell from his lips. After his departure, which seemed to me all too soon, I became the disciple of Master Alberic."

67. John of Salisbury, *Metalogicon,* bk. 1, ch. 9, 31.

68. Ibid., Prologue, 5.

69. Ibid., bk. 1, ch. 1, 9.

70. Ibid., bk. 4, ch. 25, 241.

The significance of logic lies in its overall role in philosophy. "Among the various branches of philosophy," John explains, "logic has two prerogatives: it has both the honor of coming first and the distinction of serving as an efficacious instrument throughout the whole body [of philosophy]."[71] Those who actually do philosophy, the natural and moral philosophers, "can construct their principles only by the forms of proof supplied by logicians." If they succeed without the use of logic, "their success is due to luck, rather than to science," for "logic is 'rational' [philosophy]." Without logic, even a very bright person "will be greatly handicapped in philosophical pursuits if he is without a rational system whereby he may accomplish his purpose."[72] John explains further that logic came into existence because

> there was [evident] need of a science to discriminate between what is true and what is false, and to show which reasoning really adheres to the path of valid argumentative proof, and which [merely] has the [external] appearance of truth, or, in other words, which reasoning warrants assent, and which should be held in suspicion. Otherwise, it would be impossible to ascertain the truth by reasoning.[73]

Not only does John pay tribute to logic as an essential and vital part of philosophy, but he also presents a vivid sense of its appeal as a subject of study in the twelfth century. Logic is

> such an important part of philosophy that it serves the other parts in much the same way as the soul does the body. On the other hand, all philosophy that lacks the vital organizing principle of logic is lifeless and helpless. It is no more than just that this art should, as it does, attract such tremendous crowds from every quarter that more men are occupied in the study of logic alone than in all the other branches of that science which regulates human acts, words, and even thoughts, if they are to be as they should be. I refer to philosophy, without which everything is bereft of sense and savor, as well as false and immoral.[74]

As important as logic was in John's estimation, he did not view it as an end in itself. "By itself," he insisted, "logic is practically useless. Only when it is associated with other studies does logic shine, and then by a virtue that is communicated by them."[75] John of Salisbury was no Peter Abelard. He

71. Ibid., bk. 2, ch. 5, 82.
72. Ibid.
73. Ibid., bk. 2, ch. 2, 76.
74. Ibid., bk. 2, ch. 6, 84.
75. Ibid., bk. 4, ch. 28, 244; see also bk. 2, chs. 10–11, 100–101.

thought that logic and reason should not be applied to the divine mysteries. Citing Ecclesiasticus 3.22, John advises his readers not to seek after things that lie beyond our reach and comprehension and not rashly to "discuss the secrets of the Divine Trinity and mysteries whose vision is reserved for eternal life."[76] Since Peter Abelard was one who did analyze the Trinity, it is likely that John had him in mind.

THEOLOGY

If John had reservations about applying logic and dialectic to the divine mysteries and to theology in general, many other twelfth-century scholars sided with Abelard and thought it appropriate, and even necessary, to apply logic and reason to the mysteries of the faith. We must now see how the new inquisitive mentality of the twelfth century affected theology.

Theology Turns to Reason: Berengar of Tours, Anselm of Canterbury, and Peter Abelard

Perhaps the most impressive sign that reason was emerging in the eleventh century as a force to be reckoned with is its adoption by some theologians as an indispensable instrument for the interpretation of the mysteries of faith. Berengar of Tours (ca. 1000–1088), who lectured on the Eucharist at Church schools in Tours around 1030 to 1040, insisted that reason should be applied to matters of faith because reason is a gift of God. Evidence was more important than authority in theology. Berengar's views appear in a treatise titled "A Rejoinder to Lanfranc" *(Rescriptum contra Lanfrannum),* which Berengar wrote around 1065 against Lanfranc of Bec, who had attacked his interpretation of the Eucharist.[77] In his reply to Lanfranc, Berengar used parts of the old logic that involved the theories of predicables and categories.[78] He believed that reason could not support the view that accidents can exist apart from a substance. But the Eucharist assumes that the accidents of the bread do not subsist in a substance after the consecration. That is, the bread no longer exists, so that its accidents can no longer inhere in it. Because he judged it unreasonable, Berengar denied that the accidents of the bread could exist independently of their substance. He

76. Ibid., bk. 4, ch. 41, 272.
77. The issues between Berengar and Lanfranc and the roles they played in theology are described by Toivo Holopainen, *Dialectic and Theology in the Eleventh Century* (Leiden: E. J. Brill, 1996). On Lanfranc, see Chapter 3, 44–76; on Berengar, Chapter 4, 77–118.
78. Ibid., 92–93.

therefore assumed that the form of the bread continued to exist after it was consecrated and that another, second, form, which is actually the body of Christ, is added to the form of the bread. Thus did Berengar deny the act of transubstantiation.[79]

In his dispute with Lanfranc, Berengar clearly indicates his strong conviction that reason is a proper instrument for doing theology. "I say," he declares against Lanfranc,

> you do not hesitate to write of me that I desert sacred authorities; through divine mercy it will become manifest that this is a false accusation and not the truth, as soon as the moment comes when sacred authorities are to be brought forward out of the necessity of using that point of departure; nevertheless, it is incomparably superior to act by reason in the apprehension of truth; because this is evident, no one will deny it except a person blinded by madness.[80]

Berengar declares his sense of shame at those who rely on authority in matters where reason is the ultimate authority.[81] He regards the use of reason as a virtually religious obligation, as is evident when he declares that

> [i]t is clearly the property of a great heart to have recourse to dialectic in all things, because to have recourse to dialectic is to have recourse to reason; and he who refuses this recourse, since it is in reason that he is made in the image of God, abandons his glory, and cannot be renewed from day to day in the image of God.[82]

Thus, Berengar believed that dialectic, which was the embodiment of reason, was a powerful and legitimate tool that should be used in all matters where it was relevant, including the Christian faith. Berengar, however, did not oppose the use of authoritative Christian writings, which he regarded as the proper repository of Church doctrine. But in order to discover the genuine and original meaning of those writings, they had to be read intelligently with the aid of reason. Thus, Berengar may be regarded as "a representative of the Augustinian programme of faith in search of understanding: he applies his reason to revealed doctrine, as it is conveyed by the sacred authorities, not in order to demolish it but in order to arrive at a coherent interpretation of it as a whole."[83]

79. See Etienne Gilson, *History of Christian Philosophy in the Middle Ages* (London: Sheed and Ward, 1955), 615, n. 41.
80. Translated by Holopainen, *Dialectic and Theology*, 109.
81. Ibid., 109, and n. 86 for the Latin text.
82. Ibid., 116, and n. 110 for the Latin text.
83. Ibid., 118.

Roscelin of Compiègne (ca. 1050–ca. 1125) was alleged to have applied logic and nominalism to the Trinity, inquiring "why the Son was incarnate and not the Father or the Holy Spirit, and how, if God is one, it was possible for one Person to be incarnate and not the other two."[84] It was this kind of boldness – something new in the eleventh century[85] – that alarmed traditional theologians, who feared the uses to which reason, in the form of logic, or dialectic, could be put. Lanfranc of Bec (ca. 1010–1089) opposed Berengar and regarded his use of dialectic as an abuse of the discipline. Berengar, however, has been characterized as the first "to stir up a major theological controversy in which all parties used the dialectical method, thus contributing greatly to its extension."[86]

Although many probably viewed this phenomenon with approval, some in the eleventh century found it abhorrent, as did Saint Peter Damian (1007–1072), who rejected the liberal arts as useless and objected to the application of logic to any aspect of the faith. To show his contempt for logic, he allowed that God could undo an historical event by willing that it had not happened. Although this would produce a contradiction, and strike at the heart of logic, Peter accepted it as an indication of the superiority of faith over logic.[87]

The application of reason to theological problems eventually won widespread support in the twelfth century because the basis for such a dramatic move had been clearly laid in the eleventh century. Saint Anselm of Canterbury (1033–1109), who, after being prior, and then abbot, of the abbey of Bec, became archbishop of Canterbury in 1093, was one of those most responsible for the new emphasis on reason in theology. Like many

84. This is a summary of Roscelin's views by Evans, *Old Arts and New Theology* (Oxford: Clarendon Press, 1980), 198.

85. Ibid., 2.

86. David Knowles, *The Evolution of Medieval Thought* (Baltimore: Helicon Press, 1962), 95–96.

87. Copleston, *History of Philosophy*, Vol. 2: *Medieval Philosophy: Augustine to Scotus* (Westminster, MD: The Newman Press, 1957), 145–146. The issue Peter Damian raised about whether God could undo the past was destined to become a major theological focal point in the thirteenth and fourteenth centuries. As William Courtenay put it: "The question of whether God has the power to make a past thing never to have been was considered by most theologians in the thirteenth and fourteenth centuries to be an eminently debatable and fruitful topic." See William J. Courtenay, "John of Mirecourt and Gregory of Rimini on Whether God can Undo the Past," in *Recherches de théologie ancienne et médiévale* 40 (Louvain, 1973), 165; reprinted in William J. Courtenay, *Covenant and Causality in Medieval Thought: Studies in Philosophy, Theology and Economic Practice* (London: Variorum Reprints, 1984), VIIIb.

who followed this path, Anselm insisted that belief was an indispensable prerequisite for understanding faith, an Augustinian view which he proclaimed in his *Proslogium* with these words:

I do not attempt, O Lord, to penetrate Thy profoundity, for I deem my intellect in no way sufficient thereunto, but I desire to understand in some degree Thy truth, which my heart believes and loves. For I do not seek to understand, in order that I may believe; but I believe, that I may understand. For I believe this too, that unless I believed, I should not understand.[88]

In order to understand, however, Anselm used reason extensively in his theology. He thought it important to explain rationally what he believed. In this spirit, Anselm's "confidence in reason's power of interpretation is unlimited," so that he "did not shrink from the task of proving the necessity of the Trinity and the Incarnation."[89] His three proofs of God's existence in the *Monologium* are based solely on reason, as is his subsequent simpler proof, the famous ontological proof of the existence of God, in his *Proslogium.*

The *Monologium* is a remarkable treatise. Anselm explains in the opening lines that he wrote it at the request of his fellow monks, who asked that he produce a meditation for them that did not draw upon Scripture, but was based solely on reason:

Certain brethren have often and earnestly entreated me to put in writing some thought that I had offered them in familiar conversation, regarding meditation on the Being of God, and on some other topics connected with this subject, under the form of a meditation on these themes. It is in accordance with their wish, rather than with my ability, that they have prescribed such a form for the writing of this meditation; in order that nothing in Scripture should be urged on the authority of Scripture itself, but that whatever the conclusion of independent investigation should declare to be true, in an unadorned style, with common proofs and with a simple argument, be briefly enforced by the cogency of reason, and plainly expounded in the light of truth.[90]

Anselm says that he sought to meet that request to the fullest extent possible.[91] In the *Monologium,* Anselm argues as if he had never heard of the

88. Translated by Copleston, ibid., vol. 2, 156.
89. Gilson, *History of Christian Philosophy in the Middle Ages,* 129.
90. St. Anselm, *Proslogium; Monologium; An Appendix in Behalf of the Fool Guanilon; and Cur Deus Homo,* translated from the Latin by Sidney Norton Deane, with an Introduction, Bibliography, and Reprints of the Opinions of Leading Philosophers and Writers on the Ontological Argument (La Salle, IL: The Open Court Publishing Co., 1944), 35.
91. Anselm, *Monologium,* 36; Holopainen, *Dialectic and Theology,* 121.

Christian faith, an attitude he carried over into his next treatise, *Proslogium,* where he informs the reader that

I began to ask myself whether there might be found a single argument which would require no other for its proof than itself alone; and alone would suffice to demonstrate that God truly exists, and that there is a supreme good requiring nothing else, which all other beings require for their existence and well-being.[92]

To meet the challenge he had posed for himself, Anselm fashioned a famous proof for the existence of God, a proof known subsequently as the ontological argument. In the *Proslogium,* Anselm's argument appears in Chapters 2 and 3. In light of its fame and the highly rationalistic nature of the argument, I shall quote all of the two chapters that are relevant:

Chapter II

And so, Lord, do thou who dost give understanding to faith, give me, so far as thou knowest it to be profitable, to understand that thou art as we believe; and that thou art that which we believe. And, indeed, we believe that thou art a being than which nothing greater can be conceived. Or is there no such nature, since the fool hath said in his heart, there is no God? (Psalms xiv.1) But, at any rate, this very fool, when he hears of this being of which I speak – a being than which nothing greater can be conceived – understands what he hears, and what he understands is in his understanding; although he does not understand it to exist...

Hence, even the fool is convinced that something exists in the understanding, at least, than which nothing greater can be conceived. For, when he hears of this, he understands it. And whatever is understood, exists in the understanding. And assuredly that, than which nothing greater can be conceived, cannot exist in the understanding alone. For, suppose it exists in the understanding alone: then it can be conceived to exist in reality; which is greater.

Therefore, if that than which nothing greater can be conceived, exists in the understanding alone, the very being, than which nothing greater can be conceived, is one, than which a greater can be conceived. But obviously this is impossible. Hence, there is no doubt that there exists a being, than which nothing greater can be conceived, and it exists both in the understanding and in reality.

Chapter III

And it assuredly exists so truly, that it cannot be conceived not to exist. For, it is possible to conceive of a being which cannot be conceived not to exist; and this is greater than one which can be conceived not to exist. Hence, if that, than which nothing greater can be conceived, can be conceived not to exist, it is not that, than which

92. Anselm, *Proslogium,* Preface, 1.

nothing greater can be conceived. But this is an irreconcilable contradiction. There is, then, so truly a being than which nothing greater can be conceived to exist, that it cannot even be conceived not to exist; and this being thou art, O Lord, our God.[93]

A modern summary account of Anselm's argument will aid in understanding this important proof and convey something of its logical flavor:

The definition of God, in Whom all Christians believe, contains the statement that God is a Being than which no greater can exist. Even the Fool in the Psalm (Ps. 14:1), who said that there was no God, understood what was meant by God when he heard the word, and the object thus defined existed in his mind, even if he did not understand that it exists also in reality. But if this being has solely an intramental existence, then another can be thought of as having real existence also, that is, it is greater (by existence) than the one than which no greater can exist. But this is a contradiction in terms. Therefore the Being than which no greater can be conceived exists both in the mind and in reality.[94]

The proof was attacked almost as soon as it appeared. Its significance, however, may be measured by the fact that philosophers down through the centuries have argued about its meaning and validity.[95]

Anselm's application of rigorous reasoning to theology has earned him the title "father of scholasticism." He was apparently the first to treat theology in a manner sufficiently rigorous to lay the foundations for its conversion to a science by twelfth- and thirteenth-century theologians.[96] Comparing Anselm to Berengar shows how far he had come. Reason was paramount for both, but "for Berengar, the primary task of reason in theology is to function as a means of interpreting the authoritative writings of the Church. For Anselm, the primary task of reason in theology is to construct rational demonstrations for articles of faith. Because of his rational method, Anselm appears to be more of a rationalist than the schoolmaster of Tours."[97]

93. Anselm, *Proslogium,* 7–9.
94. Knowles, *Evolution of Medieval Thought,* 103.
95. For a brief summary of the history of the proof, see Knowles, *Evolution of Medieval Thought,* 103–106. For a translation of the proof and for specific reactions of philosophers, see Alvin Plantinga, ed., *The Ontological Argument: From St. Anselm to Contemporary Philosophers,* with an Introduction by Richard Taylor (Garden City, NY: Doubleday & Co., 1965).
96. See Gilson, *History of Christian Philosophy in the Middle Ages,* 139.
97. Holopainen, *Dialectic and Theology,* 132. Holopainen gives an excellent summary of Anselm's ontological proof (133–155).

By the beginning of the twelfth century, reason and its most potent formal manifestation, logic, had become significant and formidable features of the intellectual life of Western Europe. Although authorities were still respected, and often venerated, many scholars were eager to employ reason and evidence, rather than invoke authority, to resolve problems and better to understand the physical and spiritual worlds. As we shall see, this profound change in attitude toward reason and authority, which began in the eleventh century but was given a firm base in the twelfth century, must not be viewed as a rebellion against the faith and revelation. It was not an effort to replace revelation with reason. That momentous change of attitude would not become a feature of European thought until the seventeenth century.

During the twelfth century and the later Middle Ages, reason was applied to revelation and to doctrinal matters of faith in dramatic ways and, of course, there was also the whole physical world to which reason could be applied. Although we have seen preliminary stirrings in the eleventh century, the substantial beginnings of this dramatic movement occur in the twelfth century, even before the impact of the new translations of Greco-Islamic logic, science, and natural philosophy, which had been felt in Europe as early as the first quarter of the twelfth century. The new sense of rationality would have major repercussions in natural philosophy, theology, law, and medicine, in the disciplines that would form the basis of the four faculties of major universities from the thirteenth century to the end of the Middle Ages.

But where it is obvious that reason lies at the heart of a text on logic, theology was quite another problem. Peter Abelard, who was famous for his writings on logic, would also apply logic to theology and thus continue the tradition initiated by Anselm and his colleagues in the eleventh century. Judging by the reaction he caused, Abelard went far beyond.

Many have debated whether Abelard was a rationalist. Although some scholars have thought so,[98] the recent consensus holds that he was not a rationalist in the modern sense. He did not use reason to challenge the authority of the Church, or attempt to repudiate doctrinal beliefs. In a letter to Héloise, after he had been condemned for his theological views at Sens in 1141, he declared: "I will never be a philosopher, if this is to speak against St. Paul; I would not be an Aristotle, if this were to separate me from Christ."[99] Abelard believed strongly that philosophy should be used

98. For a brief description of the attitudes of historians of philosophy toward Abelard, see Knowles, *Evolution of Medieval Thought,* 122–123.

99. Translated by David Knowles, ibid., 123. Translated from *Epistolae ad Heloissam,* letter 17 in the *Patrologiae cursus completus, Series Latina,* edited by Migne, 221 vols. (Paris, 1844–1864), vol. 178, 375.

to defend the faith against heretics and unbelievers. He regarded divine authority as supreme.

But, like Anselm, he thought logic and reason were essential for understanding the faith one accepted on the basis of authority. He sought wherever possible to explain articles of faith by reason and rational argument. Indeed, he believed that defense of the faith required the use of reason. "Those who attack our faith," he declared,

> assail us above all with philosophical reasonings. It is those reasonings which we have principally enquired into and I believe that no one can fully understand them without applying himself to philosophical and especially to dialectical studies.[100]

Abelard was here echoing a sentiment that was rather common in the twelfth century. It was the idea that heretics who attacked the faith used philosophical arguments, and that it was essential for defenders of the faith to respond in kind. As we saw in the first chapter, this attitude became more widespread and intensive by the latter part of the twelfth century when the Cathar heresy swept southern Europe, and philosophical arguments were used to buttress its beliefs. When the Dominican Order was formed in the first quarter of the thirteenth century to combat the Cathars, the members of the order studied Aristotelian natural philosophy in depth to provide rational and natural arguments that could be used in their preachings against their determined heretical foes.[101] Thus, in addition to force and coercion in the form of crusades and the Inquisition, persuasion by reason held a prominent place in the struggle.

It was not just the defense of the faith that prompted Abelard to appeal to philosophical ideas. He had great confidence in the use of reason because he was, above all, a logician or dialectician. He seems to have been driven, almost compulsively, to analyze concepts, ideas, and propositions. He had to understand by the light of reason what he read and studied. It was Peter Abelard who largely shaped scholastic theology into a scientific discipline as

100. Translated by D. E. Luscombe, "Peter Abelard," in Peter Dronke, ed., *A History of Twelfth-Century Western Philosophy* (Cambridge: Cambridge University Press, 1988), Chapter 10, 294, n. 57.

101. This subject is treated by Roger French and Andrew Cunningham, *Before Science: the Invention of the Friars' Natural Philosophy* (Aldershot Hants, Eng.: Scolar Press, 1996), Chapters 5–8, 99–201; especially 150–160.

it was done in the late Middle Ages.[102] Although the word *theologia* is Greek in origin, Abelard employed it for the discipline of theology, to which he gave a new significance and direction:

For theology to him, and then to generations of others, was a matter of *dialectical argumentation,* not of insight gained by meditation, nor the decisions of episcopal or other authoritative sources. This 'theology' was therefore quite different from the *sapientia* of the monasteries. For in this 'theology' mystery and revealed truth were to be investigated by the test of reason. This 'theology' was a new, God-centred subject, for which the seven liberal arts – and especially logic – were to be essential bases. Theology was the application of *scientia* to the understanding of the nature of God and of the Christian religion. Pagan learning had thus been brought back as essential for understanding the divine.[103]

Because he was quintessentially a logician and dialectician, Abelard applied logic to the truths and dogmas of the faith. He accepted, as he had to, that logic was subordinate to faith, but he treated the truths of faith as if they were a series of propositions in need of logical analysis.[104] Although he would not challenge the authority of the Church, he was prepared to push reason as far as it would go. Indeed, he regarded it as a duty to try to understand faith by the use of reason, as is evident in his attempt to demonstrate rationally (in *Introduction to Theology,* bk. 3) the existence of God and God's unity.[105] In his autobiography, Abelard reveals his motives for writing his first work on theology, and simultaneously suggests that the sentiments he had were commonplace among his students:

I first applied myself to writing about the foundation of our faith with the aid of analogies provided by human reason, and I wrote a treatise of theology – on the divine unity and trinity – for our scholars, who were asking for human and philosophical reasons and clamouring more for what could be understood than for what could be said. They said in fact that the utterance of words was superfluous unless it were followed by understanding, and that it was ridiculous for anyone to preach to others what neither he nor those taught by him could accept into their understanding.[106]

102. See Armand Maurer, *Medieval Philosophy* (New York: Random House, 1962; reprinted 1969), 59. Although Abelard shaped it, St. Anselm of Canterbury is usually regarded as the father of scholastic theology. See Gilson, *History of Christian Philosophy in the Middle Ages,* 139.
103. French and Cunningham, ibid., 57–58.
104. See Knowles, *Evolution of Medieval Thought,* 124.
105. For a summary of these demonstrations, see Weinberg, *A Short History of Medieval Philosophy,* 76–77.
106. Translated in Luscombe, "Peter Abelard," 293.

That Peter Abelard effected a dramatic change in the theological world is apparent from his most famous work, *Sic et Non (Yes and No)*. Peter wanted students of theology to think for themselves and to arrive at their own answers to problems. Toward this end, he formulated a number of queries in his *Sic et Non*, that allowed, and seemingly encouraged, alternative answers. Among his questions were these:

1. That faith is to be supported by human reason, *et contra*.
5. That God is not single, *et contra*.
32. That to God all things are possible, *et non*.
106. That no one can be saved without baptisms of water, *et contra*.
141. That works of mercy do not profit those without faith, *et contra*.
145. That we sin at times unwillingly, *et contra*.
154. That a lie is permissible, *et contra*.
157. That it is lawful to kill a man, *et non*.[107]

For each proposition, Abelard marshaled arguments pro and con, taking them largely from the Church Fathers. The reader could see at a glance that the Fathers were in disagreement, often contradicting one another. Abelard, however, provided no answers, and thus left each question unresolved, a risky procedure because it tended to undermine confidence in a host of problems that were central to Church doctrine and dogma. After all, if the Church Fathers disagreed on so many relevant issues, what was one to think? At the conclusion of his prologue to the *Sic et Non*, Abelard provides an explanation:

I present here a collection of statements of the Holy Fathers in the order in which I have remembered them. The discrepancies which these texts seem to contain raise certain questions which should present a challenge to my young readers to summon up all their zeal to establish the truth and in doing so to gain increased perspicacity. For the prime source of wisdom has been defined as continuous and penetrating enquiry. The most brilliant of all philosophers, Aristotle, encouraged his students to undertake this task with every ounce of their curiosity. In the section on the category of relation he says: 'It is foolish to make confident statements about these matters if one does not devote a lot of time to them. It is useful practice to question every detail.'[108] By raising questions we begin to enquire, and by

107. These eight propositions are extracted from a list of twelve translated by Charles Homer Haskins, *The Renaissance of the Twelfth Century* (Cleveland/New York: Meridian Books, World Publishing Co., 1964; first published 1927), 354–355.

108. In the revised Oxford translation (1984) of J. L. Ackrill, Aristotle says (*Categories*, ch. 7, 8b. 23–24): "It is perhaps hard to make firm statements on such questions without having examined them many times. Still, to have gone through the various difficulties is not unprofitable." The Latin text Abelard used came to him from Boethius's translation.

enquiring we attain the truth, and, as the Truth has in fact said: 'Seek, and ye shall find; knock, and it shall be opened unto you.' He demonstrated this to us by His own moral example when He was found at the age of twelve 'sitting in the midst of the doctors both hearing them and asking them questions.' He who is the Light itself, the full and perfect wisdom of God, desired by his questioning to give his disciples an example before He became a model for teachers in His preaching. When, therefore, I adduce passages from the scriptures it should spur and incite my readers to enquire into the truth and the greater the authority of these passages, the more earnest this enquiry should be.[109]

In this passage, Abelard tells us that he wished to encourage enquiry and thereby attempt to establish the truth among the conflicting opinions of the Church Fathers. For it is "by raising questions we begin to enquire, and by enquiring we attain the truth." In the same Prologue, Abelard discusses how to compare what different writers say and how to account for the meaning of a word, which may take on different meanings in different contexts. It is essential to interpret troublesome texts and passages and to determine, to the extent possible, whether they are the result of miscopying or poor translation, or due to the reader's ignorance. If, after careful analysis, two authorities still disagree, it is necessary to weigh one against the other in the larger scheme of things. Nothing is exempted from this rigorous inspection except the Bible and those pronouncements that the Church has accepted as true. All other authorities and texts are open to criticism and analysis in an effort to arrive at the truth.[110]

Although Abelard was not the first to call for rigorous analytic inspection of the corpus of Christian authors that had shaped Church doctrine,[111] he laid emphasis on human reason as had not been done before. His confidence in reason to determine the outcome of disagreements and conflicts was great indeed. What such theologians as Berengar of Tours and Anselm of Canterbury had begun in the eleventh century, Abelard and other masters continued and intensified in the twelfth century, laying the foundation

109. The translation is by Anders Piltz, *The World of Medieval Learning,* translated into English by David Jones (Totowa, NJ: Barnes & Noble Books, 1981; first published in Swedish, 1978), 82. The translation was made from the version in Migne, *Patrologia Latina,* vol. 178, col. 1349. In *Peter Abailard Sic Et Non, A Critical Edition* by Blanche Boyer and Richard McKeon (Chicago: The University of Chicago Press, 1976, 1977), see Prologue, 103–104, lines 330–346.

110. On these aspects of Abelard's method, see Piltz, *The World of Medieval Learning,* 81.

111. In the eleventh century, Bernold of Constance (d. 1100), using the "old logic" as a tool, established similar rules for the exegesis of the Bible. See Piltz, ibid., 74–75.

for what was to come in subsequent centuries. Peter Abelard's exhortation in the Prologue of his *Sic et Non,* that "by raising questions we begin to enquire, and by enquiring we attain the truth," became characteristic of scholastic methodology, which, however, departed from Abelard by not only raising the questions but also answering them. Theologians had been moving toward widespread application of dialectics to theology since the eleventh century. Indeed, "faith was being fashioned into a science."[112]

The Reaction to the Rationalization of Theology

During the first few decades of the twelfth century, theology was well on its way to becoming a profession. Theologians were now called "master" *(magister),* as we see in the titles of Master Manegold of Lautenbach,[113] who was one of the first, Master Ivo of Chartres, Master Anselm of Laon, Master Abelard, Master Hugh of Saint-Victor, and the most famous of all: Master Peter Lombard, whose name, as we shall see, became virtually synonymous with theology.[114] The new masters sought to bring order out of the mass of often-contradictory comments that they had inherited from the Church fathers and saints. They began to give their own interpretations of revelation. Soon they began to cite one another's opinions in an authoritative manner, thus adding another layer of authority to that of the historical authority of the Bible and the Fathers. As one churchman put it: "If new things please you, then look into the writings of Master Hugh or [Master] Bernard or Master Gilbert or Master Peter Comestor where you will find no lack of roses and lilies."[115] The new breed of theological masters sought to analyze ideas and to speculate about them. In short, they ushered in scholastic theology, which laid emphasis on the use of reason.

It is ironic that as the new theology emerged, the old traditional theology, sometimes referred to as "monastic theology," had its supporters and developed alongside it in the twelfth century. Monastic theologians emphasized knowing God as directly as possible, and they also emphasized

112. See M.-D. Chenu, *Nature, Man, and Society in the Twelfth Century: Essays on New Theological Perspectives in the Latin West,* with a preface by Etienne Gilson. Selected, edited, and translated by Jerome Taylor and Lester K. Little (Chicago: The University of Chicago Press, 1968; originally published in French in 1957), Chapter 8 ("The Masters of the Theological 'Science'"), 279–280.

113. Although he was a "Master," Manegold opposed the application of dialectic to the faith. See Gilson, *History of Christian Philosophy in the Middle Ages,* 615, n. 41.

114. Chenu, *Nature, Man, and Society in the Twelfth Century,* 274.

115. Chenu, ibid., 275. In note 13, Chenu gives the Latin text.

love of God and how the soul comes to know God.[116] They stressed contemplation rather than analysis. The most striking difference lay in their respective attitudes toward the liberal arts and especially to logic and dialectic. Monastic theologians were suspicious of what the new theology, and the new attitude toward learning, represented. In short, they were fearful of, and hostile to, the use of reason and all that came to be associated with it. Reason involved endless probing and poking around. It was the work of the curious and prideful. These were dangerous and worrisome tendencies. Rupert of Deutz (1070–ca. 1129), a well-known monastic theologian of his day, put it forcefully:

> Shamefully, they dared to examine the secrets of God in the Scriptures in a presumptuous way, motivated by curiosity and not by love. As a result they became heretics. God has decreed that the proud are not to be admitted to the sight of divinity and truth.[117]

The most famous of the traditional, "monastic" theologians was Bernard of Clairvaux (1090–1153), who laid great emphasis on the personal experience of ecstasy. He claimed that his soul had had ecstatic unions with God and that such experiences were not communicable to others.[118]

A major difference between traditional theologians, such as Rupert of Deutz, Bernard of Clairvaux, William of Saint Thierry, and many others, and the new theologians, such as Anselm of Laon and Peter Abelard, lay in the fact that the former did not regard the newly developed liberal arts as essential to the study of the Bible, whereas the latter thought that the secular arts, especially dialectic or logic, were indeed essential because they could reveal much that had been overlooked or missed by traditional Biblical commentators.[119] It often seemed as if the new theologians assumed that the use of reason would lead to faith, whereas the traditionalists held that only faith could produce proper reasoning. The failure to understand each other on such vital matters underlies the assault by Bernard of Clairvaux against Peter Abelard, which resulted in the condem-

116. For a detailed description of monastic theology, see B. P. Gaybaa, *Aspects of the Medieval History of Theology 12th to 14th Centuries* (Pretoria: University of South Africa, 1988), Chapter 1, 7–65.
117. Cited by Gaybaa, ibid., 43. The passage is from Rupert of Deutz's *De Trinitate et operibus ejus libri XLII* in Migne, *Patrologia Latina* 167: 199–1828. The translated passage is from cols. 1084–1085.
118. See Gilson, *History of Christian Philosophy in the Middle Ages*, 166.
119. G. R. Evans, *Old Arts and New Theology*, 58.

nation of Peter's works in 1140. Once Bernard was aroused against Peter – and Bernard's friend, William of Saint Thierry, was instrumental in inciting Bernard's wrath – he sent numerous letters to Church authorities, including Pope Innocent II, warning about the dangers that Abelard's ideas posed to the Church. In his formal letter to Pope Innocent II, Bernard listed 19 heretical beliefs ostensibly held by Abelard, who was condemned by the bishops gathered at the Council of Sens on June 1 (or June 2), 1140.[120] In his letter to Pope Innocent II, Bernard declares:

In fine, to say a lot in a few words, our theologian [Abelard] lays down with Arius that there are degrees and grades in the Trinity; with Pelagius he prefers free will to grace; and like Nestorius he divides Christ by excluding the human nature he assumed from association with the trinity.[121]

Bernard repeated these accusations of heresy against Abelard in a number of other letters,[122] and added many more charges.[123] Bernard was convinced, however, that Abelard's heresies derived from an excessive reliance on reason, as he explains in a letter to a Cardinal Haimeric:

He has defiled the Church; he has infected with his own blight the minds of simple people. He tries to explore with his reason what the devout mind grasps at once with a vigorous faith. Faith believes, it does not dispute. But this man, apparently holding God suspect, will not believe anything until he has first examined it with his reason.[124]

Although the two antagonists were reconciled before the death of Abelard in 1142, their approaches to faith and theology were irreconcilable. Their mutually hostile attitudes encapsulate the intellectual struggle that went on through much of the twelfth century. Indeed, in 1177 or 1178, Walter of St. Victor (d. 1180) wrote a treatise titled *Against the Four Labyrinths [or Minotaurs] of France (Contra quatuor Labyrinthos Franciae),* which was an attack upon Peter Abelard, Peter Lombard (see next paragraph), Peter of Poitiers, and Gilbert of la Porrée, all of whom

120. For a concise account of the condemnation, see Marenbon, *The Philosophy of Peter Abelard* (Cambridge: Cambridge University Press, 1997), 26–32.

121. *The Life and Letters of St. Bernard of Clairvaux,* newly translated by Bruno Scott James (London: Burns Oates, 1953), letter 242, p. 323.

122. *Life and Letters,* letters 240 (p. 321); 243 (p. 324); 244 (p. 325); 249 (p. 329).

123. For a particularly heavy barrage, see letter 238 ("To the Bishops and Cardinals in Curia"), p. 316, par. 2.

124. *Life and Letters,* letter 249, p. 328.

he believed guilty of heresies of various kinds because they applied philosophy and dialectic to the faith.[125] By the end of that century, the issue was no longer in doubt. The new learning that poured into Europe from Islam and the Byzantine Empire proved decisive. It was nourishment for the reasoned approach to learning – both secular and theological. Bernard and his like-minded colleagues were simply overwhelmed and bypassed, although an analogous battle would be fought during the thirteenth century over the relationship between the faith and Aristotelian natural philosophy. Only this time it was not about the propriety of applying natural philosophy and logic to theology – that battle was long over – but about the role of God in natural philosophy, an issue that was quickly resolved.

The New Theology: The "Four Books of Sentences" of Peter Lombard

One of the great events in the history of theology occurred just after Bernard of Clairvaux passed from the scene. Sometime between 1155 and 1158, Peter Lombard (ca. 1095–1160) completed his *Four Books of Sentences,* a theological treatise that would become the basic textbook in theology until the end of the seventeenth century, a period of approximately five centuries.[126] Peter brought to fruition and culmination the trends in speculative theology that had been in process during the first half of the twelfth century. As part of the professionalization of theology in this period, theologians had to devise a pedagogical strategy for the teaching of theology. In achieving this objective, they invented "systematic theology," which was centered on collections of opinions, or *sententiae,* from which we get the term *sentences.* These opinions were drawn from the Church Fathers and other sources. The new "systematic theology" sought to organize these opinions logically. Although a large body of theological literature existed before the twelfth century, prior to that time

> no Latin theologian had developed a full-scale theological system, with a place for everything and everything in its place, in a work that went well beyond the bare

125. See Gerard Verbeke, "Philosophy and Heresy: Some Conflicts Between Reason and Faith," in W. Lourdaux and D. Verhelst, *The Concept of Heresy in the Middle Ages (11th–13th C.);* Proceedings of the International Conference, Louvain, May 13–16, 1973 (Leuven: Leuven University Press; The Hague: Martinus Nijhoff, 1976), 173–174.

126. For a study of Peter Lombard's life and thought, see Marcia Colish, *Peter Lombard,* 2 vols. (Leiden: E. J. Brill, 1994). For the dating of 1155–1158, see 25.

essentials, that treated theology as a wholesale and coherent intellectual activity, and that, at the same time, imparted the principles of theological reasoning and theological research to professionals in the making.[127]

Systematized collections of sentences, or opinions, were written in the twelfth century prior to Peter Lombard's collection,[128] but it was the latter's work that triumphed over all others and eventually became the textbook of the theological schools in the universities that emerged around 1200. In the prologue of his *Four Books of Sentences* (for the content of each book, see Chapter 6), Peter asserts that philosophizing is a futile exhibition of learning and that reasoning should play a secondary role in theology. But this was probably a conciliatory gesture to traditionalists, because his work testifies to the untrustworthiness of these remarks.[129] Peter Lombard used philosophy and logic extensively to clarify and explain terms and propositions. He was also critical of his sources, willing to criticize revered authorities, especially St. Augustine. With Peter Lombard's *Sentences,* theology was poised to become a science, although the formal arguments to justify its scientific status would not come until the thirteenth century.

As if to herald the coming status of theology, two works at the end of the twelfth century treated theology as if it were a science, indeed, an axiomatized science: One is by Alan of Lille (ca. 1128–1202), the other by Nicholas of Amiens (second half of twelfth century). Together, these two treatises constitute a kind of axiomatic theology. The first of these, by Alan of Lille, was probably written in the 1170s and titled *Theological Rules (Regulae Theologicae).* Alan took as his model Boethius's *De hebdomadibus* (see the section on Boethius in this Chapter), in which Boethius declared at the beginning that he would use rules employed by mathematicians. As we saw, however, Boethius left it to his readers to use the appropriate axiom in the right place. Some scholars in the twelfth century – for example, Clarenbaldus of Arras and Gilbert of Poitiers – tried to apply Boethius's axioms to the appropriate places.[130] Alan of Lille, however, presents a series of rules – 134 to be precise – that would prove helpful for the arts as applied

127. Colish, ibid., 34.

128. For a description of many of these, see Colish, *Peter Lombard,* Chapter 2: "The Theological Enterprise," 35–77. See also Colish, "Systematic Theology and Theological Renewal in the Twelfth Century," *Journal of Medieval and Renaissance Studies* 18, no. 2 (1988), 135–156.

129. Colish, *Peter Lombard,* 84–85.

130. See Evans, *Alan of Lille: The Frontiers of Theology in the Later Twelfth Century* (Cambridge: Cambridge University Press, 1983), 74–75.

to theology. They are not mathematical, but they are axiomatic rules. It is a kind of "theological grammar, to show in detail how the laws of the *artes* must be modified if they are to be used in theology."[131] Although Alan considered Boethius's *De hebdomadibus* as his model, his treatise is closer to the *Liber de causis,* a work that was derived from Arabic sources and is ultimately traceable to Proclus's *Elements of Theology.*[132] The system is axiomatic and deductive because Alan derives propositions from axioms.

Toward the end of the twelfth century (between 1187 and 1191), Nicholas of Amiens composed *On the Art of the Catholic Faith (De arte catholicae fidei),* a title that Nicholas devised because, as he explained, it was "composed in the manner of an art," and therefore "contains definitions, distinctions and propositions, proving what is proposed by a sequence of arguments which conform to rules."[133] Nicholas took a geometrical treatise, Euclid's *Elements,* as the model for his treatise. In the fashion of Euclid's geometric method, Nicholas employs definitions (which he calls *descriptiones*), postulates (which he calls *petitiones*), and common notions, or axioms (which he calls *communes conceptiones*). Here is how he describes them:

> The descriptions [that is, definitions] are put forward for this reason: that it might be clear in what sense the words appropriate to this art should be used.... [The postulates] are so called because, although they are not able to be proved through other statements – being like maxims, even if not as obvious as they – I nevertheless 'postulate' *(peto)* that they should be accepted in order that what follows should be approved ... [The common notions] are so called because they are so obvious that the mind having once heard them immediately understands them to be true.[134]

From these definitions, postulates, and axioms, Nicholas constructs a series of theorems, the fifth of which illustrates his method: "The composition of form and matter is the cause of the substance." The proof is as follows:

131. Ibid., 79.

132. Charles Burnett, "Scientific Speculations" in Peter Dronke, ed., *A History of Twelfth-Century Western Philosophy* (Cambridge: Cambridge University Press, 1988), 165.

133. Translated by Burnett, "Scientific Speculations," 164. The *De arte catholicae fidei* has also been attributed to Alan of Lille. For evidence in favor of Nicholas of Amiens's authorship, see Evans, *Alan of Lille,* Appendix I ("A Note on the Authorship of the 'De arte catholicae fidei'"), 172–187.

134. Translated by Burnett, "Scientific Speculations," 164–165. I have added the words within the first pair of brackets. The idea of applying a mathematical form to theology may go back to Boethius's *De hebdomadibus,* where, as we saw, Boethius organized a theological problem in a mathematical format. The *De hebdomadibus* was well known in the twelfth century. See Burnett, ibid., 163. Nicholas may have derived the idea of an axiomatic, mathematical format from Boethius, but the form he followed derives from Euclid's *Elements.*

For indeed substance consists of matter and form, consequently matter and form are the cause of the composition of substance (by the first postulate: there is a component cause of all composition). Again, neither the form can exist unless it is composed with matter, nor can matter exist unless it is composed with form (as has been proved in the fourth theorem); therefore form and matter have actual existence because of their composition; therefore their composition is the cause of their existence. But their existence is the cause of the substance; therefore (by the first theorem) the composition of the form with matter is the cause of the substance (since all that is cause of a cause is cause of its effect).[135]

Why did Nicholas choose to organize a treatise on theology in the manner of a geometrical work? Because of his firm belief that authority and Scripture alone would not convince unbelievers of the truth of the faith. They would be immune to such appeals. Therefore, he declares:

I have carefully put in order the probable reasons in favor of our faith, – reasons of a kind that a clear-sighted mind could scarcely reject, so that those who refuse to believe the prophecies and Scripture find themselves brought to it at least by human reasons.[136]

In these axiomatized theological works, we see how far twelfth-century scholars had come in their desire to organize theological ideas in a rigorous and rational manner. The application of reason to theology became characteristic of twelfth-century thought. By the end of that century, the genie of reason had been loosed from its bottle, never again to reenter. Indeed, the West was on a path that would eventually exalt reason over revelation. This was the ultimate consequence of what had been started by St. Anselm and other like-minded theologians, even though none of them had ever considered exalting reason over faith. "Yet once reason was separated from faith for analytical purposes, the two began to be separated for other purposes as well. It was eventually taken for granted that reason is capable of functioning by itself, and ultimately this came to mean functioning without any fundamental religious beliefs whatever."[137] This state of affairs would occur after the Middle Ages, in the period we have come to call "The Age of Reason."

135. The translation is by Etienne Gilson, *History of Christian Philosophy in the Middle Ages*, 177. Gilson attributes the work to Alan of Lille, "which," he declares, "historical research first ascribed to Nicholas of Amiens" (ibid., 176). The pendulum seems to have swung back to Nicholas of Amiens.

136. Gilson, *History of Christian Philosophy*, 176.

137. Berman, *Law and Revolution: The Formation of the Western Legal Tradition* (Cambridge, MA: Harvard University Press, 1983), 197–198.

NATURAL PHILOSOPHY

During much of the twelfth century, logic was the major tool for those who wished to apply reason to theology. By the end of the century, however, Aristotle's natural philosophy was becoming readily available and, despite some difficulties, would transform theology by the latter part of the thirteenth century. It could do this because, as we shall see (Chapter 5), natural philosophy was a quintessentially rational discipline. In the first chapter, we saw that prior to the influx of Aristotle's natural books, natural philosophy was essentially Platonic. Although the level of knowledge of natural philosophy was quite modest in the twelfth century, a number of Platonic natural philosophers revealed a genuine desire to understand and appreciate nature in rational terms.[138]

Adelard of Bath (ca. 1080–1142)

The most significant of this group is Adelard of Bath, who was important both as a translator and natural philosopher. He traveled widely, visiting France, Salerno, Sicily, Syria, and perhaps Palestine. In the course of his travels, he learned Arabic and eventually translated scientific works from Arabic to Latin, most notably Euclid's *Elements* and the astronomical tables of the Islamic mathematician and astronomer, al-Khwarizmi.[139] As one who traveled through Arabic lands and became a translator from Arabic to Latin, Adelard was undoubtedly influenced by aspects of Islamic science, although specific influences of the latter are difficult to detect. Among Adelard's most relevant and widely read works is a treatise on natural philosophy titled *Natural Questions (Quaestiones Naturales),* probably written around 1116.[140] The format of the *Natural Questions* involves a dialogue in the form of questions and answers between Adelard and his unnamed nephew, the latter raising the questions, the former replying.

The content of Adelard's natural philosophy is not especially noteworthy, but his attitude toward authority and reason, and his approach to learning,

138. For a discussion of the nature of Aristotelian natural philosophy, see the section "What is natural philosophy?" in Chapter 5.
139. For a brief description of Adelard's life and works, see Marshall Clagett, "Adelard of Bath," in *Dictionary of Scientific Biography,* vol. 1 (1970), 61–64. Clagett believes (p. 63) that "Adelard should be considered, along with Gerard of Cremona and William of Moerbeke, as one of the pivotal figures in the conversion of Greek and Arabic learning into Latin."
140. See Louise Cochrane, *Adelard of Bath: The First English Scientist* (London: British Museum Press, 1994), 50. The work was also translated into Hebrew.

are. In a discussion about animals, Adelard informs his nephew that he learned one thing "under the guidance of reason, from Arabic teachers," presumably, in the light of what follows, not to trust authority. "You, captivated by a show of authority," he informs his nephew sarcastically,

> are led around by a halter. For what should we call authority but a halter? Indeed, just as brute animals are led about by a halter wherever you please, and are not told where or why, but see the rope by which they are held and follow it alone, thus the authority of writers leads many of you, caught and bound by animal-like credulity, into danger. Whence some men, usurping the name of authority for themselves, have employed great license in writing, to such an extent that they do not hesitate to present the false as true to such animal-like men. For why not fill up sheets of paper, and why not write on the back too, when you usually have such readers today who require no rational explanation and put their trust only in the ancient name of a title?[141]

It is not likely that an equally powerful statement in opposition to mindless obedience to authority can be found in all of the later Middle Ages.

Adelard's nephew asks why the Creator gave to animals the means of defending themselves by providing them with attachments like horns, tusks, and claws, but did not furnish man with similar weapons. Although man is dearer to the Creator than all the other animals, Adelard replies that God, nevertheless, chose not to give us weapons for defense, or the means for swift flight. Instead, He gave us something "which is much better, and more worthy, reason I mean, by which he so far excels the brutes that by means of it he can tame them, put bits in their mouths, and train them to perform various tasks. You see, therefore, by how much the gift of reason excels bodily defenses."[142]

Prior to the twelfth century, and even in the twelfth century, scholars often interpreted the physical world in theological terms. They viewed the world "as a kind of shadow and a symbol of divine power and providence."[143] Thus, Hildegard of Bingen (1098–1179), the famous nun who

141. Translated by Richard C. Dales in Dales, ed., *The Scientific Achievement of the Middle Ages* (Philadelphia: University of Pennsylvania Press, 1973), 41. The translation is from the Latin edition by Martin Müller, *Die Quaestiones Naturales des Adelardus von Bath* (Münster I. W., 1934), in *Beiträge zur Geschichte der Philosophie und Theologie des Mittelalters,* vol. 31, 2. The passage occurs in Chapter 6 of the *Natural Questions.*

142. Translated by Dales, *The Scientific Achievement of the Middle Ages,* 43–44, from Chapter 15 of Adelard's *Natural Questions.*

143. From Alistair Crombie, "Science," in A. L. Poole, ed., *Medieval England,* 2 vols. (Oxford, 1958), vol. 2, Chapter 18, 579. Also cited in Cochrane, *Adelard of Bath,* 44–45.

wrote of her visions and also intermingled religion and natural philosophy in some of her works, claimed to have heard a voice from heaven, which said: "God who established all things by His will, created them to make His name known and honored, not only moreover showing in the same what are visible and temporal, but also manifesting in them what are invisible and eternal."[144] Adelard of Bath, and many others in the twelfth century, adopted a radically different attitude, one that eventually triumphed over the old theological approach. "I will detract nothing from God," Adelard declares, "for whatever is is from Him." But "we must listen to the very limits of human knowledge and only when this utterly breaks down should we refer things to God."[145]

Adelard uttered a similar sentiment near the beginning of his treatise. In response to his nephew's query about why plants rise from the earth, and the nephew's conviction that this should be attributed to "the wonderful operation of the wonderful divine will," Adelard replies that it is certainly "the will of the Creator that plants should rise from the earth. But this thing is not without a reason," which prompts Adelard to offer a naturalistic explanation based on the four elements.[146] A few chapters later, the nephew sees a flaw in Adelard's argument about the nourishment of plants, arguing that if air is a vital element in the sustenance of plants, the air ought to continue to sustain a plant even after it had been pulled out of the earth. "But," says the nephew, "because air cannot furnish this to them, despite the fact that they desire air, your whole explanation is destroyed, and the accomplishment of all things must rather be ascribed to God." Adelard replies with an appeal to reason:

> I take nothing away from God, for whatever exists is from Him and because of Him. But the natural order does not exist confusedly and without rational arrangement, and human reason should be listened to concerning those things it treats of. But when it completely fails, then the matter should be referred to God. Therefore,

144. From Hildegard's treatise titled *Scivias,* translated in Lynn Thorndike, *A History of Magic and Experimental Science,* 8 vols. (New York: Columbia University Press, 1923–1958), vol. 2, 137.
145. From Adelard's *Natural Questions,* Chapter 4, as cited in Cochrane, *Adelard of Bath,* 45. Cochrane's translations from Adelard's *Natural Questions* are from Hermann Gollancz, *Dodi ve-Nechdi ("Uncle and Nephew"). The Work of Berachya Hanakdan, now edited from MSS. at Munich and Oxford, with an English Translation, Introduction etc. to which is added the first English Translation from the Latin of Adelard of Bath's "Quaestiones Naturales"* (London: Humphrey Milford Oxford University Press, 1920), 96.
146. Translated by Dales, *The Scientific Achievement of the Middle Ages,* 39, from Chapter 1 of Adelard's *Natural Questions.*

since we have not yet completely lost the use of our minds, let us return to reason.[147]

Only when reason was incapable of resolving a question, or of determining a cause, did Adelard believe it proper to invoke God as an explanation of an effect. Otherwise, we might well ponder why God endowed humans with the ability to use their reason. Adelard harshly condemns those who have usurped authority for themselves and who write for readers "who require no rational explanation" and put their trust in authorities:

For they do not understand that reason has been given to each person so that he might discern the true from the false, using reason as the chief judge. For if reason were not the universal judge, it would have been given to each of us in vain. It would be sufficient that it were given to one (or a few at most), and the rest would be content with their authority and decisions. Further, those very people who are called authorities only secured the trust of their successors because they followed reason; and whoever is ignorant of reason or ignores it is deservedly considered to be blind. I will cut short this discussion of the fact that in my judgment authority should be avoided. But I do assert this, that first we ought to seek the reason for anything, and then if we find an authority it may be added. Authority alone cannot make a philosopher believe anything, nor should it be adduced for this purpose.[148]

Adelard's emphasis on the use of reason is rather remarkable. His message is clear. He firmly believed that God was the creator of the world, and that God provided the world with a rational structure and a capacity to operate by its own laws. In this well-ordered world, natural philosophers must always seek a rational explanation for phenomena. They must search for a natural cause and not resort to God, the ultimate cause of all things, unless the secondary cause seems unattainable.

William of Conches (d. after 1154)

William of Conches, Adelard's contemporary, who, like Adelard, "followed the guidance of reason,"[149] was a teacher in France but gave it up because he became disillusioned by the lack of standards and by what he viewed as the pandering tactics of his teaching colleagues. William wrote on natural philosophy, presenting his views in a gloss on Plato's *Timaeus* and in a work

147. Translated by Dales, ibid., 40, from Chapter 4 of Adelard's *Natural Questions.*

148. Translated by Dales, ibid., 41–42, from Chapter 6 of Adelard's *Natural Questions.*

149. Thorndike, *A History of Magic and Experimental Science,* vol. 2, 50.

titled *Dragmaticon,* which was a revision of an earlier work titled *Philosophy of the World (Philosophia Mundi).* In the earlier treatise, William made some assertions about the Trinity that drew the wrathful attention of William of Saint Thierry, who complained by letter to Bernard of Clairvaux, just as he had done with Peter Abelard. As with Abelard, William of Saint Thierry accused William of Conches of holding heretical opinions about the Trinity, and that William "stupidly and haughtily ridicules history of divine authority."[150] William took cognizance of this attack in his *Dragmaticon* and corrected errors he had made earlier in his *Philosophy of the World (Philosophia Mundi),* requesting his readers to make appropriate corrections, especially about his views on the Trinity. He withdrew his assertion that in the Trinity the Father represents power and the Holy Spirit will. In doing so, he acknowledged that there was no Scriptural authority for such claims.

But in the same *Dragmaticon,* William of Conches asserted his independence of patristic authority in matters relevant to natural philosophy when he declared:

> It is not lawful to speak against any matter concerning either the Catholic faith or Church regulations, nor be aroused in opposition to men [such as] the Venerable Bede or other holy Fathers; however, in those matters concerning philosophy, if they err in any respect, it is permissible to differ from them. For even though they were greater men than we are, yet they were men."[151]

William thought it improper to invoke God's omnipotence as an explanation for natural phenomena. Like all natural philosophers in the Middle Ages, William of Conches believed that God was the ultimate cause of everything, but, like Adelard of Bath, he believed that God had empowered nature to produce effects and that one should therefore seek the cause of those effects in nature. He also rejected the idea that Scripture was of use in natural philosophy. Some priests believed only what they found in the Bible. William explained that

150. Thorndike, ibid., vol. 2, 59–60. For a good account of William's approach to natural philosophy and theology, see E. J. Dijksterhuis, *The Mechanization of the World Picture,* translated by C. Dikshoorn (Oxford: Clarendon Press, 1961; first published in Dutch in 1950), 119–123; and for a perceptive analysis of William's role in the emergence of natural philosophy in the twelfth century, see Joan Cadden, "Science and Rhetoric in the Middle Ages: The Natural Philosophy of William of Conches," *Journal of the History of Ideas* 56 (Jan. 1995), 1–24.

151. Translated from *Dragmaticon,* 65, 66, by Tina Stiefel, *The Intellectual Revolution in Twelfth-Century Europe* (New York: St. Martin's Press, 1985), 79; see 91–92, n. 4, for the Latin text.

when modern priests hear this, they ridicule it immediately because they do not find it in the Bible. They don't realise that the authors of truth are silent on matters of natural philosophy, not because these matters are against the faith, but because they have little to do with the strengthening of such faith, which is what those authors are concerned with. But modern priests do not want us to inquire into anything that isn't in the Scriptures, only to believe simply, like peasants.[152]

Adelard of Bath and William of Conches represent a new kind of natural philosopher who emerged in the twelfth century, one who laid heavy emphasis on the use of reason and would ostensibly use only those traditional authorities that passed the test of reason. They had what we might plausibly call a "scientific temperament." But their Platonic natural philosophy was erected on a meager base of knowledge that was already in process of being supplanted by the new natural philosophy that was entering Western Europe by way of translations from Arabic and Greek. At the core of the new natural philosophy were the natural books of Aristotle, which would form the basis of natural philosophy for the next five hundred years. The characteristic feature of Aristotle's works is an emphasis on reason. Because of Aristotle's unrivaled place in Western thought, and the central role of natural philosophy, reason would come to play a vital role in late-medieval intellectual life, as will be seen in Chapter 3.

But reason was not applied only to theology and natural philosophy. Law and medicine were also beginning to be shaped by it in the twelfth century. These two disciplines became significant university subjects with formidable faculties from the thirteenth century on. Medicine was reshaped and law was given an institutional base. In this chapter, I shall describe the revolutionary transformation of law in the eleventh and twelfth centuries; in the next chapter, I shall briefly discuss medicine and law as university subjects.

LAW

By the late eleventh and early twelfth centuries, as church and state grew stronger in Western Europe, each recognized the need for a cohesive and coherent legal system that could cope with the myriad problems that con-

152. Translated from William's *Gloss on Boethius* by Stiefel, ibid., 86; 93, n. 25 for the Latin text. John Newell, "Rationalism in the School of Chartres," *Vivarium* 21 (1983), 108–126, is largely an account of William of Conch's attitude toward reason.

fronted a dynamic society growing ever more complex. Civil and ecclesiastical law had, of course, existed for centuries. Even during the transformation of the Roman Empire, and during the lowest points of urban life in Western Europe, law, both secular and ecclesiastical, existed in the various parts of Europe. Prior to the late eleventh century, secular, or civil law, was rooted in feudal and local custom, while ecclesiastical law was also localized and customary. Over many centuries, ecclesiastical law came to consist of a mass of often contradictory and conflicting decrees issued by popes, Church councils, and Christian emperors. To add to the mix, and confusion, there were also Scriptural and patristic literature. From the fifth to twelfth centuries, many collections of Church law were compiled on the basis of this body of confused ecclesiastical law, which ranged over many subjects that were often in conflict.[153] Governmental laws were local and provincial and were administered not by properly trained judges but by laymen, such as kings and assorted nobles. There were no law schools; no professional jurists; no hierarchical system of courts; and, above all, no properly developed legal treatises that embodied general principles of law.

By the late eleventh and twelfth centuries, the realization dawned that some rationalization of both civil and ecclesiastical law was desperately needed if church and state were to carry out their respective functions. As Haskins has put it, the twelfth century "was an age of political consolidation, creating a demand for some 'common law' wider in its application then mere local custom and based on principles of more general validity."[154]

As ecclesiastical and secular authorities became more centralized, law became more systematized and more widespread. In the course of the twelfth and thirteenth centuries,

> there emerged a class of professional jurists, including professional judges and practicing lawyers. Intellectually, western Europe experienced at the same time the creation of its first law schools, the writing of its first legal treatises, the conscious ordering of the huge mass of inherited legal materials, and the development of the concept of law as an autonomous, integrated, developing body of legal principles and procedures.[155]

Although both civil and canon law developed simultaneously and independently, they exerted a mutual influence.

153. For an impressive list of collections, see Roger E. Reynolds, "Law Canon: To Gratian," in Strayer, ed., *Dictionary of the Middle Ages*, vol. 7, 395–413.

154. Charles Homer Haskins, *The Renaissance of the 12th Century*, 207.

155. Berman, *Law and Revolution*, 86.

Civil Law

In the opening lines of a chapter titled "The Origin of Western Legal Science in the European Universities," Harold Berman, says that "Maitland called the twelfth century 'a legal century.' It was more than that: it was *the* legal century, the century in which the Western legal tradition was formed."[156] What made this possible was the discovery, near the end of the eleventh century, of a manuscript copy of the *Digest* (also called the *Pandects*), which formed the most important part of the *Corpus luris Civilis,* the code of ancient Roman law that Emperor Justinian (who ruled from 527 to 565) had ordered drawn up during 533–534.[157] The *Digest* was the fundamental authoritative commentary on the Roman law. It was apparently lost and unused between 603 and 1076, in which year the *Digest* was cited in a legal case.[158] As Berman describes it,

> [t]he Digest was a vast conglomeration of the opinions of Roman jurists concerning thousands of legal propositions relating not only to property, wills, contracts, torts, and other branches of what is today called civil law, but also to criminal law, constitutional law, and other branches of law governing the Roman citizen.... The Digest was not a code in the modern sense; it did not attempt to provide a complete, self-contained, internally consistent, systematically arranged set of legal concepts, principles, and rules.[159]

Prior to the rediscovery of the *Digest,* Roman law had survived in customary law and in various summaries and collections, but Roman law in the sense of a legal science was virtually absent in Western Europe. Not only was Roman law embedded in the absent *Digest,* but the latter was itself a difficult text to comprehend, largely because the authors who wrote it did so from different points of view and also because they failed to remove many contradictions that had entered the text.[160] Because the

156. Berman, ibid., 120. Frederick William Maitland (1850–1906), a professor of English law at the University of Cambridge, was one of the greatest historians of medieval English law. For a brief sketch of his life and work, see Norman F. Cantor, *Inventing the Middle Ages* (New York: William Morrow and Company, 1991), Chapter 3, 48–78.

157. In addition to the *Digest,* the *Corpus iuris civilis* included the *Code,* the *Institutes,* and the *Novels.* The *Corpus iuris civilis* has been said to "rank with the Bible in the list of the most important books in the history of Western civilization." Quoted from Charles Donahue, Jr., "Law, Civil – Corpus luris, Revival and Spread," in Strayer, ed., *Dictionary of the Middle Ages,* vol. 7 (1986), 418.

158. Haskins, *Renaissance of the 12th Century,* 198–199.

159. Berman, *Law and Revolution,* 128.

160. Donahue, "Law, Civil," *Dictionary of the Middle Ages,* vol. 7 (1986), 419.

Europe of the late eleventh and twelfth centuries was very different from that of Justinian's Roman Empire of the sixth century, one might properly ask how the legal material of the *Digest* could prove of any use to the legal needs of European society more than five centuries later. The answer lies in the attitude that European legal scholars adopted toward the Roman law in the late eleventh and twelfth centuries. They viewed it "as the law applicable at all times and in all places."[161] For them, it was "the *true* law, the ideal law, the embodiment of reason."[162] Their task was to adapt it to the needs of their age.

The *Digest* was of great importance because it presented many actual cases, which were followed by the opinions of various jurists. Beginning in the twelfth century, systematization was conferred upon the *Digest* by legal commentators in the West, mostly at newly established universities, such as the University of Bologna, one of the oldest in Christendom. It was European legal scholars who expanded the particular cases in the *Digest* into general legal principles. The method by which that systematization was accomplished was by the application of analysis and synthesis, or, what is often called the scholastic method. This involved glossing the cases in the *Digest* word by word, or line by line, and also making careful distinctions, moving from the most general to the more particular. As part of this method, teachers also posed questions *(questiones)* arising out of the cases. Using the scholastic method with its various techniques, legal scholars analyzed Roman law and reconciled its contradictions, reconciliations that could then be applied to contemporary law.

The rational systematization of the Roman civil law has been traditionally identified with the achievements of a few famous scholars. The most significant name associated with the civil law is that of Irnerius (or Guarnerius) (ca. 1055–ca. 1130), who, beginning around 1087, taught the whole *Corpus luris Civilis* at Bologna, focusing mostly on the *Digest.* Irnerius proceeded by glossing the text. His successors did likewise and were known as glossators. Around 1250, more than a century after Irnerius, a glossator named Odofredus explains how he will proceed in his lectures:

> Concerning the method of teaching the following order was kept by ancient and modern doctors and especially by my own master, which method I shall observe: First, I shall give you summaries of each title before I proceed to the text; second, I shall give you as clear and explicit a statement as I can of the purport of each law

161. Berman, *Law and Revolution,* 122.
162. Berman, ibid., 123.

[included in the title]; third, I shall read the text with a view to correcting it; fourth, I shall briefly repeat the contents of the law; fifth, I shall solve apparent contradictions, adding any general principles of law [to be extracted from the passage], commonly called 'Brocardica,' and any distinctions or subtle and useful problems (*quaestiones*) arising out of the law with their solutions, as far as the Divine Providence shall enable me.[163]

The method Odofredus describes goes back to the twelfth century. We can see how systematically Roman law was approached, and especially how the basic scholastic technique was applied, when Odofredus says that he will "solve apparent contradictions" and add whatever else has to be added to render the law intelligible, such as distinctions, subtleties, and problems. And, not to be ignored, is Odofredus's further intention to provide solutions to difficulties that might arise from the application of "distinctions, subtleties, and problems." Odofredus's aim, as it was for all teachers of the law, was to explain everything, or as much as was possible.

Thus was the Roman law that was inherited from the Justinian code gradually expanded into a universal secular law. Indeed, this had already been achieved in the twelfth century. The development of a universal secular law based on interpretations of the Roman law by legal scholars at the law schools that had developed in Europe, most notably in Italy (especially in Bologna), was a monument to the application of reason for the advancement and benefit of society. What had been produced was "a scientific jurisprudence and a jurisprudential method."[164]

Canon Law

The mass of the Church's often-inconsistent legal documents buried in collections spread around Europe had become quite large by the twelfth century.[165] A major step toward the unification and systematization of these conflicting collections occurred around 1140, when a Bolognese monk named Gratian published an aptly named treatise titled *A Concordance of Discordant Canons (Concordia discordantium canonum),* which came to be called the *Decretum.* The *Concordance,* or *Decretum,* a work of more than

163. Haskins, *Renaissance of the Twelfth Century,* 203.

164. Ernst H. Kantorowicz, "Kingship Under the Impact of Scientific Jurisprudence," in Marshall Clagett, Gaines Post, and Robert Reynolds, eds., *Twelfth-Century Europe and the Foundations of Modern Society* (Madison: The University of Wisconsin Press, 1966), 105, n. 2.

165. For a description of many of these collections, see Reynolds, "Law Canon: To Gratian," in *Dictionary of the Middle Ages,* vol. 7, 395–413.

1,400 printed pages in a modern nineteenth-century edition, has been characterized as "the first comprehensive and systematic legal treatise in the history of the West, and perhaps in the history of mankind."[166] It was truly revolutionary.[167] Prior to its appearance, two collections of canon law had been compiled in the eleventh century that did not follow a chronological order; they actually sought to organize the laws into categories under such headings as baptism, the Eucharist, excommunication, witches, perjury, and numerous other groupings.[168] What makes Gratian's treatise noteworthy is that he followed the highly organized, rationalistic, scholastic method. According to Stanley Chodorow, Gratian "stated a proposition or asked a question, set out canons supporting different positions on the matter, and reconciled conflicts in a dictum that served also to enunciate a doctrine."[169] He followed a similar method to that which Peter Abelard used in his *Sic et Non,* except that where Peter deliberately chose not to reconcile the conflicts, Gratian did. Thus, Gratian did in the 1140s for canon law what Peter Lombard did in the 1150s for theology: He brought order and organization into a vital discipline.

Gratian divided the book into two parts, the first covering subject themes in an orderly manner, including definitions of different types of divine and human law, whereas in the second part, he presented 36 hypothetical cases that were useful for teaching canon law because they included questions that Gratian analyzed. He

> gathered more canonical material than any of his predecessors, and had gone beyond them in a decisive way by organizing it into a comprehensive and coherent statement of church law. By applying to the canons the rational techniques of the theological school of Peter Abelard, he had created a new canonical jurisprudence.[170]

One of the most significant contributions of Gratian was his reappraisal of customary law, which was the law by which most communities lived. Gratian, and the many canon lawyers who followed him, set up criteria for repudiating customary laws that fell short of proper standards. Mere custom, powerful

166. Berman, *Law and Revolution,* 143.
167. Stanley Chodorow, "Law, Canon: After Gratian," in *Dictionary of the Middle Ages,* vol. 7 (1986), 413.
168. The two authors were Burchard, Bishop of Worms, whose work appeared around 1012 and Ivo, Bishop of Chartres, whose treatise was issued in 1095. See Berman, *Law and Revolution,* 143–144.
169. Chodorow, "Law, Canon: After Gratian," *Dictionary of the Middle Ages,* vol. 7, 413.
170. Chodorow, ibid.

though it was, would no longer suffice. Among other criteria, a law had to meet the test of reasonableness, a monumentally significant advance and a tribute to the role that canon law would play in the ongoing emphasis on reason. The scholastic technique, as we shall see more fully in the discussion of natural philosophy in the later Middle Ages, is characterized by a series of questions in each of which pros and cons by different authorities are enunciated and, after careful argumentation, a conclusion is reached. Often a third, or new, opinion emerges. This was especially true for law.

Typical of the questions Gratian posed and resolved was whether priests should read profane, that is, pagan, literature.[171] Gratian first presents arguments against the use of profane literature, citing passages from the Fourth Carthaginian Council ("A Bishop should not read the books of the heathen"), a few passages from St. Jerome, one from Rabanus Maurus, and another from Origen, who "understands by the flies and frogs with which the Egyptians were smitten, the empty garrulousness of the dialecticians and their sophistical arguments."[172]

Then, with reference to all that has been said against reading profane literature, Gratian says:

> From all which instances it is gathered that knowledge of profane literature is not to be sought after by churchmen.
> But, on the other hand one reads that Moses and Daniel were learned in all the wisdom of the Egyptians and Chaldaeans.[173]

Gratian presents next the case for priests to read profane literature, saying first:

> One reads also that God ordered the sons of Israel to spoil the Egyptians of their gold and silver; the moral interpretation of this teaches that should we find in the poets either the gold of wisdom or the silver of eloquence, we should turn it to the profit of useful learning.[174]

Gratian then draws on other Christian authors – St. Ambrose, Venerable Bede, St. Jerome, Pope Eugene, St. Augustine, and Pope Clement – who

171. The question is translated and explained in Arthur O. Norton, *Readings in the History of Education: Mediaeval Universities* (Cambridge, MA: Harvard University Press, 1909), 60–75. Norton's translation was made from Distinction 37 of the Lyons edition of 1580. It is also summarized in Berman, *Law and Revolution,* 147.

172. Norton, *Readings in the History of Education,* 64.

173. Norton, ibid.

174. Norton, ibid.

called for the faithful to read profane literature so that they might better understand the faith and defend it, and even turn the profane authorities against themselves. Occasionally, Gratian even includes his own commentary, as when he chooses to elaborate on the passage included from Venerable Bede, asking, if Bede's arguments in favor of reading profane literature are sound, "why are those [writings] forbidden to be read which, it is shown so reasonably should be read?" Gratian mentions those who read profane literature and poetry for pleasure, but reserves his praise for those who read such works "to add to their knowledge, in order that through reading the errors of the heathen they may denounce them, and that they may turn to the service of sacred and devout learning the useful things they find therein. Such are praiseworthy in adding to their learning profane literature."[175]

In his conclusion, Gratian declares that it "is evident from the authorities already quoted ignorance ought to be odious to priests." He ends the question with a passage from Augustine, where the latter excuses ignorance in someone who could not find any way of learning, but the person "who having the means of knowledge did not use them," cannot be forgiven.[176]

Thus did Gratian bring order from conflicting opinions about the Church and the faith. In this instance, he resolved the dilemma

> by stating that anyone (and not only priests) ought to learn profane knowledge not for pleasure but for instruction, in order that what is found therein may be turned to the use of sacred learning. Thus Gratian used general principles and general concepts to synthesize opposing doctrines – not only to determine which of two opposing doctrines was wrong, but also to bring a new third doctrine out of the conflict."[177]

By the end of the twelfth century, civil and canon law were transformed and set upon a new course. Although they would continue to be changed, they were established in the universities of Western Europe as disciplines where reason was systematically applied to laws that were now intended to be universal in scope, while also attempting to meet the needs of the new merchant class and the guilds of workers and craftsmen that emerged in the commercial life of a now vibrant society.

We saw that a momentous change in the way theology was done occurred in the twelfth century. Theology became a profession, and its

175. Norton, ibid., 68–70.
176. Norton, ibid., 75.
177. Berman, *Law and Revolution*, 147.

characteristic feature was the application of reason to God and the mysteries of the faith. Although there had been resistance by some powerful figures, most notably St. Bernard of Clairvaux, the new professionalism had triumphed by the end of the twelfth century. As will be seen in the next chapter, theology would undergo further changes in the late Middle Ages, while natural philosophy and medicine would be totally transformed by a massive influx of new learning from outside of Western Europe. The steps in this dramatic process and the role of reason in the intellectual life of medieval universities must now be described.

3

REASON TAKES HOLD

Aristotle and the Medieval University

FOR MOST OF THE TWELFTH CENTURY, LEARNING IN WESTERN Europe was based on treatises that had originated within a Latin tradition that was comprised largely of handbooks, compendia, and encyclopedic works. Much of this knowledge had roots in Greek treatises written during the Hellenistic period.[1] Some of this encyclopedic knowledge reached Latin authors, who either knew enough Greek to understand it or had access to translations.

THE LATIN TRADITION OF LEARNING IN THE EARLY MIDDLE AGES PRIOR TO THE INFLUX OF NEW TRANSLATIONS

Two Roman authors of the first century A.D. made significant contributions. Seneca (d. A.D. 68) titled his most famous treatise *Natural Questions* and filled it with information about geographical and meteorological phenomena into which he interjected morals drawn from nature, a feature that made his book popular with Christians. Pliny the Elder (A.D. 23/24–79), Seneca's contemporary, wrote a massive encyclopedic treatise in 37 books titled *Natural History.* Almost anything was grist for Pliny's insatiable information mill. These two treatises and the numerous extant handbooks in Greek and Latin formed a basis for subsequent authors who wrote for a Latin-speaking audience. As we saw in Chapter 1, between the fourth and eighth centuries, such authors as Macrobius, Chalcidius, Martianus Capella, Boethius, Cassiodorus, Isidore of Seville, and Venerable Bede pro-

1. For this summary of available learning in the early Middle Ages, I draw upon my earlier work, *The Foundations of Modern Science in the Middle Ages* (Cambridge: Cambridge University Press, 1996), 12–17.

vided the means for understanding the cosmic structure and operation of the world. For natural philosophy in the form of cosmology and cosmic reflections, they relied primarily on the *Timaeus* of Plato, in a partial Latin translation and commentary by Chalcidius, to which was added bits and pieces of encyclopedic knowledge.

Underlying the meager natural philosophy of the early Middle Ages were the seven liberal arts – grammar, rhetoric, and dialectic or logic, called the *trivium;* and the four scientific disciplines: geometry, arithmetic, music, and astronomy, which came to be known as the *quadrivium.*[2] The Latin encyclopedists provided the substantive character of the seven liberal arts that endured throughout the later Middle Ages. Martianus Capella wrote a treatise on the seven liberal arts, the *Marriage of Philology and Mercury,* in which he describes each of the arts. In addition to the great contributions he made in logic (see Chapter 2), Boethius wrote treatises on each of the four quadrivial sciences, two of which – *Music* and *Arithmetic* – were extant throughout the Middle Ages and widely used. Cassiodorus argued that the liberal arts should form the basis of a Christian education. By the end of the seventh century, they served that function as well as they could.

But the fare that the seven liberal arts, natural philosophy, and medicine provided was meager. Treatises written in the eleventh and twelfth centuries were based on this slight body of secular literature. Despite these serious limitations, we saw that scholars in the twelfth century sought to understand nature and its operations and were eager to apply their God-given reason to mundane problems, as well as to those of the faith.[3] European scholars in the twelfth century had come to see the power and utility of reason, but had little to which they might apply it. Of the two traditional divisions of the seven liberal arts, the trivium, or linguistic component, and the quadrivium, or scientific component, scholars in the early Middle Ages had more success with the trivium than the quadrivium. Euclid's *Elements* were virtually unknown and there were no treatises in mathematical astronomy. Natural philosophy was at the level of Adelard of Bath's *Natural Questions,* which reveal a scholar who desperately wanted to analyze natural problems, but had virtually no base of natural knowledge on which to draw. Unlike natural philosophers, theologians and legal scholars in the twelfth century had a dif-

2. For the seven liberal arts, I also use Grant, ibid., 14–17.

3. Based on his study of rationalism at Chartres in the twelfth century, Newell ("Rationalism at the School of Chartres" [*Vivarium* 21 (1983)], 126) rightly concludes that "the Middle Ages did not have to wait for the rediscovery of Aristotle to have confidence in man's rational capabilities."

ferent problem: They had to organize and apply an already existing body of literature that had become chaotic, a task they did rather spectacularly.

In science and natural philosophy, however, things were quite otherwise. Without an influx of new treatises in the sciences and natural philosophy, twelfth-century European Christian scholars would have been left with an urge to use their reason, but little on which to exercise it. Given their energy and drive, they might have fashioned a body of science and natural philosophy over the next few centuries. But that would have delayed the science that eventually emerged in seventeenth-century Western Europe. Fortunately, European scholars were never put to that formidable test. A science and natural philosophy of momentous proportions lay ready at hand in the civilizations of Islam and Byzantium. Indeed, if we are to comprehend the emergence of science in Europe, we must first describe the massive introduction of new knowledge into Western Europe during the twelfth and thirteenth centuries. It was the end product of one of the most amazing occurrences in the history of humanity: the transmission of knowledge through three civilizations: Greek, Islamic, and West European, in that order; and by means of three distinct languages: Greek, Arabic, and Latin. My focus will be on the final phase of this extraordinary process: the passage of knowledge from Greek and Arabic into Latin.

THE TRANSLATIONS

With the capture of Toledo from Muslim forces in 1085 and the conquest of Sicily in 1091, Western Europeans acquired important centers of learning.[4] The expansion of Christendom coincided with the beginnings of those activities in northern Europe that were described in Chapter 2. Europeans had long been aware that Islam had a much higher level of knowledge and learning than they had. Although some contact, and a few translations from Arabic to Latin, had been made before, the reconquest finally brought European scholars into direct contact with that learning.

Over the course of the twelfth century, numerous European scholars sought to make the new learning available in the Latin language. To do this, they went to locales where these previously unknown literary treasures could be found. Thus it was that they went to Spain, especially to Toledo, and to Sicily for treatises in Arabic, and to Italy and Sicily for

4. The brief summary I present here is based on my earlier version in Grant, *The Foundations of Modern Science in the Middle Ages*, 22–32.

works in Greek. Since Greek and Latin are cognate languages, the most direct and simplest way was to translate from Greek to Latin. Translating from Arabic posed much greater problems. Not only is Arabic linguistically far removed from Latin, but the history of many Greek treatises in Arabic was also complex. The original Greek might have been rendered into other languages before it reached its Latin state. Obviously, the shortest route was Greek to Arabic to Latin. But many works followed a different path. Thus a given Greek work might have been translated successively into Syriac and Arabic, and then into Latin; or into Syriac, Arabic, Spanish and then Latin; or into Syriac, Arabic, Hebrew, and Latin. Chances for distortions were obviously much greater in translations from Arabic to Latin than from Greek to Latin. But, whether an original work was in Greek or Arabic, reasonably accurate translations could only be made where both the translating expertise and appropriate treatises were available. Judging from the numerous translations that were made into the Latin language, there was an intense and widespread desire to obtain as much of the new knowledge as possible.

The new knowledge was almost exclusively from the domains of science and natural philosophy. The humanities and literature played almost no role. Within the domains of science and natural philosophy, treatises became available on logic, mathematics, astronomy, optics, mechanics, natural philosophy, and medicine, as well as works on astrology, magic, and alchemy. Among the numerous translators, two were spectacular contributors: Gerard of Cremona (d. 1187), who translated exclusively from Arabic to Latin in the second half of the twelfth century, and William of Moerbeke (ca. 1220–1286), who translated exclusively from Greek to Latin in the latter half of the thirteenth century. Each of these translators alone could have altered the course of Western intellectual life. Gerard translated some 70 works and Moerbeke approximately 50. Gerard translated the basic works of Aristotle, the *Almagest* of Ptolemy, Euclid's *Elements,* the *Algebra* of al-Khwarizmi, and other mathematical works. He also translated numerous treatises by Islamic authors that significantly supplemented the corpus of Greek science and natural philosophy. By translating major medical texts of Galen, Avicenna, and Rhazes (al-Razi), Gerard raised the level of medicine and gave it a new foundation. William of Moerbeke translated most of the works of Aristotle, as well as many commentaries on Aristotle's works by Greek commentators in late antiquity, such as Alexander of Aphrodisias, John Philoponus, Simplicius, and Themistius. In 1269, he translated nearly all of the scientific works of Archimedes.

The combined translations of Gerard of Cremona and William of Moerbeke represented an enormous aggregation of works in which reason and rationality were prime factors. Western Europe had never seen anything like it and would never see its like again.

The Works of Aristotle

Numerous works vital for the advance of science were now available in the West. The impact of Euclid's *Elements* and Ptolemy's *Almagest* alone were capable of transforming the basis of science. It was as if the West had left a barren desert and moved to a richly watered oasis. But of the authors from the Greco-Islamic tradition of science and natural philosophy, none surpasses or equals Aristotle (384–322 B.C.), the great Greek philosopher and scientist.

The sheer number, range, and intrinsic importance of his works makes this claim indisputable. A count of the treatises listed in the table of contents of the Oxford English translation of Aristotle's works published by the Princeton University Press (1984) yields a count of 47 works, of which 16 are spurious or of doubtful authorship. Of the 31 authentic treatises, we find a number that laid the initial foundations for a variety of disciplinary studies. Aristotle is the acknowledged founder of two disciplines: logic (*Categories, On Interpretation, Topics, Sophistical Refutations,* and *Prior Analytics*) and biology *(Parts of Animals, History of Animals, Generation of Animals, Movement of Animals,* and *Progression of Animals).* He formulated a system of natural philosophy to describe the operations and structure of the world (in his *Metaphysics, Physics, On the Heavens, On the Soul, On Generation and Corruption, Meteorology,* and the *Small Works on Natural Things [Parva naturalia])* that was found serviceable for approximately 2,000 years until abandoned in the seventeenth century. The first book of Aristotle's *Metaphysics* represents the first history of philosophy and the first history of science, while his *Posterior Analytics* is the first extant treatise on the philosophy, or methodology, of science. He also wrote seminal treatises on government *(Politics),* ethics, *(Eudemian Ethics* and *Nicomachean Ethics),* rhetoric *(Rhetoric),* and literary criticism *(Poetics).*

Scholars estimate that approximately one-fifth of Aristotle's works has survived, but that this one-fifth is a good representative sample of his overall output. What is extraordinary, however, is the fact that

> [m]ost of what has survived was never intended to be read; for it is likely that the treatises we possess were in origin Aristotle's lecture-notes – they were texts which he tinkered with over a period of years and kept for his own use, not for that of a

reading public. Moreover, many of the works we now read as continuous treatises were probably not given by Aristotle as continuous lecture-courses. Our *Metaphysics,* for example, consists of a number of separate tracts which were first collected under one cover by Andronicus of Rhodes, who produced an edition of Aristotle's works in the first century BC.[5]

It was the edition of Aristotle's works by Andronicus of Rhodes that passed down to subsequent generations of scholars and students in the civilizations of Byzantium, Islam, and the Latin West.[6] Scholars in these civilizations were reading Aristotle's private notes and not his polished treatises. They were rough drafts in which books and chapters were not always in the proper order. They are filled with difficulties. According to Jonathan Barnes: "There are abrupt transitions, inelegant repetitions, careless allusions. Paragraphs of continuous exposition are set among staccato jottings. The language is spare and sinewy."[7] Judged by his extant writings, Aristotle has always been regarded as a dull, pedantic author by comparison to Plato, whose elegance and polish place his dialogues among the literary masterpieces of the Western world. Ironically, Aristotle's published works, that is, those works that were deliberately released for a wider audience, were regarded, in the ancient world, as of high literary quality. Unfortunately, for Aristotle, those works did not survive. Only his extensive private notes and jottings have come down to us and shaped our views of him.

Given the truth of these assessments of Aristotle's works, why did they play such an overriding role in Islamic and Western thought? Why did Aristotle's rough works have a greater impact on Islamic and Latin thought than Plato's eminently more readable polished dialogues? In part, the forms of their respective works play a role. Plato's dialogues were always regarded as more difficult to render into other languages than was Aristotle's prose. Moreover, not until the twentieth century did it become known that Aristotle's works were his own private lecture notes. Until then, Aristotle's readers regarded his works as complete and final treatises on the subjects he covered. They had no idea of their true nature. Indeed, until the early

5. Barnes, *Aristotle* (Oxford: Oxford University Press, 1982), 3. For a more detailed and illuminating account of the fate of the treatises attributed to Aristotle, see Scott L. Montgomery, *Science in Translation: Movements of Knowledge Through Cultures and Time* (Chicago: The University of Chicago Press, 2000), 5–10.
6. For a brief account of the story of Adronicus's edition of Aristotle's works, see G. E. R. Lloyd, *Aristotle: The Growth and Structure of His Thought* (Cambridge: Cambridge University Press, 1968), 13–14.
7. Barnes, *Aristotle,* 3.

twentieth century, Aristotle's readers implicitly assumed that he had written all of his works at the same time, as if his thought had never evolved or developed, so that any contradictions or inconsistencies were only apparent, not real. The perennial challenge for students of Aristotle's thought was to reconcile apparent discrepancies – that is, to explain them away.

But the most important reason for Aristotle's dominance in the areas of logic, science, and natural philosophy over a period of 2,000 years is the sheer range of topics he covered and, for the most part, the comprehensive and authoritative manner in which he treated those topics. There was nothing to compare with it. The habit and custom developed of expressing one's own views on any of the topics Aristotle wrote about by writing a commentary on the work in which Aristotle treated the topic.

From the standpoint of reason, Aristotle's works were destined to have the greatest impact on Western thought. It was not only Aristotle's treatises that influenced Western Europe, but also a whole complex of literature and learning that grew up around them to form what we conveniently call "Aristotelianism." For along with Aristotle's own works came many commentaries on those works by Greek commentators of late antiquity and by Islamic authors, especially Avicenna (ibn Sina) and Averroës (ibn Rushd), who wrote during the ninth to twelfth centuries. Aristotle's influence extended beyond his own works and the commentaries thereon. His ideas also influenced other authors whose works entered Europe, especially *The Introduction to Astronomy* by Abu Ma'shar, which is really an astrological treatise, but one in which numerous ideas of Aristotle are summarized. It was through Abu Ma'shar's *Introduction to Astronomy* that Aristotle's influence may be said to have first entered Europe in a meaningful way.

Aristotle is the most important figure in science and natural philosophy during the Latin Middle Ages. This is undoubtedly attributable to the fact that his works in logic and natural philosophy were made the basis of an education in the arts faculties of medieval universities for almost five centuries. A vivid testimony to Aristotle's hold on the Western mind may be gleaned from the fact that approximately 2,000 manuscripts of his works have been identified. If 2,000 manuscripts survived, one can readily imagine thousands more that did not. Aristotle was easily the most influential of all authors studied and read in medieval universities. To understand what university students and scholars were exposed to for many centuries, we must first see what they received from Aristotle by way of an attitude toward reason and reasoned argument, and then see what medieval natural philosophers did with Aristotle's logic and natural philosophy.

ARISTOTLE'S LEGACY TO THE MIDDLE AGES

We have seen generally why Aristotle's works were so important for the history of science and natural philosophy. For this study, however, Aristotle's importance lies in the way he elevated reason and thought to the highest level of activity in the universe, and, even more than that, the way he actually used reason to understand and resolve problems and to organize his thoughts.[8] Aristotle believed profoundly in the rationality of the universe, a rationality that he believed was built into its very fabric, from the top down. He assumed that God was an immaterial, immobile substance who was associated with the outermost mobile sphere of the universe, the sphere of the fixed stars. Despite absolute immobility, God, the unmoved mover, causes the movement of the outermost sphere. In doing this, God himself remains immobile and imperturbable, blissfully unaware that there is a world that he somehow affects. The unmoved mover engages in only one activity: It thinks, and thinks only about itself, because it is the only object worthy of its own attention.[9] Thus does the noblest entity in the universe engage in thought, reasoned thought.

But Aristotle also wrote about human reason, elaborating on how it operates independently of the body, even concluding that we have an active and a passive reason, the former immortal, the latter perishing with the body.[10] Aristotle regarded only the reasoning part of the human soul as immortal and held that humans were like the gods only in their ability to reason. With his lofty sense of the role of reason, Aristotle declares that "for man ... life according to intellect [i.e., reason] is best and pleasantest, since intellect more than anything else *is* man."[11] Although Aristotle and the Greeks seemed to lack a specific term for "reason,"[12] there can be no doubt that Aristotle, at least, had firm ideas about it.

8. Michael Frede declares that "Aristotle is the paradigm of an extreme rationalist. All strictly speaking known truths are truths of reason, seen to be true by reason, either immediately or mediated by deduction." At the conclusion, Frede asserts that for Aristotle, the acquisition of reason "is made possible primarily by perception and experience." See Frede, "Aristotle's Rationalism," in Michael Frede and Gisela Striker, eds., *Rationality in Greek Thought* (Oxford: Clarendon Press, 1996), 158 and 173.
9. Aristotle, *Metaphysics* 12.7.1072b. 14–30.
10. See Aristotle, *On the Soul (De anima),* Chapters 4 and 5.
11. *Nicomachean Ethics* 10.7.1178a.6. I have added the bracketed term and equated "intellect" with "reason." This brief passage is perhaps as close as we get to "Man is a rational animal," a sentiment that is often attributed to Aristotle, but which I have not found in his works.
12. According to Max Black, "Aristotle and other Attic thinkers had no single word or phrase equivalent to our 'rational' or 'rationality'. Thus Aristotle frequently uses the three words, *nous, logos,* and *dianoia* in discussing rationality." See Max Black, "Ambiguities of Rationality," in Newton Garver and Peter H. Hare, eds., *Naturalism and Rationality* (Buffalo, NY: Prometheus Books, 1986), 29–30.

But it is not only an abstract elevation of reason to the highest level of human activity that makes Aristotle so important for the subsequent emphasis on rationality in Western thought. It is, rather, the way he actually used reason to approach and solve problems; it was his incessant striving for objectivity and his perpetual efforts to arrive at truth by the careful consideration of evidence and alternative arguments. Aristotle was instinctively a rationalistic thinker whose treatises are models of reasoned argumentation. What Aristotle conveyed to his readers in the Latin Middle Ages was a way of approaching problems and an overall system of the world that was embedded in his various works. Although his system of the world was of the greatest importance to medieval scholars, who saw the world through his eyes, it is Aristotle's method of investigating problems that taught medieval natural philosophers how to use reason in their inquiries about the operations of the physical world. So dominant were his works that they could hardly avoid serving as models for his university followers.

To understand things properly, Aristotle thought it advisable to describe the opinions of others, usually his predecessors, and by so doing to identify important problems.[13] This is the historical dimension of Aristotle's approach. In the first book of the *Metaphysics,* he presents the opinions of numerous pre-Socratic philosophers and of Plato on the principles that constitute all things. After his lengthy summary, Aristotle says that all told, the early philosophers named all the causes of things, but did so "vaguely; and though in a sense they have all been described before, in a sense they have not been described at all. For the earliest philosophy is, on all subjects, like one who lisps, since in its beginnings, it is but a child."[14]

Aristotle sees his mission as one in which he brings order and clarity to what has been inadequately explained by his predecessors. In *Parts of Animals,* Aristotle takes up the problem of classification of animals and declares: "Some writers propose to reach the ultimate forms of animal life by dividing the genus into two differences. But this method is often difficult, and often impracticable."[15] He then spends the next three chapters showing why the method of dichotomy is unsatisfactory for classifying animals and then declares, "Having laid this foundation, let us pass on to our next topic."[16] At the beginning of his treatise on *On Generation and Corruption,* Aristotle says that he will "study coming-to-be and passing

13. For a substantial list of these, see Lloyd, *Aristotle,* 284–285.
14. *Metaphysics* 1.10.993a.12–16, trans. W. D. Ross.
15. *Parts of Animals* 1.2.642b.5–7.
16. *Parts of Animals* 1.4.644b.20.

away" and the processes associated with it, such as "growth and alteration. We must inquire what each of them is: and whether alteration has the same nature as coming-to-be, or whether to these different names there correspond two different processes with distinct natures."[17] To gain insight into these problems, Aristotle launches into an historical inquiry, explaining:

> On this question, indeed, the early philosophers are divided. Some of them assert that the so-called unqualified coming-to-be is alteration, while others maintain that alteration and coming-to-be are distinct.[18]

Aristotle then describes the opinions of Empedocles, Anaxagoras, and Leucippus (with whom he also associates Democritus) and then, in the second chapter, declares:

> In general, no one except Democritus has applied himself to any of these matters in a more than superficial way. Democritus, however, does seem not only to have thought about all the problems, but also to be distinguished from the outset by his method. For, as we are saying, none of the other philosophers made any definite statement about growth, except such as any amateur might have made.[19]

In his works on logic, Aristotle makes no mention of his predecessors because he found nothing relevant in their works. "As regards our present inquiry [into modes of argument]," he says in the *Sophistical Refutations,* "it is not that one part of it had been worked out before, and another not, but rather that it did not exist at all."[20]

In lieu of an historical context, Aristotle often approaches problems as puzzles *(aporia)* to be solved, where it is important to identify difficulties and to present possible solutions.[21] We see this in its most dramatic form in his *Metaphysics,* where in the opening paragraph of the first chapter of the third book, Aristotle explains his procedure:

> We must, with a view to the science which we are seeking, first recount the subjects that should be first discussed. These include both the other opinions that some have held on certain points, and any points besides these that happen to have been overlooked. For those who wish to get clear of difficulties it is advantageous to state

17. *On Generation and Corruption* 1.1.314a.4–6.
18. *On Generation and Corruption* 1.1.314a.7–8.
19. *On Generation and Corruption* 1.2.315a.31–315b.2.
20. *Sophistical Refutations* 183b.34–36; translated by Lloyd, *Aristotle,* 285.
21. See Barnes, "Life and Work," in Jonathan Barnes, ed., *The Cambridge Companion to Aristotle* (Cambridge: Cambridge University Press, 1995), 23.

> the difficulties well; for the subsequent free play of thought implies the solution of the previous difficulties, and it is not possible to untie a knot which one does not know. But the difficulty of our thinking points to a knot in the object; for in so far as our thought is in difficulties, it is in like case with those who are tied up; for in either case it is impossible to go forward. Therefore one should have surveyed all the difficulties beforehand, both for the reasons we have stated and because people who inquire without first stating the difficulties are like those who do not know where they have to go; besides, a man does not otherwise know even whether he has found what he is looking for or not; for the end is not clear to such a man, while to him who has first discussed the difficulties it is clear. Further, he who has heard all the contending arguments, as if they were parties to a case, must be in a better position for judging.[22]

For the remainder of the chapter, Aristotle lays out question after question that he will consider in subsequent chapters, in the manner described.

In the opening chapter of his *Politics,* Aristotle emphasizes the usefulness of resolving compounded entities into their constituent elements. "As in other departments of science," he declares,

> so in politics, the compound should always be resolved into the simple elements or least parts of the whole. We must therefore look at the elements of which the state is composed, in order that we may see in what the different kinds of rule differ from one another, and whether any scientific result can be attained about each one of them.[23]

Later in the *Politics,* we see the questioning manner in which Aristotle approached political and governmental questions, using the same approach he used in all his other investigations. "Having determined these questions," he declares that

> we have next to consider whether there is only one form of government or many, and if many, what they are, and how many, and what are the differences between them.
>
> A constitution is the arrangement of magistracies in a state, especially of the highest of all. The government is everywhere sovereign in the state, and the constitution is in fact the government. For example, in democracies the people are supreme, but in oligarchies, the few; and, therefore, we say that these two constitutions also are different: and so in other cases.
>
> First, let us consider what is the purpose of a state, and how many forms of rule there are by which human society is regulated.... [24]

22. Aristotle, *Metaphysics,* bk. 3, ch. 1, 995a.24–995b.4.

23. Aristotle, *Politics* 1.1.1252a.17–21.

24. Aristotle, *Politics* 3.6.1278b.6–16.

This is typical of Aristotle's methodology and procedure. He usually establishes a basis for proceeding, setting out alternatives and defining terms and concepts. Thus, in his treatise *On the Heavens,* he declares that not every body is either light or heavy, but having said that, he realizes that he has not yet defined these terms and promptly states:

> We must explain in what sense we are using the words 'heavy' and 'light', sufficiently, at least, for our present purposes: we can examine the terms more precisely later, when we come to consider their essential nature. Let us then apply the term 'heavy' to that which naturally moves toward the centre, and 'light' to that which moves naturally away from the centre. The heaviest thing will be that which sinks to the bottom of all things that move downward, and the lightest that which rises to the surface of everything that moves upward.[25]

Aristotle clearly had the "scientific temperament." He sought to be objective and detached and gives the appearance of wanting to examine all relevant evidence for every problem. Above all, however, he sought to arrive at true generalizations and categorizations. Indeed, he always seems to divide and categorize things, whether dealing with knowledge, animals, or inanimate objects and their changes and motions. His division and analysis of knowledge played a monumental role in the history of thought and the use of reason.

Aristotle regarded all knowledge as belonging to one of three categories: practical, productive, and theoretical.[26] Practical knowledge guided conduct and was concerned with ethical and political behavior (Aristotle devoted the *Ethics* and *Politics* to this aspect of knowledge). Productive knowledge was directed toward making useful, as well as beautiful and artistic things (toward this end, Aristotle wrote his *Rhetoric* and *Poetics*). By contrast, Aristotle regarded theoretical knowledge as the highest form of knowledge, because it was knowledge for its own sake. He divided it into three parts: *physics* (φυσικη) (or natural philosophy, and less commonly, natural science, as it was sometimes called in the Middle Ages), which treats of physical bodies that have a separate existence and are capable of motion, and are therefore subject to change; *mathematics,* which treats of immovable objects, such as points, lines, surfaces, and volumes, that are abstracted from physical bodies, and therefore do not have separate existence; and, finally, there is *metaphysics,* also called *theology,* or *first philoso-*

25. Aristotle, *On the Heavens,* 1.3.269b.20–26.
26. Aristotle classifies knowledge and the sciences in *Metaphysics,* bk. 6, ch. 1. See also Jonathan Barnes, *Aristotle,* 23–24.

phy, which Aristotle regarded as the most exalted discipline because it was concerned with immaterial substances that were unchangeable and wholly separate from matter (to the minds of Aristotle and his medieval followers, the less change an entity suffered, the more perfect and noble it was). Within this class of beings is the supreme immaterial substance, God, also known as the "Unmoved Mover" or "Prime Mover."

Although Aristotle often invoked God and gods, he did not do so for reasons of piety. "Aristotle's gods are too abstract, remote and impersonal," argues a perceptive modern student of Aristotle, "to be regarded as the objects of a religious man's worship. Rather, we might connect Aristotle's remarks about the divinity of the universe with the sense of wonderment which nature and its works produced in him."[27] Aristotle's sense of wonderment about the world, his belief in the worthiness of impartial inquiry into natural phenomena, and his concern for empirical data derived by the senses appear in a famous passage from his biological treatise, *Parts of Animals,* where he says:

Of substances constituted by nature some are ungenerated, imperishable, and eternal,[28] while others are subject to generation and decay.[29] The former are excellent and divine, but less accessible to knowledge. The evidence that might throw light on them, and on the problems which we long to solve respecting them, is furnished but scantily by sensation; whereas respecting perishable plants and animals we have abundant information, living as we do in their midst, and ample data may be collected concerning all their various kinds, if only we are willing to take sufficient pains. Both departments, however, have their special charm. The scanty conceptions to which we can attain of celestial things give us, from their excellence, more pleasure than all our knowledge of the world in which we live; just as a half glimpse of persons that we love is more delightful than an accurate view of things, whatever their number and dimensions. On the other hand, in certitude and in completeness our knowledge of terrestrial things has the advantage. Moreover, their greater nearness and affinity to us balances somewhat the loftier interest of the heavenly things that are the objects of the higher philosophy. Having already treated of the celestial world as far as our conjectures could reach, we proceed to treat of animals, without omitting, to the best of our ability any member of the kingdom, however ignoble. For if some have no graces to charm the sense, yet nature, which fashioned them, gives amazing pleasure in their study to all who can trace links of causation,

27. Barnes, *Aristotle,* 64–65.
28. Aristotle is here alluding to the incorruptible celestial ether from which the planets, stars, and orbs are made.
29. In Aristotle's cosmos, all animate and inanimate entities that exist in the sublunar region are subject to generation and corruption.

and are inclined to philosophy. Indeed, it would be strange if mimic representations of them were attractive, because they disclose the mimetic skill of the painter or sculptor, and the original realities themselves were not more interesting, to all at any rate who have eyes to discern the causes. We therefore must not recoil with childish aversion from the examination of the humbler animals. Every realm of nature is marvellous: and as Heraclitus, when the strangers who came to visit him found him warming himself at the furnace in the kitchen and hesitated to go in, is reported to have bidden them not to be afraid to enter, as even in that kitchen divinities[30] were present, so we should venture on the study of every kind of animal without distaste; for each and all will reveal to us something natural and something beautiful. Absence of haphazard and conduciveness of everything to an end are to be found in nature's works in the highest degree, and the end for which those works are put together and produced is a form of the beautiful.[31]

Aristotle did not include logic within the classification of theoretical knowledge because he regarded logic as an essential instrument, or *organon,* that an educated person ought to have in order to know what is capable of demonstration and what is not. Aristotle thought it essential to acquire such knowledge before undertaking the study and analysis of the sciences.[32] Since Aristotle's division of knowledge was generally adopted in the Middle Ages, the application of reason to the theoretical sciences in that period involved logic, theology, mathematics, and natural philosophy, and the exact sciences, which were often called the *middle sciences* (for example, astronomy and optics), because they were derived from the application of mathematics to different aspects of natural philosophy and therefore lay somewhere in the middle between pure mathematics and natural philosophy.

Although medieval scholars believed that Aristotle had arrived at a faithful interpretation of the physical nature of the world, most of the principles and conclusions he derived have been subsequently shown to be inapplicable or inadequate, and have been abandoned. But this in no way diminishes the measure of Aristotle's worth and value for the history of medieval and early modern thought. To gain some idea of his contributions to the Middle Ages, we must understand what it was like for Aristotle and his medieval followers

30. That is, gods.
31. Aristotle, *Parts of Animals* 1.5.644b.21–645a.25. In this magnificent passage, Aristotle also reveals his bias for a hierarchical universe, where the celestial region is nobler than the terrestrial region, and for a teleological universe, one in which all activities are for an end or purpose.
32. Aristotle speaks of the study of logic in *Metaphysics* 4.3.1005b.3–5; 1006a.6–10; and in the *Nicomachean Ethics* 1.3.1094b.23–26.

to engage in scientific inquiry. Aspects of science that moderns take for granted did not form part of the methodology of medieval science. Careful, methodical observation of phenomena was lacking, as were carefully controlled experiments. Also missing was the systematic application of mathematics to physical phenomena. Although significant instances of these fundamental activities of modern science can be found in the late Middle Ages, they were sporadic and episodic, never routine and systematic.

In a culture such as that of the Middle Ages, in which the powerful tools for scientific research and inquiry just described were largely absent, how could nature be interpreted and analyzed in order to come to some understanding of a world that would otherwise be unknowable and inexplicable? The most powerful weapon available was human reason, as Aristotle employed it in those of his works that were so familiar to university scholars and students. In studying Aristotle, they could directly witness the way in which Aristotle employed reason to construct principles and generalizations about the world based on a modicum of observation that was often highly selective in order to justify this or that generalization.[33] For the most part, however, Aristotle relied on a priori reasoning to form a picture of the structure and operation of the world. Logic and reason served as the primary factors to determine the way the world had to be.

Medieval Latin scholars eagerly embraced Aristotle's methodology and his approach to the physical world, while adding important ideas about the cosmos from Christian faith and theology. In the grand, historical development of science, the conscious application of reason to the natural world was the first step in the process that would eventually produce modern science. Without the systematic use of reason, science would be impossible. In the ancient and medieval worlds, Aristotle's works represented the epitome of reason. His ideas, attitudes, and methods permeated and dominated the thought of Western Europe between 1200 and 1650, and perhaps even to 1700. They did so because Aristotle's works had been made the basis of a medieval university education. Virtually all students studied his works on logic, natural philosophy, and metaphysics. Aristotle's ideas were thus rooted in the curricula of medieval universities

33. Aristotle was much more empirically minded in his biological works, which include much observational data. He relied much less on observation in the works that had their greatest impact on the Middle Ages, namely, the corpus of natural books that included the *Physics, On the Heavens (De caelo), Meteorology, On Generation and Corruption,* and *On the Soul (De anima).*

where they became the common intellectual property of all educated Europeans. The reason and reasoned arguments that were characteristic features of Aristotle's works became commonplace in Western Europe. What, then, was the medieval university, where Aristotle reigned and where reason was institutionalized?

THE MEDIEVAL UNIVERSITY

Few institutions have had a greater impact on Western society than the university.[34] Universities as we know them today may differ radically from their sister institutions of the thirteenth century, but there is considerable resemblance and an unbroken history that firmly links them. If reason and rationality had a home during the late Middle Ages, it was surely in the university. The fact of its existence and the formal structure that it assumed, however, owe a great deal to the concept of a corporation, an institutional legal development that was unique to the West.[35]

The Formation of Europe's Corporate Spirit

Although the concept of a corporation was derived from Roman law, it was only in the twelfth century that it began to receive full and systematic development. From the late eleventh to the thirteenth centuries, corporation law was developed within the Roman Catholic Church, an outgrowth of the newly rationalized canon law. In Roman law as exemplified by the Justinian Code, it was the imperial authority that conferred the privileges of a corporation on a public entity.[36] As Berman explains,

> the church rejected the Roman view that apart from public corporations (the public treasury, the cities, churches) only collegia recognized as corporations by the imperial authority were to have the privileges and liberties of corporations. In contrast under canon law any group of persons which had the requisite structure and purpose – for example, an almshouse or a hospital or a body of students, as well as a bishopric, or indeed, the Church Universal – constituted a corporation without special permission of a higher authority.[37]

34. For more extensive coverage, see Grant, *The Foundations of Modern Science in the Middle Ages*, Chapter 3, 33–53.
35. See Toby Huff, *The Rise of Early Modern Science: Islam, China, and the West* (Cambridge: Cambridge University Press, 1993), 120.
36. For medieval corporations, I rely heavily on Berman, *Law and Revolution*, 215–221.
37. Berman, ibid., 219.

During the Middle Ages, it was commonplace for similar-minded and similar-oriented individuals to come together to form corporations. They came from all walks of life: business, Church, education, and the professions. Merchants of a certain kind, wool merchants, for example, would organize themselves into a guild, or corporation, as would craftsmen, such as weavers, millers, bakers, and masons. The most common Latin term designating such corporate organizations was *universitas* (less often, *collegium* or *corpus*). "In the end," as Toby Huff explains, "it was but an accident of history that the Latin universitas (corporation or whole body) came to refer exclusively to the places of higher learning that retain the name universities."[38]

The advantages of a corporation were many. It was a fictional entity upon which numerous legal rights were conferred. Any debt owed by the corporation was not also owed by its individual members. A corporation could own property, draw up contracts, and engage in court actions by suing, or being sued. It was also a prototype of representative government, since the members of a corporation elected their own officers, who could act on their behalf. Indeed, the actions taken by corporate officers were decided by a majority of the members in accordance with an old Roman maxim that "What touches all should be considered and approved by all" *(Quod omnes tangit omnibus tractari et approbari debet).*[39] The elected officer (or officers) had the right to represent the corporation in a court of law, or before church and state (the state could be a kingdom, duchy, or municipality). The objectives of corporations was to protect the interests of their constituent members. To achieve this, corporations had to promulgate laws and statutes, which the members were legally required to obey.[40] Hence, they became law-making bodies. The members gave up certain rights in order to acquire the protection of their respective corporations.

The corporation, or *universitas,* is of crucial significance for this study because the universities that had emerged by 1200 – Bologna, Paris, and Oxford – were self-governing corporations with numerous rights and privileges. It was these corporate entities, these universities, that would institutionalize reason and reasoned discourse in European society.

As we saw in the first chapter, in the system of education that prevailed in the twelfth century, masters usually taught in what were called cathedral schools, schools that were associated with a cathedral church that was the

38. Huff, *The Rise of Early Modern Science,* 135.

39. See Berman, *Law and Revolution,* 608, n. 54, and Huff, *The Rise of Early Modern Science,* 135. On this, Huff and Berman draw upon Gaines Post, *Studies in Medieval Legal Thought: Public Law and the State, 1100–1322* (Princeton: Princeton University Press, 1964), Chapter 4.

40. Berman, *Law and Revolution,* 219.

residence of a bishop. Here teachers and students congregated. Students were taught whatever was customary and available in the seven liberal arts, with logic as the most basic subject. Cathedral schools were located in urban centers that had been growing and developing for a few centuries. It was at such schools in a variety of cities that Peter Abelard and his many colleagues taught. Students and masters moved from school to school, the students seeking the most suitable master, the masters seeking to draw as many students as possible and to receive appropriate fees. Masters and students were often strangers in the cities in which they taught, and were taught, and usually lacked rights and privileges. Without some form of organization, teachers and students were at a disadvantage in negotiations with Church and municipal authorities on teaching and living conditions.

Under such circumstances, masters and students could see advantages in organizing themselves into corporations, in the manner of merchants and craftsmen. By the end of the twelfth century, such organizations were already in existence in Paris, Bologna, and Oxford. They were called "universities" and could be an association, or "university of masters" *(universitas magistrorum),* or "university of students" *(universitas scholarium),* or a "university of masters and students" *(universitas magistrorum et scholarium).* The University of Paris, perhaps the most important university in medieval Europe, was a university of masters, since it was the teaching masters who organized the university as an institution. Each subdivision of a university was itself a corporation. At Paris, each of the four faculties of the complete university – arts, theology, medicine, and law – was a corporation with rights and privileges, as was each of the four nations, or subdivisions, of the arts faculty of the University of Paris, namely the French, Norman, Picard and English-German nations. Each corporation of teaching masters was self-governing and controlled the affairs of its own corporate domain, making rules for curriculum, examinations, admission of new masters to its faculty, and the granting of degrees. By the end of the Middle Ages, more than 60 universities were in existence, having been recognized by popes and secular rulers, who regarded them as prestigious institutions.[41]

The Curriculum of Medieval Universities

While the concept of a corporation was vital for the formation of medieval universities, the translations that had been ongoing through most of the

41. For a list to 1500, see Jacques Verger, "Patterns," in Hilde de Ridder-Symoens, ed., *A History of the University in Europe,* Vol. 1: *Universities in the Middle Ages* (Cambridge: Cambridge University Press, 1992), Chapter 2, 62–65.

twelfth century also played a monumental role in their emergence and development. By 1200, a body of literature in science and natural philosophy had come into existence in Western Europe. Based largely on translations from Greek and Arabic into Latin, the new literature represented a knowledge explosion. The seven liberal arts were expanded with treatises never before known, especially in the sciences, and by treatises that were more complete versions of works already known. But the major portion of this new literature was relevant to natural philosophy, a subject that had not been formally taught and was known only in a rudimentary form, centered on part of Plato's *Timaeus*. The newly translated natural philosophy was, of course, Aristotle's natural philosophy, along with the commentaries on those treatises and works that incorporated Aristotle's ideas. Medical works also formed part of the new legacy.

Here was a body of literature that was ready-made to serve as a curriculum for the newly emerging universities. But of the four faculties that comprised any complete medieval university, the translated material was directly relevant to only two: arts and medicine. The law and theology faculties had their own literatures that were initially largely independent of the translations. It was not long, however, before theology would become heavily dependent on Aristotle's works in natural philosophy, as we shall see in Chapter 6.

Aristotle's natural philosophy became the curriculum for the faculty of arts, which relegated the old seven liberal arts to introductory status and made natural philosophy the primary subject of an arts education. The new arts curriculum was particularly important because it was required of all students matriculating for a bachelor of arts degree. Because the arts degree was a prerequisite for matriculation in the higher faculties of theology, medicine, and law, virtually all students at a medieval university took the natural philosophy curriculum, which was primarily Aristotle's natural philosophy. It was the curriculum they all shared. Hence, its great significance. A study of the natural philosophy that was taught and discussed at medieval universities reveals a highly rational enterprise. Since Aristotelian natural philosophy was the basic curriculum for all students for some five centuries, it is obvious that natural philosophy, with its rationalistic characteristics, was institutionalized in the universities. Educated individuals in medieval society were thus exposed to a curriculum that emphasized a reasoned approach to problems about the physical world.

The Analytical Character of Learning at Medieval Universities

Medieval university education in the arts faculties was usually concerned with the seven liberal arts and three categories of philosophy: natural, moral, and metaphysical, known as the "three philosophies." In the trivium of the seven liberal arts, students were exposed to logic (dialectic), grammar, and rhetoric; within the quadrivium, they learned arithmetic, geometry, astronomy, and music. Thus, not only did they learn logic, the very embodiment of reason, but they also studied inherently rational disciplines, such as arithmetic and geometry, which they could apply to astronomy, as well as to other exact sciences, such as optics and statics. Among the philosophical disciplines, natural philosophy was taught to both undergraduates and to students matriculating for a master of arts degree. Moral philosophy and metaphysics were subjects taught primarily to graduate students in the arts. But reason was also applied to the three philosophies. In all of these disciplines, Aristotle provided the model of procedure, although in many instances medieval scholars abandoned some of his solutions, or simply went beyond him. Over the duration of a four-year university course leading to the bachelor's degree, or a six-year course that included a master of arts degree, university students were exposed to the subject areas just mentioned.

A medieval university education in arts was primarily an education in logic, natural philosophy, and the exact sciences, where reason functioned as the most important tool of interpretation and analysis. In the absence of courses in literature and history and other humanities subjects, the medieval university offered an education that was overwhelmingly oriented toward analytical subjects: logic, science, mathematics, and natural philosophy. Modern university students face nothing comparable. The incredible array of course offerings and majors available in a large, modern university makes it possible for students to avoid, almost completely, the rigors of analytical courses. This was not possible in the late Middle Ages, when the curriculum was much the same for all students, and was overwhelmingly analytical and rational. Never before had an analytical curriculum of such extent and range been implemented anywhere.

Because the curriculum was basically Aristotelian in its methodology and objectives, there was little interest in seeking out new knowledge and making new discoveries. The purpose of medieval education was, rather, to discover eternal truths about the world, to explain why things are as they are and will remain much as they are. The idea was to learn about the structure and operation of the world, not, however, for the purpose of manipulating it, or accumulating knowledge for the advancement of the human race. Such objectives and ideals were still in the distant future and would not emerge until the sixteenth and seventeenth centuries, and even later. The purpose of medieval learning was to know about these things so that we might understand how God's creation works, or, in the tradition of the ancient Greeks, and especially of Aristotle, to know about these things because knowledge, especially theoretical knowledge, is a noble end in itself.

The rationalistic curriculum that would achieve these ideals and begin Europe's long march toward the Age of Reason was first institutionalized in medieval universities. Nothing like it had ever been done before. And yet, despite advancements made by various individuals in logic and in other disciplines, universities were dedicated to an essentially conservative curriculum that prevailed until the seventeenth century. It is ironic that as the universities maintained and defended an ossified curriculum, the rest of Western European society was advancing on many fronts at an amazing pace (for some sense of these dramatic advances, see Chapter 7, the section on "Contemporary Attitudes toward 'medieval' and 'Middle Ages'"). Before the curriculum was discarded as too narrow and obsolete, the medieval universitiy had established itself as the locus for reasoned discourse. Universities became places where society expected to see the application of reason to problems of nature, law, theology, and medicine, the major subjects of study.

To appreciate and understand how reason operated at the medieval university, it is necessary to examine the literary formats that were used to organize the various subject matters; and to sample briefly a few treatises that served as vehicles for disseminating learned medieval views about the physical and spiritual worlds.

Scholastic Literature: Form and Example

Medieval university education was based upon authoritative texts. These were inherited from the ancient world – for example, from Aristotle, Euclid, Ptolemy, Hippocrates, and Galen; from Islamic authors – from Avicenna (ibn Sina), Averroës (ibn Rushd), Rhazes (al-Razi), and Alhazen (ibn al-Haytham); or they were produced by scholastic authors who began to produce authoritative Latin works in various fields as early as the twelfth century, even before the appearance of universities, as the works by Gratian in law and Peter Lombard in theology attest. In presenting their opinions on various subjects, teaching masters drew their material from such standard authoritative university textbooks, especially from Aristotle's works on natural philosophy. The aim of every teaching scholar was to organize the presentation of opinions on a specific theme in a manner that would be regarded as objective and definitive. The literature that was developed to embody these goals is itself a good illustration of the emphasis on rationality that is characteristic of medieval learning.

During the late Middle Ages, scholars could choose to write about a subject in a variety of ways. They could write a straightforward treatise *(tractatus)* on a given theme. For example, a number of authors wrote treatises on proportions in which they presented theorems about mathematical proportionality and then applied those theorems to problems in natural philosophy. In the fourteenth century, Thomas Bradwardine and Nicole Oresme wrote treatises in this genre, to which they assigned titles such as *Treatise on Proportions (or Ratios),* or some variant thereof.[42] Most treatises in the exact sciences were thematic tractates, as, for example, Roger Bacon's *Perspectiva* and John Pecham's *Communis perspectiva,* which were concerned with optics, or perspective, as it was usually called in the Middle Ages. Many medical works also fall into this category, as does *The Anatomy of Mundinus* by Mondino de' Luzzi (ca. 1265–1326). In theology, Thomas Aquinas's great theological treatise, *Summa theologiae,* is a comprehensive study of theology independent of any text.

Rather than write thematic treatises, however, medieval authors, especially in natural philosophy and theology, more frequently chose to present their opinions by way of a commentary on an authoritative text, a practice that gave rise to a vast commentary literature in natural philosophy and theology. At first, university teachers lectured by reading the required texts and explaining, or glossing, terms and expressions that they deemed obscure or unclear. The written counterpart to this lecturing technique was the glossing of terms between the lines of text or in the margins. Other techniques followed upon the mere glossing of a text. One approach, which could be presented either as a lecture or written text, or both, was to intermingle a summary, or paraphrase, of a portion of text with an explanation and commentary on it, a technique that Albertus Magnus (Albert the Great) used. Another technique was to read, or write, a passage from the required text and then explain its meaning. In the process of explaining the meaning of the passage, a teacher might also cite the opinions of other commentators. In the written commentaries, a section of text would be presented, followed by its commentary, which in turn was followed by the next segment of text, followed by its commentary, and so on sequentially through the entire treatise.

42. Bradwardine called his treatise *Treatise on Proportions, or On the Proportions of Speeds in Motions*: see H. Lamar Crosby, ed. and trans., *Thomas of Bradwardine His "Tractatus de proportionibus," Its Significance for the Development of Mathematical Physics* (Madison: The University of Wisconsin Press, 1955). Oresme titled his treatise *On Ratios of Ratios (De proportionibus proportionum).* See *Nicole Oresme "De proportionibus proportionum" and "Ad pauca respicientes,"* edited with Introductions, English Translations, and Critical Notes by Edward Grant (Madison/Milwaukee: The University of Wisconsin Press, 1966). As can be seen, I have translated the Latin term *proportio* as "ratio," whereas H. Lamar Crosby has rendered it as "proportion."

This form was most evident in the Aristotelian commentaries by Averroës, the Muslim commentator who exercised a great influence on the West after the translation of his numerous Aristotelian commentaries in the thirteenth century. Averroës's format was used by such eminent scholastic authors as Walter Burley and Nicole Oresme. Some authors chose not to include the full text of each section of Aristotle's work, but merely cited the first few words (cue-words) of each section, added their commentary, and then cited the cue-words of the next section, and so on, as did Roger Bacon and Thomas Aquinas.[43] The commentaries on authoritative texts described thus far were all intended to explicate as clearly as possible the meaning and significance of the text itself. The words, in the form of ideas and thoughts, of the original author were the focus of this approach.

By the thirteenth century, however, another method had evolved in natural philosophy that would eventually prove the dominant form of medieval commentary on an authoritative text. Instead of commenting on the text by summarizing it, or setting it out verbatim, the practice developed of raising questions about themes and ideas in the text.[44] This was initially done in the classroom and then in the written texts from which masters lectured and from which students then studied. At first, questions were posed and discussed at the end of the commentary, but in time they completely displaced the commentary. The result was the emergence of the question form of scholastic literature, a form that became almost synonymous with what has come to be known as the scholastic method.[45]

The question form of literature is historically linked to the disputations that were held in the medieval universities and which formed part of the educational process. That process was founded on the "ordinary disputation" *(disputatio ordinaria)* in which a master posed a question and assigned students to defend the affirmative and negative sides. The master was then responsible for "determining," or resolving, the question. Because they were frequently exposed to disputations, medieval students gained a good sense of how to marshal arguments on both sides of a question. At the end of their fourth year, students were required to determine a question by them-

43. See *Commentary on Aristotle's "Physics" by St. Thomas Aquinas,* translated by Blackwell, Spath, and Thirlkel, Introduction by Vernon J. Bourke (New Haven: Yale University Press, 1963), xvi–xvii.

44. Questions were already used in theology in the twelfth century; see Monika Asztalos, "The Faculty of Theology" in Ridder-Symoens, ed., *Universities in the Middle Ages,* 410.

45. For a more detailed presentation, see Grant, *The Foundations of Modern Science,* 40–41, 127–131.

selves after having heard all the affirmative and negative arguments. If this requirement was completed to the satisfaction of the master and his colleagues, the student was granted the Bachelor of Arts degree.

The scholastic questions *(questiones)* literature was based on the structure of a medieval disputation. The question *(questio)* mode of explaining a text was in vogue for approximately four centuries during which the format remained remarkably stable. In Chapter 5, on natural philosophy, a complete question will be presented to show the rational and scientific character of the genre. Here it is sufficient to outline the structure of the typical question.

Every question began with an enunciation of the problem, usually asking whether (*utrum* was the Latin term) this or that is the case. For example, "whether there could be an infinite dimension," or "whether the earth always is at rest in the center of the universe."[46] As in a disputation, arguments were presented for or against the enunciated thesis. If the author offered a series of affirmative arguments, anywhere from 1 to 10, or even more, he would usually end up defending a version of the negative side. Or, the reverse might obtain: The author presents a sequence of negative arguments, from which it could usually be inferred that he would ultimately defend the affirmative side. These initial arguments were called the "principal arguments" *(rationes principales)*. They were followed by a statement of the opposite position, which might take the form "Aristotle says the opposite," or "Aristotle determines the opposite," or "The Commentator [Averroës] affirms the opposite," and so on. After representing the opposite opinion, the author might then explain his understanding of the question, raise doubts about it, and even define ambiguous terms in the question. The author was now ready to express his own opinions, usually by way of distinct, numbered conclusions. When this task was completed, the author took the final step: a point by point response to each of the principal arguments enunciated at the outset of the question.

What follows is a schematized outline of a typical medieval question:

1. The statement of the question;
2. principal arguments *(rationes principales)*, usually representing alternatives opposed to the author's position;

46. The first question is from Albert of Saxony's *Questions on the Physics*, bk. 3, qu. 11 (in Patar's edition it is bk. 3, qu. 12, 557–564); the second is from John Buridan's *Questions on De caelo*, bk. 2, qu. 22. Translations of the enunciations of these questions appear in Edward Grant, *A Source Book in Medieval Science* (Cambridge, MA: Harvard University Press, 1974), 201, 205.

3. opposite opinion *(oppositum,* or *sed contra),* a version of which the author will defend. In support of this opinion, the author often cites a major authority, often Aristotle himself; or cites from a commentary on a work of Aristotle; or invokes a theological authority in a theological treatise, such as a *Commentary on the Sentences of Peter Lombard;*
4. qualifications, or doubts, about the question, or about some of its terms [optional];
5. body of argument (author's opinions by way of a sequence of conclusions);
6. brief response to refute each principal argument.[47]

With the completion of this brief sketch of the form and structure of medieval scholastic literature, we must next turn our attention to the substantive ways in which reason was employed in the literature of the faculties of arts and theology. In Chapters 4 and 5, I shall consider logic and natural philosophy, respectively, which were the primary subjects of the arts faculty; and in Chapter 6, consider theology, which was, of course, the subject matter of the faculties of theology. I concentrate on these two faculties because they were the ones most intimately involved in the relationship between reason and revelation, a relationship that established the boundary conditions for the use of reason in the Middle Ages. The constraints placed upon reason in the Middle Ages came from beliefs about the faith as embodied in revelation. Since the faith, and therefore revealed truth, was interpreted solely by professional theologians who taught, or were trained, in theological faculties, and since reason was regarded as the legitimate domain of natural philosophers and logicians who taught in the arts faculties, we can see that any conflict between reason and revelation would directly involve the faculties of theology and arts. While the classic struggle between reason and revelation begins in the Middle Ages, it continues on into the seventeenth and eighteenth centuries, when many will assert the complete independence of reason from revelation, and even go beyond and judge revelation by the criteria of reason.

Although, for reasons already stated, the faculties of medicine and law are not central to the major objectives of this volume, we must briefly consider them before turning to the faculties of arts and theology. As university subjects, medicine and law were of great importance for the establishment of reasoned argumentation, largely because each discipline used reason as

47. For a similar description of the question format but specifically tailored for a theological treatise, see Gilson, *History of Christian Philosophy in the Middle Ages* (London: Sheed and Ward, 1955), 328.

the basic means of understanding its subject matter. At this point, therefore, I shall give brief summaries of the role reason and rationality played in each of these major university disciplines, taking medicine first.

Medicine and the Medical Faculties in the Medieval Universities

Although faculties of medicine were established in most medieval universities, three of them were especially important: Montpellier, Paris, and Bologna. Probably no more than half of medical practitioners during the Middle Ages were university trained. Given the state of medical knowledge in the Middle Ages, it is unlikely that a university-trained physician was any more successful than one who never attended a university and was self-taught, or simply a charlatan.[48] It is probable, however, that all things being equal, the university-trained physician had more prestige than one not so trained. For the university-trained physician was not just learned in medical doctrine derived from various authoritative university texts, but he was also likely to be knowledgeable about natural philosophy, since most university-educated physicians had bachelor of arts degrees and were therefore familiar with Aristotelian natural philosophy. Faculties of medicine were drawn to arts faculties, no doubt, because "venerable truisms, repeated by university medical writers, asserted that all the liberal arts and natural philosophy were necessary for medicine."[49] Most who wrote on medicine "had extensive training in logic, astrology, and natural philosophy."[50] Indeed, some physicians were outstanding natural philosophers, as, for example, Peter of Abano (d. ca. 1316). Most physicians were in agreement that the liberal arts were vital for the study of medicine. Of all arts subjects, logic was regarded by many as the most important for medical purposes.[51] Some

48. See Nancy Siraisi, "The Faculty of Medicine," in H. de Ridder-Symoens, ed., *A History of the University in Europe,* Vol. 1: *Universities in the Middle Ages* (Cambridge: Cambridge University Press, 1992), 361–363.

49. Siraisi, ibid., 375.

50. Siraisi, ibid. To gain some idea of what they studied, see Chapters 4 (logic) and 5 (natural philosophy) of this study. Astrology was also studied because it was an integral part of medical prognostication during the Middle Ages. See Roger French, "Astrology in Medical Practice," in Luis García-Ballester, Roger French, Jon Arrizabalaga, and Andrew Cunningham, eds., *Practical Medicine from Salerno to the Black Death* (Cambridge: Cambridge University Press, 1994), 30–59.

51. See Pearl Kibre, "Logic and Medicine in Fourteenth Century Paris," in A. Maierù and A. Paravicini Bagliani, eds., *Studi sul XIV secolo in memoria di Anneliese Maier* (Rome: Edizioni di storia e letteratura, 1981), 415–416.

physicians at the University of Padua even sought to make medicine more rigorously scientific by approaching it as a mathematical discipline.[52]

As with the faculties of arts and theology, medicine was taught from authoritative text books, from Greek or Islamic sources. The most notable authoritative texts from Greek sources came from Galen, Dioscorides, and Hippocrates; and from Islamic sources, the most important were Avcienna (his *Canon of Medicine* was widely used), Haly Abbas, al-Razi (Rhazes), Averroës, and Albucasis, the last named having produced a famous book on surgery.[53] Latin physicians also contributed works to the corpus of medical authorities.

Medicine was both a practical discipline, or an art, that involved healing the body, and it was also a theoretical discipline that sought to determine a corpus of unchanging truths. The theoretical aspects of medicine made it a science *(scientia).* That science gained greater prestige from Aristotle's conception of a science and the ways scientific knowledge was acquired.[54] Not only medicine was influenced by Aristotle's natural philosophy and his methodological conceptions in the *Posterior Analytics* and *Metaphysics;* the medical faculties were also influenced by the way natural philosophy had been taught by the scholastic method (see Chapter 5). They utilized the method of disputation and often proceeded by organizing commentaries on medical texts into a series of probing questions, each of which was highly structured.

On its theoretical side, medieval medicine was in many ways akin to natural philosophy in its approach. As Nancy Siraisi explains:

> The introduction of scholastic methods of disputation and reconciliation into medicine led to the useful result that learned physicians isolated clearly the discrepancies among ancient authors (most notably between Galen and Aristotle). But scholastic medicine like scholastic natural philosophy, is criticized not only for the premium it set on honing disputational skills, but also for providing only tenuous connections between theory and the data of experience. Yet it is hard to see how learned physicians could have developed any system of relating theory and experience different from that of the contemporary Aristotelian natural philosophers whose ideas about the nature of, and means of arriving at, scientific knowledge they shared. Moreover, it seems likely that their experience of medical practice

52. See Brian Lawn, *The Rise and Decline of the Scholastic 'Quaestio Disputata' With Special Emphasis on its Use in the Teaching of Medicine and Science* (Leiden: E. J. Brill, 1993), 76.
53. For a list of texts, see Siraisi, "The Faculty of Medicine," 377–379.
54. For more on Aristotle's importance for medicine, see García-Ballester, "Introduction," *Practical Medicine from Salerno to the Black Death,* 11.

familiarized professors of medicine to a greater extent than their colleagues in philosophy with the need to take account of the empirical and the particular. Certainly, learned physicians were very conscious of the dual nature of medicine as craft and learning and devoted considerable discussion to the ambiguous relations between its two aspects.[55]

Thus, while the nature of medical practice compelled attention to empirical data, medicine sought to be as scientific and theoretical as possible, which meant that it strove to emulate natural philosophy. The emphasis was clearly on reasoned and logical arguments. The question form of literature was commonly used in the medical faculties in the thirteenth and fourteenth centuries.[56] But the rational character of medieval medicine and of European society as a whole is nowhere better exemplified than in the acceptance of human dissection in postmortems to determine the cause of death. This occurred in Italy in the late thirteenth century, in a way that is little understood. But the practice soon became established.

Bernard Tornius (1452–1497)[57] conveys a good idea of a postmortem, which he performed in the second half of the fifteenth century. Although the interpretations presented are perhaps strange and alien, Tornius displays a scientific attitude and approach in his work, as we see in what follows:

Worshipful Judge, I grieve over thy sad lot, for to lose one's offspring is hard, harder to lose a son, and hardest [to lose him] by a disease not yet fully understood by doctors. But for the sake of the other children, I think that to have seen his internal organs will be of the greatest utility. Now, therefore, I will not hesitate to state as briefly as I can what we have seen and draw my honest conclusion and adduce remedies which in my judgment are advantageous.

In the first place, the belly appeared quite swollen, although the abdomen was thin. But after dividing according to rule the abdomen and peritoneum, we saw the intestines and the bladder, which was turgid and full of urine. Removing further the colon and caecum, there appeared in them more gross wind than filth. Then when the ileum and jejunum and duodenum were removed, two worms were found, quite large and white, showing phlegm rather than any other humor. After the intestines had been cut off from the mesentery, since nothing notable was found therein, seeing that the bladder was turgid, I had it cut open and a great

55. Siraisi, "The Faculty of Medicine," 384.

56. For a detailed discussion of the questions literature in medicine, see Lawn, *The Rise and Decline of the Scholastic 'Quaestio Disputata,'* Chapter 7 ("Medical *Quaestiones Disputate* c. 1250–1450"), 66–84.

57. Tornius "taught medicine at Pisa from 1478 until his death in 1497." Michael McVaugh's annotation in Grant, *A Source Book in Medieval Science,* 740, n. 1. (See next note.)

quantity of urine appeared, although before he died, as they reported, he discharged a large amount of urine. Afterwards we examined the liver, which was marked with certain spots like ulcers and somewhat swollen about the beginning of the chilic [i.e., portal] vein. But what is more remarkable, there appeared around the source of the emulgent veins in the hollow of the chilic vein an evident obstruction by which the whole cavity was filled with viscous humor for the space of a thickness of a finger, beyond which humor no blood was seen beneath, while the emulgent veins were full of blood, quite watery in character, and the swollen kidneys were also full of this sort of blood, or perhaps of much urinal wateriness admixed with it. Moreover, the ascending chilis [vena cava] had the branch to the heart filled with much blood, and the heart was much swollen, and so the auricles too appeared swollen beyond measure. When these were cut open a great part of the blood came out, and so almost all the blood was found near the heart. But the vein which carries the nourishing blood to the lungs was also full of similarly viscous humor and seemed wholly free from blood. Having seen this much, I did not examine further concerning anything else, since the cause of his death was apparent in my judgment.

From these facts, I infer, first, that this lad had contracted a great oppilation [obstruction] either from birth or in course of time, and it is safe to assume that matter of this sort was accumulated by gradual congestion rather than brought by a deflux from another member.

Second, I infer that those worms were generated after the beginning of his principal illness and were in no way the cause of his death.

Third, I infer that when transmission of blood through the chilic vein and the pulmonary vein was prevented, ebullition and fever resulted. And because in that blood there was much phlegm, that fever was like a phlegmatic [quotidian] one in many of its accidents, though from the manner of its oncoming and development it seemed like a double tertian. For every third day it came on worse in the night, as those present reported and I infer clearly from his restlessness and perceived from his pulse.

I infer, fourth, that those spots of the liver were generated after the oppilation.

Fifth and last, I infer that any son of yours of the same constitution is to be preserved to his twelfth year with the usual medicines which I will mention in closing.[58]

Tomius continues on with instructions on how to protect the remaining children and promises frequent visitations.

58. Translated by Lynn Thorndike, "A Fifteenth-Century Autopsy," in Thorndike, *Science and Thought in the Fifteenth Century: Studies in the History of Medicine and Surgery, Natural and Mathematical Science, Philosophy and Politics* (New York: Columbia University Press, 1929), 126–128. Thorndike's translation is reprinted with annotations by Michael McVaugh in Grant, *A Source Book in Medieval Science,* 740–742. Thorndike (p. 126) suggests that "in modern phraseology, the autopsy seems to have revealed that the boy suffered from multifold metastatic abscesses of the liver, the result of septicemia or pyleophlebitis."

Shortly after the beginning of postmortems, the dissection of humans was introduced into medical schools where it became institutionalized for the teaching of anatomy. This is no minor achievement. Most societies forbade cutting the dead human body.[59] Except for Egypt, human dissection had been forbidden in the ancient world. By the second century A.D., it was also forbidden in Egypt. It was never allowed in the Islamic world. Astonishingly, its introduction into the West at the end of the thirteenth century went unopposed by the Church.

That postmortems were soon transformed into anatomical dissections at the medieval medical schools is made evident by a famous anatomical textbook – the first ever – that was published in 1316 by Mondino de' Luzzi (ca. 1265–1326), professor of anatomy at the University of Bologna. Mondino based his text on his own anatomical investigations. In his introduction, he explains that his purpose is

> to give, among other topics, some of that knowledge of the human body and of the parts thereof which doth come of anatomy. In doing this I shall not look to style but shall merely seek to convey such knowledge as the chirurgical usage of the subject doth demand.
>
> Having placed the body of one that hath died from beheading or hanging[60] in the supine position, we must first gain an idea of the whole, and then of the parts. For all our knowledge doth begin from what is known. For though the known is oft vague and though our knowledge of the whole is of a surety vaguer than that of the parts, we yet begin with a general consideration of the whole.[61]

Dissection of cadavers at medical schools was an irregular occurrence. But it was done solely for teaching purposes, and not as research to advance knowledge of the human body. Nevertheless, the integration of human dissection as a normal part of a medical education was of monumental importance. Without the firm beginnings in human dissection initiated in the Middle Ages, the significant progress that was to come from the great anatomists of the sixteenth and seventeenth centuries – for example, Leonardo da Vinci (1452–1519), Andreas Vesalius (1514–1564), Gabriele Falloppio (1523–1562), and Marcello Malpighi (1628–1694) – could not

59. They were less sensitive about cutting the living human body, which occurred frequently in warfare and torture.

60. Most dissections were performed on the bodies of executed criminals.

61. See Grant, *A Source Book in Medieval Science,* 730. The translation is from *The Fasciculo di Medicina. Venice, 1493,* with an Introduction, etc., by Charles Singer (Florence: R. Lier & Co., 1925).

have occurred.[62] That a European, Christian society overcame the long-standing fear of, and deep hostility to, human dissection is a tribute to the rational character of medieval medicine and to those unknown individuals who made it possible.

Law in the Medieval Universities

Gratian and others had already systematized and rationalized canon law in the twelfth century while others were doing the same for civil law; that is, they applied the scholastic method and adapted Roman law to contemporary society. With the advent of the universities, the law faculties taught canon and civil law, using Gratian's *Decretum* for the former and the Roman law for the latter. Some law faculties taught only canon law, as did the law faculty of the University of Paris, which was forbidden to teach civil law. As the scholastic method took deeper root in the universities for the subjects of natural philosophy and theology, law was also significantly affected. A good indication of this is the elaborate terminology that was developed to distinguish the different kinds of commentaries and to identify various divisions within a given text. Logic also influenced law, as is obvious from the introduction into legal analyses of *insolubilia* (unsolvables; see Chapter 4 for this term) and *distinctiones* or *divisiones*.[63] There was also a term *(Quaestiones legitimae)* to indicate that a legal analyst was seeking to reconcile apparent or real contradictions in a legal text. Just as in the faculties of arts and theology, disputed questions were debated among students who were supervised by a master who resolved, or determined, the questions.

It is vital to realize that the legal profession, which was formed in the Middle Ages and existed primarily in the universities, "produced a science of laws."[64] In his excellent study of the formation of the Western legal tradition, Harold Berman distinguished nine "principal social characteristics of Western legal science in its formative period, especially as they were influenced by the universities."[65] University law schools disseminated through all of Western

62. I have relied here on Grant, *The Foundations of Modern Science in the Middle Ages,* 204–205. For a concise but penetrating discussion of medieval dissection, see also David C. Lindberg, *The Beginnings of Western Science: The European Scientific Tradition in Philosophical, Religious, and Institutional Context, 600 B.C. to A.D. 1450* (Chicago: University of Chicago Press, 1992), 342–345.

63. See Antonio García y Garcia, "The Faculties of Law," in Hilde de Ridder-Symoens, ed., *A History of the University in Europe,* Vol. 1: *Universities in the Middle Ages* (Cambridge: Cambridge University Press, 1992), Chapter 12, 395. For the numerous terms used, see 394–397.

64. Berman, *Law and Revolution,* 160.

65. Ibid., 161. The nine characteristics appear on 161–163.

Europe "a transnational terminology and method."[66] By the methods they used (his third point), "European universities ... made possible the construction of legal systems out of preexisting diverse and contradictory customs and laws." Indeed, the emphasis on "harmonizing contradictions" made possible the synthesis of various kinds of law. In a fifth point, Berman observes that law was taught alongside "theology, medicine, and the liberal arts," which used the scholastic method, so that the "law student could not help knowing that his profession was an integral part of the intellectual life of his time." In perhaps his most important point – the eighth – Berman declares that the "Western universities raised the analysis of law to the level of a science, as that word was understood in the twelfth to fifteenth centuries, by conceptualizing legal institutions and systematizing law as an integrated body of knowledge, so that the validity of legal rules could be demonstrated by their consistency with the system as a whole."

Although nationalism has eroded the transnational character of medieval law, Berman sees "something of the Bologna tradition, and something of the scholastic dialectic, survive nine centuries later – even in the law schools of America."[67] While nationalism may have produced a "decline of unity and common purpose in Western civilization as a whole," and thereby adversely affected the transnational status of law as it existed in the Middle Ages, Berman saw a hopeful sign in the "economic, scientific, and cultural interdependence" that was emerging "on both a regional and worldwide basis."[68] With the advent of the global economy and global interdependence, the medieval ideal of law as a transnational phenomenon may yet reemerge as a powerful force in human civilization.

Although I have, regrettably, given short shrift to medicine and law, I hope that I have, nonetheless, said enough about them to convince readers that they were highly rational intellectual disciplines within the university system of the Middle Ages. But, important as medicine and law are, they are not my primary concern, because they were not involved in the central issue that concerned reason in the Middle Ages: the relationship of reason to revelation. The subjects that are most relevant to this issue are logic, natural philosophy, and theology, the first two of which were taught in faculties of arts, the last of which was taught in faculties of theology. Since it was regarded as an essential tool for the two other disciplines, let us begin with logic.

66. This is the second of Berman's points (p. 162).
67. Berman, *Law and Revolution,* 163.
68. Ibid., vi.

4

REASON IN ACTION

Logic in the Faculty of Arts

THE FACULTY OF ARTS OF ANY MEDIEVAL UNIVERSITY HAD MORE students and more teaching masters than any of the three other higher faculties: theology, medicine, and law. This was necessarily true because the bachelor of arts and master of arts degrees were prerequisites for entry into the higher faculties. Therefore, all students began their careers in the arts faculty. By virtue of the subjects taught, the faculty of arts was the primary repository of reason in any medieval university. This is evident from the range of courses taught: astronomy, mathematics, optics, logic, and natural philosophy.[1] All were inherently analytical subjects except natural philosophy, which was nevertheless taught as if it were analytical.

THE OLD AND NEW LOGIC

Although logic was a basic subject, it was always regarded as an instrument for the critical study of all other areas of learning. We have already seen the role it played in the twelfth century. In the thirteenth century, Peter of Spain reiterated the central role that logic was accorded in the twelfth century when he declared that "[l]ogic is the art which provides the route to the principles of all methods, and hence logic ought to come first in the acquisition of the sciences."[2] With the translations of Aristotle's previously unknown logical works, which were added to the old logic, logic was given a substantial foundational role in the curriculum of the medieval university.

1. For a discussion of the courses in the arts faculty, see Gordon Leff, "The *Trivium* and the Three Philosophies," in H. de Ridder-Symoens, ed., *A History of the University in Europe,* Vol. 1: *Universities in the Middle Ages* (Cambridge: Cambridge University Press, 1992), Chapter 10.1, 307–336, and John North, "The *Quadrivium,*" in H. de Ridder-Symoens, ed., Chapter 10.2, 337–359.
2. Translated from Peter's *Summulae Logicales* by Alexander Broadie, *Introduction to Medieval Logic* (Oxford: Clarendon Press, 1993), 2.

The "old logic," as we saw, had been based on Boethius's Latin translations of Aristotle's *Categories* and *On Interpretation*, along with Boethius's commentaries and original treatises. However, Boethius had also translated most of Aristotle's logic, including the *Prior Analytics*, *Topics*, and *Sophistical Refutations*, although these translations were unknown and virtually unused from the sixth to the twelfth centuries. Sometime after 1120, they reemerged and, along with James of Venice's translation from the Greek of Aristotle's *Posterior Analytics*, formed the new logic, which, by 1159, was known to John of Salisbury, who mentioned them in his *Metalogicon*.[3] The "new logic," which came to be called "modern logic" *(logica moderna)*, replaced the "old logic" derived from Boethius (see Chapter 2). New textbooks on logic were written that reshaped Aristotle's logic and gave it a purely medieval flavor. Perhaps the most famous of these texts was the *Summulae logicales* of Peter of Spain (d. 1277), who later became Pope John XXI. Written in the 1230s, Peter's treatise became the standard logical text into the seventeenth century, by which time it had gone through some 166 editions.

To have a proper appreciation of the role of reason in the Middle Ages, one should realize that all undergraduates in the arts faculties of medieval universities studied logic. They did so because almost all scholars believed what Peter of Spain had declared in the opening words of his treatise just quoted, namely, that logic was essential for the study of all other disciplines. This meant that virtually all students who obtained the Master of Arts degree and continued their education in the professional schools of theology, law, and medicine had been trained in the basics of formal logic. Since logic was regarded as a tool that could, and should, be applied to all disciplines in which reasoning and argumentation were involved, physicians, theologians, and lawyers found occasions in which to use formal logic.

Six treatises by Aristotle formed the basic core of the new logic that emerged in the twelfth century. These are his *Categories*, in which Aristotle identifies 10 basic kinds of entities into which things may be divided, including substance, quantity, quality, relation, and place (the 10 categories are concerned as much with metaphysics as with logic);[4] *On Interpretation* (*De interpretatione* in Latin, although, during the Middle Ages, the treatise

3. See Bernard G. Dod, "Aristoteles latinus," in Norman Kretzmann, Anthony Kenny, and Jan Pinborg, eds., *The Cambridge History of Later Medieval Philosophy from the Rediscovery of Aristotle to the Disintegration of Scholasticism 1100–1600* (Cambridge: Cambridge University Press, 1982), Chapter 2, 46.

4. For more on Aristotle's *Categories*, see Lloyd, *Aristotle: The Growth and Structure of His Thought* (Cambridge: Cambridge University Press, 1968), 112–116.

was regularly cited as *Peri hermeneias,* a transliteration of the Greek title); *Prior Analytics,* which contains Aristotle's theory of the syllogism; *Posterior Analytics,* in which Aristotle presents his theory of scientific demonstration; *Topics,* where Aristotle studies nondemonstrative reasoning and shows how to argue effectively; and the *Sophistical Refutations,* a work in which Aristotle collected types of fallacies and analyzed them.[5]

There is little doubt that of these treatises, the *Sophistical Refutations* (or *Sophistic Refutations*) played the most significant and innovative role in the development of medieval logical thought. Beginning in the twelfth century – it appeared in Latin translation around 1120 – its influence on the course of medieval logic was great:

> Unlike the *Posterior Analytics,* which took a lot of getting used to, the *Sophistic Refutations* was relatively easy to get into and to understand. There was nothing especially obscure about it. And unlike the *Prior Analytics,* there was obviously a lot of work that remained to be done. Moreover, the discovery and avoiding of fallacies was *very* important in theological matters, where you had to keep straight what you were saying about the Trinity, and about the two natures but one person in Christ, and so on. In short, the *Sophistic Refutations* was tailor-made for the twelfth century to go to work on. And that is exactly what happened.
>
> The *Sophistic Refutations,* and the study of fallacy that it generated, produced a whole new logical literature. There was, for instance, the *sophismata* literature – as we find illustrated in Buridan's *Sophismata*.... And the theory of "supposition" ... developed out of the study of fallacies.
>
> In fact, whole new kinds of treatises came to be written on what were eventually called "the properties of terms" – semantic properties that were important in the study of fallacies. These treatises, and the logic contained in them, are the peculiarly mediaeval contributions to logic. It is primarily on these topics that mediaeval logicians exercised their best ingenuity.[6]

FORMS OF LITERATURE IN LOGIC

Treatises on logic took a variety of forms. The commentary and the questions forms of literature that were predominant in natural philosophy had counterparts in the literature on logic. Commentaries on logic were largely expository and explanatory, and did not add much to the sum total of

5. In the description of Aristotle's logical treatises, I follow Paul Vincent Spade, *Thoughts, Words, and Things: An Introduction to Late Mediaeval Logic and Semantic Theory,* Version 1.1 on Spade's Web site (copyright, 1998), Chapter 2 ("Thumbnail Sketch of the History of Logic to the End of the Middle Ages"), 11–12.
6. Spade, *Thoughts, Words, and Things,* Chapter 2, 39.

knowledge about logic. The questions format described in Chapter 3 found favor with some teaching masters. For example, in the fourteenth century, John Buridan wrote questions on topics drawn from a few logical treatises and embraced them with the umbrella title *Brief Questions on the Ancient Art (Quaestiones [breves] super Artem veterem).* He also wrote questions on Aristotle's logical treatises, including *Questions on the Prior Analytics (Quaestiones super librum Priorum)* and *Questions on the Book of the Topics (Quaestiones super librum Topicorum).* Like Buridan, Albert of Saxony also wrote *Brief Questions on the Ancient Art (Quaestiones super Artem veterem),* as well as *Questions on the Books of the Posterior Analytics (Quaestiones super libros Posteriorum).*[7]

Most logical treatises, however, do not fall into the genres of commentaries or questions. These formats were not customarily used for teaching and scholarship in logic, a subject that was usually taught and written about either as a whole or thematically. Works in the first category were compendia, or introductions, which covered the whole field of logic and often included the word *summa* (*summae* in the plural) or *summulae* (little *summas*) in the title. Examples are Peter of Spain's *Summulae logicales* (written in the 1230s),[8] the most popular logical textbook in the Middle Ages, and William of Sherwood's *Introduction to Logic* (composed in the mid–thirteenth century).[9]

A brief description of the contents of William of Sherwood's *Introduction to Logic* will convey a good idea of the range of topics that an undergraduate was expected to learn and master. William divided his book into six chapters, five of which correspond to five of Aristotle's logical treatises.[10] The first chapter, which corresponds to Aristotle's *On Interpretation,* treats statements. The "science of discourse," says William,

> has three parts: grammar, which teaches one how to speak correctly; rhetoric, which teaches one how to speak elegantly; and logic, which teaches one how to

7. Charles H. Lohr, "Medieval Latin Aristotle Commentaries" in *Traditio* 23 (1967), Authors A-F, 348–349 (Albert of Saxony), and 26 (1970) Authors: Jacobus-Johannes Juff, 164–166 (Buridan).
8. Peter of Spain became Pope John XXI (he died in 1277) and illustrates the intimate connection between university learning in logic, natural philosophy, and theology, on the one hand, and the Church on the other.
9. William of Sherwood's name also takes the form William of Shyreswood, or William of Shirwood. He died in 1249.
10. See *William of Sherwood's "Introduction to Logic,"* translated with an introduction and notes by Norman Kretzmann (Minneapolis: University of Minnesota Press, 1966), 16.

speak truly. Logic is principally concerned with the syllogism, the understanding of which requires an understanding of the proposition; and, because every proposition is made up of terms, an understanding of the term is necessary.[11]

Under statements, William considers 28 subtopics, including "statements and propositions," "the parts of a statement," "the quality of categorical statements," "the laws of opposition," "hypothetical statements," "assertoric and modal statements," "the six modes," and "the quantity of modal statements." In the second chapter, William treats "The Predicables," which he defines as "what *can be* said of something else."[12] Under this rubric, William includes sections on genus, species, interrelations of genera and species, differentia, property, and accident. The second chapter corresponds to Aristotle's *Categories:* the third corresponds to Aristotle's *Prior Analytics* and is devoted to the syllogism with its moods and figures.[13] It contains one of the oldest versions of the famous medieval mnemonic verse form that was intended as an aid in remembering the syllogistic figures and moods.[14] The fourth chapter treats dialectical reasoning based on the dialectical syllogism, which "produces opinion on the basis of probable [premisses]."[15] This chapter corresponds to Aristotle's *Topics.* Properties of terms are taken up in the fifth chapter, which includes the four subdivisions of signification, supposition, copulation, and appellation. This chapter does not correspond to any of Aristotle's works and represents an aspect of logic that is distinctively medieval. The topic of the final, and sixth chapter, is sophistical reasoning, under which William considers such themes as equivocation, ignorance regarding refutation, begging the original issue, treating what is not a cause as a cause, and treating more than one question as one. This final chapter is based upon Aristotle's *Sophistical Refutations.* The first four chapters of William's *Introduction to Logic* were drawn from the old logic, while the last two chapters represent what medieval logicians came to call

11. Ibid., 21.
12. Ibid., 51.
13. William defines figure and mood in the syllogism in section 5 of Chapter 3 (see *William of Sherwood's "Introduction to Logic,"* 60–61).
14. The verses are:

Barbara celarent darii ferio baralipton
Celantes dabitis fapesmo frisesomorum
Cesare camestres festino baroco
Darapti felapton disamis datisi bocardo ferison.

See *Sherwood's Introduction to Logic,* 66.
15. *Sherwood's "Introduction to Logic,"* 69.

modern logic *(logica moderna)*, much of which was formulated in the late Middle Ages. Sherwood's *Introduction* included the "property of terms" in Chapter 5 and the treatment of fallacies in Chapter 6.[16]

For the logicians who developed the *logica moderna,* the most important thematic topic was, as we saw, fallacies, a concern that took its origin from Aristotle's *Sophistical Refutations,* which was itself a collection of fallacies. A new logical literature on fallacies emerged. Treatises were often titled after the type of fallacy or problem that was the primary subject matter of the treatise. For example, in the thirteenth century, both Peter of Spain and William of Sherwood wrote treatises titled *Syncategoremata,* which were concerned with syncategorematic terms, that is, "words that cannot serve by themselves as subjects or predicates of categorical propositions.... For example, 'and', 'if', 'every', 'because', 'insofar', and 'under'."[17] In fact, syncategorematic terms do not signify anything. Thus, "prepositions, adverbs, conjunctions, and the quantifiers 'all' and 'some'"[18] are syncategorematic terms. Problems with syncategorematic terms led logicians to consider propositions that included such terms. Such a proposition was considered an exponible proposition, which was defined as "a proposition which has an obscure sense requiring exposition owing to its inclusion of a syncategorematic term or of a term which implicitly involves a syncategorematic term."[19] Treatises, and parts of treatises, were devoted to the problem of exponibles.

In all this, we see one of the characteristic features of medieval logic: the multiplication of themes and topics with unfamiliar, even bizarre, sounding names, that have long disappeared from the subject of logic as it has been taught from the nineteenth century to the present. Indeed, the names began to disappear in the sixteenth and seventeenth centuries as scholastic logic was

16. Ibid., 18–19.

17. See the article "Syncategoremata" by P. V. S. (Paul Vincent Spade) in Robert Audi, ed., *The Cambridge Dictionary of Philosophy* (Cambridge University Press, 1995), 783. Also see Philotheus Boehner, *Medieval Logic: An Outline of its Development from 1250–c. 1400* (Manchester: Manchester University Press, 1952), 7–9 for a discussion of treatises on syncategoremata. Categorematic terms are words that can serve as subjects or predicates of categorical propositions. A categorical proposition is one that is "composed of just three parts, subject, copula, and predicate, for example: 'A man is reading.'" See Alexander Broadie, *Introduction to Medieval Logic,* 25.

18. See Curtis Wilson, *William Heytesbury: Medieval Logic and the Rise of Mathematical Physics* (Madison: The University of Wisconsin Press, 1956), 11.

19. Wilson, ibid., 11. Broadie (*Introduction to Medieval Logic,* 3) says that exponible terms, that is, terms "which include 'only', 'except', 'in so far as', 'begins', 'ceases', and 'differs' – terms thought to be obscure in various ways and thus in need of exposition (hence the word 'exponible') – were of great interest to logicians and philosophers."

ridiculed and attacked. Gone are such themes and terms as "exponibles" and "syncategorematic," to which we may add others, such as "obligations" *(obligationes)*, "insolubles" *(insolubilia)* and its subdivision "impossibles" *(impossibilia)*, "suppositions" *(suppositiones)*, "copulation" *(copulatio)*, "relative terms" *(de relativis)*, "ampliation" *(ampliatio)*, "restriction" *(restrictio)*, "appellation" *(appellatio)*, "consequences" *(consequentiae)*, and "sophisms" *(sophismata)*.[20] Among these themes, obligations was significant because it had the structure of a familiar school disputation wherein an opponent seeks to lead a respondent into a contradiction and the respondent tries to avoid it.[21] To convey something of the flavor of medieval logic, it will be useful to say something about a few of the more important topics.

THE SOPHISM

Sophisms of all kinds were basic to medieval logic. In 1335, William of Heytesbury (fl. 1350), a Fellow of Merton College, Oxford University, wrote a treatise titled *Rules for Solving Sophisms,* and John Buridan (ca. 1300–d. after 1358), perhaps a little later, at the University of Paris, wrote a treatise titled *Sophisms (Sophismata)*. These treatises reflect the fact that a whole genre of logic literature was devoted to sophisms. But what is a sophism? It is a fallacy in an argument. According to Curtis Wilson:

> The term "sophism" was used in a rather broad sense in medieval times. It was applied to a proposition supported by an invalid argument which appeared to be valid, or by a valid argument which for some reason appeared to be invalid; to a proposition supported by a valid argument of which the premisses were false although seeming to be true, or of which the premisses were true although seeming to be false; to a proposition which, on the basis of different arguments could be as plausibly affirmed as it could be denied. Essential characteristics of a sophistical argument were its subtlety, its lack of accord with common sense, its seeming to be what it was not.[22]

Or, as Philotheus Boehner expressed it, a sophism "is usually an ambiguous or faulty proposition which requires certain distinctions before the correct logical sense can be obtained and false interpretations rejected. Hence a sophisma

20. For illustrations of most of these themes, see Boehner, *Medieval Logic,* Part I: "Elements of Scholastic Logic," 1–18. Many of these terms are described in the section "Other Themes on Medieval Logic" in this chapter.
21. See Broadie, *Introduction to Medieval Logic,* 2–3.
22. Wilson, *William Heytesbury,* 4.

may be aptly described as a proposition which from a logical viewpoint presents certain difficulties in virtue of its ambiguous or faulty formulation."[23]

John Murdoch reports the description of sophisms by a late medieval logician as "deceptive statements," each part of which can be plausibly argued. He adds that

> most *sophismata* were submitted to both proof and disproof *(probatur* vs. *improbatur)* and after that a decision or *solutio* (usually) made as to whether the proof is to be preferred (in which case the *sophisma* is true) or the disproof is to stand (in which case the *sophisma* is false). However, this definition leaves unsaid one outstanding characteristic of almost all such medieval *sophismata:* namely, that their "deceptiveness" consists in the fact that they are usually extremely bizarre statements, that they assert something that is, to all intents and purposes, counterintuitive.[24]

The literature on sophisms was important because it not only was relevant to logic but was also useful for problems in natural philosophy and theology. The sophism was routinely used in university courses on logic, and lists of them had been compiled from the thirteenth century onward.[25] Sophisms were formulated in virtually every subdivision of logic, such as signification, supposition, connotation, insolubles, ampliation, and so on.

Examples of Sophisms: John Buridan, William Heytesbury, and Richard Kilvington

What is remarkable about the large literature on logic is that much of it was intended for university students in logic courses. In the preface to his treatise, Heytesbury informs readers that his treatise is for first-year students in that subject, who were probably first-year undergraduates at the university, although at least one scholar thinks they were intended for advanced undergraduates, at least in Oxford.[26] To convey an idea of what these young students studied in their logic courses at the

23. Boehner, *Medieval Logic,* 8.
24. Murdoch, "The Involvement of Logic in Late Medieval Natural Philosophy," in Stefano Caroti, ed.; Introduction of John E. Murdoch, *Studies in Medieval Natural Philosophy* ([Florence]: Leo S. Olschki, 1989), 24–25.
25. See *John Buridan: Sophisms on Meanings and Truth,* translated and with an introduction by Theodore Kermit Scott (New York: Appleton-Century-Crofts, 1966), 3.
26. See Edith Dudley Sylla, "The Oxford Calculators," in Norman Kretzmann, Anthony Kenny, and Jan Pinborg, eds., *The Cambridge History of Later Medieval Philosophy* (Cambridge: Cambridge University Press, 1982), Chapter 27, 563.

University of Paris and at Oxford University in the fourteenth century, it will be useful to cite a few sophisms from John Buridan and Richard Kilvington, as well as to present a brief description of William Heytesbury's *Rules for Solving Sophisms.*

Because medieval logic employed terms, concepts, and examples that differ radically from anything we might encounter in modern logic, the texts quoted here, as well as the summary analysis by Curtis Wilson, may prove difficult to comprehend. We may recall R.W. Southern's words about the logical treatises of Boethius (cited in Chapter 2): "The works of Boethius are immensely difficult to understand and repellent to read."[27] Unfortunately, as the Middle Ages moved through its centuries, the difficulties of logic did not diminish, and the reading of it became no less repellent, as will be obvious from the criticisms made by Erasmus, Sir Thomas More, and Juan Luis Vives in the sixteenth century (cited in Chapter 7). But the ridicule humanists directed against the examples they selected from scholastic logic, while they may appear appropriate and justified at first glance, were technically incorrect. The examples they lampooned would have made sense to students and teachers in the fourteenth century. It was not nonsense, as humanists would have their readers believe.

And yet we can readily understand humanist disgust and frustration with medieval logic. By the time humanists arrived on the scene, it seemed "as though the logic of the late medieval period was running practically out of control. There were simply too many rules, and no assurance that new ones might not be introduced indefinitely."[28] For those of us who are not historians of that subject, the examples we encounter in the published and translated texts often seem bizarre and incomprehensible. Historians of medieval logic have penetrated the densities of logical treatises composed during the late Middle Ages. They have done laudable, even heroic, work in interpreting the subtleties, obscurities, and inherent difficulties of medieval logic. But they have yet to transmit this body of learning to a wider audience. They have not sought to make a broad range of examples intelligible to nonlogicians. Because there are no clear and straightforward examples to illustrate this or that aspect of medieval logic, with its multiplicity of rules and distinctions, the examples that I include should be viewed more as an attempt to reveal the form in which medieval logicians expressed themselves, and not as an effort to explicate substantive issues in medieval logic, which, in any event, lie beyond my capacities. Above all, however, the reader should be aware that medieval logic

27. Southern, *The Making of the Middle Ages* (New Haven: Yale University Press, 1953), 179.

28. Broadie, *Introduction to Medieval Logic,* 192.

made sense to medieval students and teachers of logic, as it does to modern logicians and historians of logic. Medieval logicians made significant contributions to their subject, advancing it far beyond Aristotle. Some of their contributions would not be duplicated until the nineteenth century.

To exemplify the kinds of sentences that scholastics used to illustrate logical points, I cite two that Juan Luis Vives presented, in the sixteenth century, to demonstrate the absurd depths to which medieval logic and scholastic logicians had descended:

1. "Varro, though he is not a man, is not a man *[hominem non esse]*, because Cicero is not Varro."
2. "A head no man has, but no man lacks a head."[29]

Alexander Broadie insists that these two sentences are by no means absurd and gives arguments to show why, arguments that cannot be repeated here.[30] In the first proposition, "what is centrally at issue is the effect upon the truth value of a proposition of the precise position in it of the negation sign, a topic of as much interest to present-day logicians as it was to the late scholastics."[31] Scholastic logicians made up rules to interpret Latin sentences on the basis of word order. Thus, where, in ordinary discourse, the sentences "Varro non est homo" and "Varro homo non est" would be understood and translated identically as "Varro is not a man," they would be interpreted differently by scholastic logicians because in the first sentence the negation sign *(non)* precedes *homo* and in the second sentence it follows it. "The result was a Latin which was to a certain extent artificial, a scientific Latin appropriate for scientific discourse."[32] One of its purposes, however, was to reduce ambiguity. Whatever its purpose and function, imposing significance on word order in a language that ignored word order infuriated humanists and was in no small measure responsible for their furious onslaught against the logicians. Broadie applies other medieval logical rules and theories to make sense of the second sentence.[33]

29. The translations are by Broadie, *Introduction to Medieval Logic,* 197. The wording of Broadie's translation differs somewhat from Rita Guerlac's translation of the same two sophisms, which will appear in Chapter 7. The substantive content of the sentences, however, remains the same.
30. For Broadie's arguments, see his *Introduction to Medieval Logic,* 197–206.
31. Broadie, ibid., 198.
32. Ibid., 200.
33. Ibid., 203–205.

Propositions in medieval logic were often constructed to illustrate the role of certain terms, such as "some," "no," "if," and so on;[34] or, to analyze a process, as, for example, how we are to understand what it means for a process to begin, or to cease, and so on. Thus, it was often the case that logicians focused their attention only on certain terms in a proposition, while ignoring the rest. Hence, medieval logicians tended to ignore the substantive content of the proposition, which was of little importance. Indeed, it seems that they avoided ordinary, content-laden propositions in favor of those that were silly sounding. The main objective was to illustrate points of logic and thereby to construct propositions that would enable them to achieve their goals.

The propositions they chose to construct were, therefore, frequently made to sound absurd and bizarre, as is evident from the following examples, drawn from fourteenth-century treatises on sophisms by Richard Kilvington and John Buridan:

Kilvington:[35]

1. Socrates is whiter than Plato begins to be white.
2. Socrates is infinitely whiter than Plato begins to be white.
5. Socrates will begin to be as white as he himself will be white.
9. Socrates will be as white as Plato will cease to be white.
14. Socrates will begin to traverse distance *A*, and Socrates will begin to have traversed distance *A*, and he will not begin to traverse distance *A* before he will begin to have traversed distance *A*.
20. Socrates will as quickly have been destroyed as he will have been generated.
37. Socrates can as quickly have the power to move stone *A* as Plato will have the power to traverse distance *C*.
45. You know this to be everything that is this.

Buridan:[36]

Sophisms from Chapter 1 (On Signification):

2. A horse is an ass.
3. God is not.
6. No man lies.

34. See ibid., 202–203.
35. These are drawn from *The Sophismata of Richard Kilvington*, Introduction, Translation, and Commentary by Norman Kretzmann and Barbara Ensign Kretzmann (Cambridge: Cambridge University Press, 1990). The numbers of the sophisms are those assigned by the Kretzmanns.
36. Buridan's sophisms from the first to seventh books are cited from *John Buridan: Sophisms on Meaning and Truth*, trans. T. K. Scott.

Sophisms from Chapter 2 (Truth Conditions for Categorical Propositions):

1. Aristotle's horse does not exist.
6. I am speaking falsely.

Sophisms from Chapter 3 (On Supposition):

2. You are an ass.
5. A name is trisyllabic.

Sophisms from Chapter 4 (On Connotation):

2. You ate raw meat today.
4. An old man will be a boy.
7. This dog is your father.

Sophisms from Chapter 5 (On Ampliation):

1. Some horse does not exist.
7. Nonbeing is known.

Sophisms from Chapter 6 (On Conventional Signification):

4. It is in our power that a man is an ass.
6. This can be true: "Man is nonman".

Sophisms from Chapter 7 (Time and Truth):

1. No spoken proposition is true.
8. All which is moved was moved previously.
9. No change is instantaneous.

Sophisms from Chapter 8 (Insolubles):[37]

3. Every man is running, therefore a donkey is running.
4. I say that a man is a donkey.
8. What Plato is saying is false.
9. What Socrates is saying is true.
11. What I am saying is false.
14. Either Socrates is sitting or the disjunction written on the wall is doubtful to Plato.
17. You are going to throw me into the water.
18. Socrates wants to eat.

How did medieval logicians view sophisms? Surprisingly, they did not regard a sophism as a "piece of fallacious or 'sophistical' reasoning," but rather as "a problem sentence or proposition, where it is possible to advance arguments both for its truth and for its falsity and we are

37. The sophisms from Buridan's eighth book are drawn from *John Buridan on Self-Reference: Chapter Eight of Buridan's 'Sophismata,'* translated with an Introduction, and a philosophical Commentary by G. E. Hughes (Cambridge: Cambridge University Press, 1982).

expected to learn something by seeing how the arguments for one side or the other can be refuted."[38]

We can see these features in sophisms drawn from Buridan's *Sophisms (Sophismata)*, a treatise in which, as we just saw, he included sophisms on the special themes of signification, supposition, connotation, insolubles, ampliation, truth conditions for categorical propositions, and time and truth. Thus, in the category of connotation,[39] in Chapter 4, Buridan offers the following as the seventh sophism:

(7) *This dog is your father.*

This is proved, because this is a father and this is yours; hence, it is your father. Similarly, pointing at a black monk it is argued that there is a white monk, because this is white and it is a monk; hence, it is a white monk. And similarly, it is argued that it is not a black monk since he is not black but white. Hence, he is not a black monk, by the argument from the whole in quantity to its parts. So also it could be argued concerning your father that he could not be your father, because he is not yours, but rather, conversely, you are his; hence, he is not your father.[40]

At the conclusion of the fourth chapter on connotation, Buridan inserts a section called "Solutions of Sophisms." Here he replies to the claims made in all the sophisms presented in the chapter on connotation. In his reply to the seventh sophism, just quoted, where he had briefly presented both sides of the argument, Buridan says:

(7) Concerning the seventh sophism, I say that this is not a valid consequence: this dog is a father, and it is yours; hence, it is your father. For the connotation of this term "your" changes, as has been said. Nor is this valid: this is white, and it is a monk; hence, it is a white monk. For in the first premiss, "white" connotes the whiteness of his body, and afterward it connotes the whiteness of his monastic garb. And for the same reason, this conclusion is not valid: the monk is not black,

38. G. E. Hughes, trans., ibid., 4.

39. Scott (*John Buridan: Sophisms on Meaning and Truth,* 43) summarizes William Ockham's distinction between absolute and connotative terms. "According to Ockham, an absolute term is one which, in a proposition, stands for some thing or things and does not consignify or refer to any other things in an indirect or secondary way, while a connotative term, when it occurs in a proposition, not only stands for some things, but also indirectly 'connotes' some other things. In the proposition 'A man is white,' the term 'man' is absolute and stands only for (presently existing) men. But the term 'white' is connotative, since it not only stands for what the subject term stands for, but also connotes whiteness as a property or number of concrete properties possessed by the objects for which it stands."

40. *John Buridan: Sophisms on Meaning and Truth,* trans. T. K. Scott, 121.

so he is not a black monk. Nor for the same reason is this valid, concerning your father: he is not yours, so he is not your father. Nor is this valid, for the same reason: you are white with reference to your teeth; therefore, you are white. Similarly, you are a good thief, so you are good. And so of others.[41]

In his section on insolubles, which is Chapter 8 of his *Sophismata,* Buridan presents a sophism that includes both the sophism and its solution. Thus, in his third sophism, Buridan declares:

The third sophism is concerned with what follows from a posited proposition. It is this: Suppose it is posited that every man is a donkey; then this follows:

3.0 *Every man is running, therefore a donkey is running.*

3.1 *Arguments in favour of the sophism:*

3.1.1 It can be proved by the following syllogism in *Darapti:*[42] 'Every man is running; by hypothesis, every man is a donkey; therefore a donkey is running'. (This is like the way we argue in a *reductio ad absurdum.* We can construct such an argument by taking something maintained by our opponent, adding to it something true, and then drawing a conclusion; and the inference itself is valid even though the conclusion may turn out to be impossible. It is like that in the present case too: the inference itself is a valid one.)

3.1.2. Things cannot be both as they were posited to be and as the premiss says they are, unless they are as the conclusion says they are. Therefore the premiss and the posited case together entail the conclusion.

3.2. *Argument for the opposite view:* It is a rule of logic that a false inference is an impossible one, and a true one is necessary. Now the mere positing of a case cannot turn an impossible proposition into a necessary one; so no matter what you care to posit or withdraw, or grant or not grant, you cannot turn a false inference into a true one. Everyone will agree that the inference 'Every man is running, therefore a donkey is running' is false, because it is not necessary; so no case that might be posited could ever make 'Every man is running' entail 'A donkey is running'.

3.3. The solution of this sophism is easy. You can state or posit or assert any proposition you like, but you can never thereby turn a necessary inference into one that is not necessary or *vice versa.* Therefore the stated sophism is false.

3.4. Nevertheless, in view of the arguments that were advanced, I should make it clear that a proposition can be posited or admitted or stated in either of two

41. Ibid., 123.

42. In the medieval mnemonic verse form to aid in remembering the three figures of the syllogism, *darapti* is the first mood in the third figure and is described by William of Sherwood as consisting of "two universal affirmatives leading directly to a particular affirmative conclusion – e.g., 'every man is an animal, every man is a substance; therefore some substance is an animal.'" See *William of Sherwood's "Introduction to Logic,"* 65.

ways: (a) It may be posited *simply,* as a proposition considered on its own; and in that case it will be irrelevant to the truth or falsity of any other propositions or inferences in the case under consideration. Or (b) it may be posited specifically *as a premiss* (or as one premiss among others) for the purpose of inferring something else; and in that case it is certainly essential to see whether the suggested conclusion does or does not follow from the posited proposition (together with any other premisses). For example, if in the present case you were to posit *simply* that every man is a donkey, the inference in the sophism would not thereby be made any more or any less valid. But if you posit as a *premiss* (i.e. for inferring a conclusion) that every man is a donkey, then I admit straight away that it does follow from this that some man is a donkey; and if you posit it as a premiss to be taken together with the other premiss that every man is running, then indeed it does follow from this that a donkey is running. And it was on that basis that the arguments proceeded.[43]

We saw that Buridan categorized his sophisms under a range of topics. In the eighth and final chapter, he considered 20 sophisms within the category of "insolubles." These largely concerned semantic paradoxes involving assertions that somehow deny their own truth. The third sophism just quoted is among them. The most famous self-referential sophism in the history of logic is the Liar paradox, a version of which Buridan considers in the eleventh sophism of the fourth book, where he analyzes the proposition: "What I am saying is false." All who had to cope with this problem faced the following dilemma: "If the proposition in question is true, then since what it asserts is that it itself is false, it seems to follow that it *is* false; but on the other hand, if it is false, then since that is just what it asserts, it seems to follow in turn that it is true. We therefore reach the apparently inescapable but quite intolerable result that it is true if and only if it is false."[44] Inspection of the variety and range of sophisms that medieval logicians confronted testifies to their desire to grapple with, and to understand, difficult and puzzling problems, as is obvious from the quoted sophisms and from the baffling nature of the Liar paradox.

By the mid–fourteenth century, sophisms usually consisted of five parts and were similar in structure to questions in natural philosophy:[45]

1. Enunciation of the sophism.
2. The arguments for one side.

43. G. E. Hughes, trans., *John Buridan on Self-Reference: Chapter Eight of Buridan's 'Sophismata,'* 39–40.
44. Hughes, ibid., 1.
45. For the structure of a question in natural philosophy, see the question from Nicole Oresme's *Questions on De celo* in Chapter 5.

3. The arguments for the other side.
4. The opinion of the author.
5. A refutation of each argument in favor of the rejected opinion.[46]

In the preface to his *Sophisma,* Richard Kilvington emphasizes the need to examine both sides of a question, and thereby captures the spirit of inquiry of those who dealt with sophisms in medieval universities. He explains:

(a) When we are able to call both sides into question, we will readily discern what is true and what is false, as Aristotle says in Book One of his *Topics.* Therefore, in order that we may more readily discern what is true and what is false, in the present work, which consists of sophismata to be thoroughly investigated, I intend, to the best of my ability, 'both' to demolish the two sides of the contradiction and also to support them by means of clear reasoning.
(b) I am led to do this by the requests of certain young men who have been pressing their case very hard. And so, wishing to give them something I have often heard them ask for, I have undertaken to make an attempt in that direction.
(c) And I will take it upon myself to begin this work first with sophismata having to do with the verb 'begins'.[47]

In turning to Heytesbury's *Rules for Solving Sophisms,* we should keep in mind that he also wrote another treatise on sophisms, titled simply *Sophismata.* In the latter, he treated 32 distinct sophisms. Almost all of these are directed to logic itself and are therefore regarded as "logical sophisms" *(sophismata logicalia).* By contrast, Heytesbury's *Rules for Solving Sophisms* is concerned with physical sophisms *(sophismata physicalia),* that is, with problems in natural philosophy. Wilson captures the intimate relationship that obtained between the two treatises and, therefore, between logic and physics, or natural philosophy, when he observes that

[t]he *Sophismata* deals with particular sophisms, while the *Regule* sets forth the principles commonly employed in the analysis of these sophisms. Surprisingly enough, the physicomathematical principles stated in the *Regule* play a major role in the analysis of the *sophismata logicalia;* on the one hand logic is used in the analysis of mathematical and physical problems, and on the other hand a kind of mathematical physics is introduced into the analysis of logical problems.[48]

46. Drawn from Hughes, *John Buridan on Self-Reference,* 4.
47. *The Sophismata of Richard Kilvington,* 1.
48. Wilson, *William Heytesbury,* 6.

In general, in the sophismata of Heytesbury – and in those of Richard Kilvington, who, as we saw, also wrote a lengthy treatise on sophismata – "one cannot separate 'physical sophismata' from logical or grammatical sophismata; physical concepts are used in traditional logical sophismata where they might not have been expected."[49]

Heytesbury divided his treatise into six chapters, each concerned with a different kind of sophism. Although it is not formally a questions treatise, Heytesbury's work is highly scholastic because, as was often done in sophismata, it raises objections to the rules proposed for solving each sophism, and then replies to each objection. Moreover, Heytesbury illustrates the rules with examples applied to particular cases.[50] What did Heytesbury include in a book he intended for first-year students of logic?

The first chapter is about insolubles and is, therefore, called *On Insolubles (De insolubilibus)*; the second chapter concerns propositions that involve the expressions "to know" *(scire)* and "to doubt" *(dubitare)* and is, therefore, called *On 'to know' and 'to doubt' (De scire et dubitare);* the third chapter is called *On relatives (De relativis)* because it is about propositions that contain relative or demonstrative pronouns. The theory of supposition plays a prominent role because Heytesbury is concerned with what terms represent, that is, the way terms "supposit," or, as Wilson puts it, "does the mode of supposition of the relative term differ from or coincide with the mode of supposition of its antecedent?"[51]

In contrast to the first three chapters, which were primarily concerned with straightforward logical problems, the last three chapters illustrate the application of logic to physical problems, that is, to natural philosophy. A problem that captured the attention of medieval natural philosophers and logicians was a discussion by Aristotle in two different books of his *Physics:* in Book VI, Chapter 5 (235b.32–236a.27) and Book VIII, Chapter 8 (263b.9–26). In the first location,[52] Aristotle asks whether, in a continuous change, there is a first instant when something actually begins and whether there is a last instant of a change when one can say that the change has been completed. Aristotle concludes that there is no first instant in a process of continuous change, but that there is an identifiable instant when the change has been completed. In the second consideration of this problem, in Book VIII, Chapter 8, Aristotle

49. Edith Sylla, "The Oxford Calculators," 547.

50. Wilson, *William Heytesbury,* 5.

51. Wilson, ibid.

52. In my summary, I follow Murdoch, "Infinity and Continuity," in Norman Kretzmann, Anthony Kenny, and Jan Pinborg, eds., *The Cambridge History of Later Medieval Philosophy* (Cambridge: Cambridge University Press, 1982), Chapter 28, 585.

reversed himself with regard to the first instant. He considers the change of something that is not-white to being white[53] and concludes, in Murdoch's words, "that the relevant first instant should be assigned to the later segment of the time interval in question; that is, there would be a first instant at which the changing subject is white." Thus did Aristotle reverse himself and argue that there is indeed a first instant in a continuous change.

In the fourteenth century, medieval natural philosophers transformed Aristotle's problem about certain aspects of the nature of a continuum into a separate genre of natural philosophy that was dedicated to a logical problem about the continuum. They called such treatises *On the first and last instant (De primo et ultimo instanti).* Walter Burley (1275– ca. 1345) wrote such a treatise in the fourteenth century. It became customary to accept Aristotle's first account and deny that there was a first instant of change. Burley argued that although there was no first instant of the process of change, there was a last instant before the change began. "One can thus say," Murdoch declares, "that any continuous change or motion is limited at its beginning by a last instant of its not-being. But the restriction to not-being means that the change or motion is *extrinsically* limited at both its ends, since if it were intrinsically limited, that would entail the existence of first or last instants belonging to the change itself, which is categorically denied."[54] In addition to rules that applied to continuous changes, in which something changed successively and therefore did not have all its parts at once, but acquired them successively, such as a continuous motion, the rules of first and last instants were also applied to permanent things, which had all of their parts at once, such as a white thing, which had all of its whiteness at the same time.

A concrete example of the use of first and last instants was furnished in a fourteenth-century treatise on natural philosophy by Blasius of Parma (ca. 1345–1416) titled *Question of Blasius of Parma On the Contact of Hard Bodies (Questio Blasij de Parma De tactu corporum durorum).*[55] In this treatise, Blasius defends the medieval and Aristotelian view that a vacuum could not occur by natural means. Blasius claims that if two plane, circular surfaces approach each other continually while remaining parallel, the air that lies

53. Murdoch explains ("Infinity and Continuity," 585, n. 57) that "Aristotle's example is of a change from being white to being not-white, but since the scholastics reversed the example, I have done so here."

54. Murdoch, ibid., 585–586.

55. For the complete title, see the bibliography. I have summarized Blasius's arguments in Grant, *Much Ado About Nothing: Theories of Space and Vacuum from the Middle Ages to the Scientific Revolution* (Cambridge: Cambridge University Press, 1981), 89–92. My citations will be to my summary account.

between the circular surfaces would become more and more rarefied as the circular surfaces approach, but the air never parts at any point to allow formation of a vacuum. Blasius argues that as long as the surfaces approach but do not meet, the air will continue to rarefy and yet will continue fully to occupy the diminishing intervening space. Hence, there is no last assignable instant in which rarefaction ceases prior to the contact of the surfaces, so that no vacuum can occur.

Of crucial importance is the interpretation placed upon the notion of contact between the surfaces. Actual contact of the surfaces must be construed as the last move in the process in which the surfaces approach each other. But the last move in which they come into contact lies outside of the process of motion of the surfaces. Indeed, the instant of contact lies outside of the process of motion and serves as an extrinsic boundary to that motion. In effect, there is no last instant in which the surfaces are separated, although there is a first instant in which they are in contact. But if there is no last instant in which the surfaces are separated, there can be no last instant in which the rarefied air departs from the intervening space that lies between the circular surfaces. Therefore, no vacuum occurs.

With the surfaces now in contact, Blasius assumes next that the surfaces can be separated while remaining parallel, and that this can occur without formation of a vacuum between the surfaces. In support of this claim, Blasius argues that there is a last instant of contact, but no first instant of separation. For if a first instant of separation exists, there would be a minimum distance of separation. But given any initial distance of separation, one can always argue that the surfaces must have been previously separated by half of that distance, and so on. It follows that there can be no initial distance of separation and, consequently, no first instant of separation. In the absence of a first instant of separation, and therefore an absence of a first distance of separation, it follows that for any instant chosen after separation, air will fully occupy the intervening space associated with that particular instant. Therefore, no vacuum can occur.

But one might argue that if an alleged first instant of separation is chosen arbitrarily, then air can enter the space between the circular surfaces. If we imagine a radius extended from the circumference of the surfaces to their centers, the air would first reach the midpoint of the radius before it reached the centers. One could then plausibly argue that a momentary vacuum must have existed at the center prior to the arrival of the "first" air as it moves through the successive positions from circumference to center. To counter such an argument, Blasius would deny that one could select a first instant of separation, without which there could be no initial entry of first air. For prior to any first instant that might be chosen, an earlier instant could be selected, and then another instant earlier than the second one, and so on. But no matter how far the process is carried, one cannot arrive at a first instant of separation. Therefore, there is no time at which a vacuum could have existed after separation.

Thus, the doctrine of first and last instants was about limits to successive and permanent things. Medieval natural philosophers found many occasions to use the logic of first and last instants. Medieval logicians were also attracted to such a concept and transferred the concern for limits in natural philosophy to logic. They began to analyze *terms and propositions* that were concerned with beginnings and endings, and thus developed a genre of logical treatise known as *De incipit et desinit (On "It Begins" and "It Ceases")*. Heytesbury made this the theme of his fourth chapter, which is devoted to propositions involving the terms "to begin" *(incipere)* and "to cease" *(desinere)* and is, therefore, titled *De incipit et desinit.*

The fifth chapter, titled *On "maximum" and "minimum" (De maximo et minimo)* involves the terms "maximum" and "minimum" and "is essentially a treatise on the setting of boundaries to the range of variable quantities of different types."[56] But the limits that concern maxima and minima are not temporal, as they were for problems of first and last instants and for "begins" *(incipit)* and "ceases" *(desinit)*. With maxima and minima, limits concern powers or capacities to do something. "Should a capacity such as Socrates' ability to lift things be limited by a maximum weight he can lift or by a minimum weight he cannot lift? What criteria can be used to decide such a question one way or another?"[57]

In his *Questions on the Physics,* book 1, question 12, John Buridan inquires about maxima and minima, asking "whether all natural beings are determined at the maximum."[58] In the first conclusion of this question, Buridan declares:

56. Wilson, *William Heytesbury,* 6.

57. Murdoch, "Infinity and Continuity," 588, n. 66. Murdoch observes (ibid.): "The problem had its origins in Aristotle's contention (*De caelo,* 1, Ch. 11, 281a.1–27) that a capacity should be defined in terms of the maximum it can accomplish, a contention to which he added the correlative information that, if some capacity can accomplish so much, it certainly can accomplish less, while if it cannot accomplish so much, it surely cannot accomplish more."

58. In bk. 1, qu. 13 of his *Questions on the Physics,* Buridan asks "whether natural beings are determined at the minimum." See *Acutissimi philosophi reverendi Magistri Johannis Buridani subtilissime questiones super octo Phisicorum libros Aristotelis diligenter recognite et revise a Magistro Johanne Dullaert de Gandavo antea nusquam impresse* (Paris, 1509); facsimile, entitled Johannes Buridanus, *Kommentar zur Aristotelischen Physik.* (Frankfurt: Minerva, 1964), fols. 16v, col. 1–17r, col. 2.

Let A be a power capable of lifting a large weight. We cannot assign a maximum weight to what A can lift. This conclusion can be proved by allowing that there is no action when the agent is equal to or less than the resistance....

Suppose that A lifts weight B and that this weight is the maximum weight that A can lift (according to our opponent); then there would have to be some excess of A to B. Let us suspend a weight C to B such that the new resistance becomes equal to A's power; it is true that A cannot lift B and C together. But since C is divisible, we can remove half, and let the other half – called D – remain attached to B; A's power exceeds the resistance of B and D and consequently A can lift it. However, B and D is greater than B; B is therefore not the maximum weight than A can lift.

One can also reason thus: Let A be a power capable of lifting weights, and B a weight whose resistance equals A's power. A cannot move B, but its power can move a smaller weight than B, for it will be greater than it by some amount; and one cannot give a weight smaller than B by an indivisible amount because a continuum is not composed of indivisibles. Hence, given any weight smaller than B, one can always give an intermediary weight larger than it and smaller than B; therefore given any weight that A's power can lift, there is a larger weight that this power can lift....[59]

There are other conclusions which are rightly deduced from the conclusions that have just been posited.

The first is as follows: One can assign a minimum to the weight that A cannot lift. It is certain, in fact, that the weight can be increased so that A can no longer lift it. It is therefore necessary that some weight mark the termination of this power; and one cannot understand that this power stops at such a weight, if it is not in one of these two ways: either his power can lift such a weight and cannot lift anything heavier – that would be the maximum weight that can be lifted (which we know to be impossible) – or his power cannot lift this weight, but can lift any lesser weight – which is our conclusion; this weight is the smallest weight which cannot be lifted, since any smaller weight can be lifted.[60]

The sixth, and final, chapter in Heytesbury's *Rules for Solving Sophisms* is about motion as Aristotle described it in his *Physics,* book 5, Chapter 1,[61] namely, motion that occurs in the categories of place, quantity, and quality, hence the title *"On the Three Predicaments [or Categories]" (De tribus predicamentis).*

59. Buridan, *Questions on the Physics,* bk. 1, qu. 12, fol. 16r, col. 1. The translation was originally made in French by Pierre Duhem, *Le Système du monde,* vol. 7, 73–74. I have used the English translation, which appears in Pierre Duhem, *Medieval Cosmology: Theories of Infinity, Place, Time, Void, and the Plurality of Worlds,* edited and translated by Roger Ariew (Chicago: University of Chicago Press, 1985), 60–61.

60. Buridan, *Questions on the Physics,* bk. 1, qu. 12, fol. 16v, col. 1.

61. See Aristotle, *Physics* 5.1.225b.5–9.

In applying logic to physical problems in Chapters 4 to 6, Heytesbury does not seek empirical verification of his results and conclusions. He operates "according to the imagination" *(secundum imaginationem);* that is, he proceeds by imagining physical circumstances and conditions without any concern whatever about their status in the real world. The abstract nature of the problems and the fact that they were all verbalized makes following the arguments a formidable undertaking better left to dedicated scholars. But whether or not one understands the twists and turns of Heytesbury's discussion, Curtis Wilson's interpretation and analysis of the three mathematically oriented chapters in Heytesbury's treatise reveals a formidable and powerful body of reasoned argument:

> Our study of the mathematical and physical content of Heytesbury's *Regule* yields the following results. In Chapter IV of the *Regule* ("De incipit et desinit"), Heytesbury shows an appreciation of the value of the limit-concept for the analysis of the instantaneous in time and motion; and by means of the logical exposition of the terms "to begin" and "to cease" he is enabled to deal accurately with simple limiting-processes, including in one case the limit of the quotient of infinitesimals. He also appears to recognize the value of the concept "infinite aggregate" for the analysis of the continuum. In Chapter V ("De maximo et minimo"), he applies the limit-concept to the bounding of the ranges of variables and aggregates. In the first part of Chapter VI, "De motu locali," he demonstrates a knowledge of the manner of calculating the distance traversed in uniformly accelerated and decelerated motions. In the second part of Chapter VI, "De augmentatione," he shows an awareness of the more obvious properties of the exponential growth function. In the third part of Chapter VI, "De alteratione," he attempts a mathematical description of intensity in space and time, with a success that is only partial, owing to false assumptions as to the nature of intensity, and owing to the function which he assigns to certain arbitrary rules of denomination.[62]

In Chapter VI, on motion, Heytesbury gives sound definitions of uniform motion and uniform acceleration, as well as a circular definition of instantaneous motion, and also makes the earliest known statement of what is known as the mean speed theorem, a fundamental theorem that Galileo made the basis of his new mechanics. Heytesbury declares that "that motion is called uniform in which an equal distance is continuously traversed with equal velocity in an equal part of time."[63] Expanding the definition of uni-

62. Wilson, *William Heytesbury*, 148.

63. Marshall Clagett has translated parts of the sixth chapter of Heytesbury's *Rules for Solving Sophisms*. For the definition of uniform motion, see Clagett, *The Science of Mechanics in the Middle Ages* (Madison: The University of Wisconsin Press, 1959), 235.

form motion, Heytesbury arrives at the definition of uniform acceleration when he pronounces that "any motion whatever is *uniformly accelerated (uniformiter intenditur)* if, in each of any equal parts of the time whatsoever, it acquires an equal increment *(latitudo)* of velocity. And such a motion is uniformly decelerated if, in each of any equal parts of the time, it loses an equal increment of velocity."[64] He defines the instantaneous motion of a point as "the distance which *would* be traversed by such a point, if it were moved uniformly over such or such a period of time at that degree of velocity with which it is moved in that assigned instant."[65] Although the definition is circular (it defines "instantaneous velocity" by a uniform speed equal to the very instantaneous velocity that is to be defined), Galileo employed it in this form and in virtually the same manner.

By an ingenious use of these definitions, Heytesbury, and other colleagues at Merton College, Oxford, derived the mean speed theorem. Heytesbury expresses it as follows:

Whether it [i.e., a latitude or increment of velocity] commences from zero degree or from some [finite] degree, every latitude, as long as it is terminated at some finite degree, and as long as it is acquired or lost uniformly, will correspond to its mean degree [of velocity]. Thus the moving body, acquiring or losing this latitude uniformly during some assigned period of time, will traverse a distance exactly equal to what it would traverse in an equal period of time if it were moved uniformly at its mean degree [of velocity].[66]

Although Heytesbury only enunciated the mean speed theorem and did not prove it, numerous arithmetic and geometric proofs of this theorem were presented during the fourteenth and fifteenth centuries (for more on this, see Chapter 5). Galileo did not greatly improve upon these definitions or the proof of the mean speed theorem, which he used in his great and famous work, *The Two New Sciences*.[67]

Richard Kilvington (d. 1361), who was probably the teacher of William Heytesbury, wrote a *Sophismata* in the 1320s at Oxford University that was to be used for teaching logic, but not, apparently, at

64. Clagett, *The Science of Mechanics*, 237.
65. Ibid., 236.
66. Ibid., 270. I have added the first bracketed phrase. For another sophism on motion from Heytesbury, see Murdoch, "The Involvement of Logic in Late Medieval Philosophy," 26.
67. For references to Galileo's works and for a discussion of Heytesbury's definitions, see Grant, essay review of Marshall Clagett (ed. and trans.), *Nicole Oresme and the Medieval Geometry of Qualities and Motions*. Reviewed in *Studies in History and Philosophy of Science* 3 (1972): 167–182.

the introductory level.[68] Kilvington included 48 sophisms in his treatise, some titles of which were mentioned earlier. Norman and Barbara Kretzmann explain that the

> terminology [in Kilvington's treatise] often is technical, and its style sometimes is highly compressed. Because unusual word order is one of the devices by which medieval logicians mark formal distinctions, at first glance some of the language used in the *Sophismata* is likely to seem unnecessarily awkward. But a closer acquaintance with the material almost always shows that the apparently awkward expressions are medieval analogues of devices of modern formal notation, such as parentheses indicating the scope of logical operators.[69]

The reason that alterations in word order can be effective in Latin is because the order of words in a Latin sentence is largely irrelevant. Because Latin is a flexional language, Latin sentences do not depend on word order for their meaning, but rather on inflections, that is, on a system of word endings in the form of declensions of nouns, pronouns, and adjectives.[70] For reasons of convenience, and under certain circumstances, logicians came to impose a word order on certain Latin sentences to alter their meanings. Thus, one order of words would signify one thing, another order of the same words would signify something else. This was a very rational procedure and seems, as the Kretzmanns suggest, to have functioned as an early version of formal, logical notation.

In the concluding sentence of the preface to his book, cited earlier, Kilvington explains that he will begin "this work first with sophismata having to do with the verb 'begins.'" In the first of these sophisms, Kilvington seeks to analyze the proposition "Socrates is whiter than Plato begins to be white" (also previously cited). In this very first sophism, Kilvington finds occasion to use the power of word order, introducing the Latin expression *in infinitum,* translated as "infinitely." When Kilvington places the term "infinitely" *(in infinitum)* at the beginning of the following sentence: "If infinitely Socrates is whiter than Plato begins to be white," the term infinitely is said to function syncategorematically and to govern the entire expression. However, when Kilvington places the term infinitely just before the term "whiter," as in the expression "If Socrates is infinitely whiter than Plato begins to be white," the term is said to function categorematically

68. *The Sophismata of Richard Kilvington,* xv.

69. Ibid. The bracketed phrase is mine.

70. See Mario Pei, *The Story of Language* (London: George Allen & Unwin Ltd., 1952), 118–119.

and to govern only whiter.[71] In both versions, the differently ordered Latin words are identical and would ordinarily be translated by the same sentence. But in the domain of medieval logic, the different order of the same words required different translations and meanings.

A striking example of the Latin word-order phenomenon of medieval logic appears in Albert of Saxony's treatise on natural philosophy. In his *Questions on Aristotle's Physics* (bk. 4, qu. 11),[72] Albert asks "whether if a vacuum did exist, a heavy body could be moved in it." In resolving this question, Albert uses two phrases, which contain the same words but follow a different order, and, therefore, convey different meanings. The first of the Latin phrases is *descenderet in infinitum velociter:*[73] the second is *in infinitum velociter descenderet.*[74] By the rules of Latin grammar and syntax, these phrases are equivalent and their differing word order irrelevant to their meaning. But not in the following citation. Here Albert relies on logical analysis where word order is crucial. In the passage with the two phrases, Albert declares:

> But you say that if a heavy simple body would "descend infinitely quickly" *(descenderet ... infinite*[75] *velociter)* in a vacuum, then a heavy simple body could be moved in a vacuum. But if it were moved, then it would be moved in time [that is, in a finite time] in a vacuum, since every motion occurs in time. Consequently, it would not be moved in an instant, which is the opposite of what Aristotle says. In replying to this, I concede the consequent, but deny the antecedent, for I did not say in the question that "it would 'descend infinitely quickly'" *(descenderet in infinitum velociter)* but I said that "infinitely quickly it would descend" *(in infinitum velociter descenderet).* The first of these statements is false and the second true; nor does the first imply the second, because in the first the words "it would descend" are taken determinately, but in the second they have merely confused supposition, because in this second statement the syncategorematic term "infinite" precedes [the words "it would descend"], while in the first statement it follows them.[76]

71. *The Sophismata of Richard Kilvington,* 149.
72. The question is translated in Grant, *A Source Book in Medieval Science,* 335–338.
73. An equivalent phrase with the same meaning is *descenderet infinite velociter,* where *infinite* replaces *in infinitum.*
74. As with the first example, an equivalent phrase would be *infinite velociter descenderet,* where *infinite* replaces *in infinitum.* In the explication of these two Latin phrases, I follow footnotes 23 and 24 on p. 338 of my translation. For the explanatory information in those two notes, I am indebted to John E. Murdoch of Harvard University.
75. Here, Albert uses "infinite" instead of "in infinitum." But they are equivalent, as we saw in the two preceding notes.
76. The translation appears in Grant, *A Source Book in Medieval Science,* 338. I have made a few changes in the translation and inserted the crucial Latin phrases in their proper places.

In this discussion, Albert, who wrote on medieval logic, employs the technical terminology of logic, using terms such as "antecedent," "consequent," "determinately," "merely confused supposition" and "syncategorematic." In the first expression, *descenderet in infinitum velociter* ("it would descend *in infinitum* quickly"), the term "infinitum"[77] follows the term it modifies, namely *descenderet* ("it would descend"), and thereby gives determinate supposition to *descenderet.* Under these circumstances, the term *infinitum* was often considered, as it is here, a categorematic infinite, which signifies that the heavy simple body descending from the concave surface of the sky would fall with a single, actual infinite speed that is greater than any other assignable speed. Since an actually infinite speed is impossible, Albert rejects this in favor of the second mode of expression, *in infinitum velociter descenderet.* Here the term "infinitum" precedes the term it modifies, namely *descenderet,* and thus gives merely confused supposition to *descenderet.* In this instance, the term *infinitum* is considered a syncategorematic infinite, which signifies that the heavy simple body will fall infinitely quickly in the sense that however large the assigned speed, the body can always descend with a yet greater speed. Thus, the syncategorematic infinite is only a potential infinite. Only under these circumstances, Albert of Saxony concluded, could a heavy, simple body move in a vacuum with a temporal speed. In this example, we see how word order could play a crucial role in assigning different meanings to the same Latin words,[78] and how logic was applied to hypothetical physical conditions.[79]

OTHER THEMES IN MEDIEVAL LOGIC

A substantial part of medieval logic dealt with the properties of terms,[80] which, as we saw, William of Sherwood had already highlighted in his *Introduction to Logic.* A number of different types of treatises, or parts of treatises, on the properties of terms were developed. Of these logical cate-

77. I use the term "infinitum" to stand for "in infinitum."
78. On determinate supposition and merely confused supposition, see John Murdoch's translation of William of Alnwick's (ca. 1270–1333) response to Henry of Harclay's (ca. 1275–1317) arguments on atomism in Grant, *A Source Book in Medieval Science,* 322, n. 17. For other examples involving word order, see Murdoch, "Infinity and Continuity," 567–568, and notes 9–11.
79. For further examples of the application of logic to natural philosophy, see Murdoch, "The Involvement of Logic in Late Medieval Natural Philosophy," 3–28, especially 24–28.
80. See Boehner, *Medieval Logic,* 9–12.

gories, the theory of supposition, which also developed from the treatment of fallacies, was extremely important.[81] Supposition theory was concerned with the way terms function, or refer, in a proposition.[82] For example, the term "horse" refers in different ways in the following propositions: in "Every horse is an animal," the term horse refers to individual horses, and is called "personal supposition"; in "Horse is a species," the term horse refers to a universal, namely species, and is called "simple supposition"; and in "Horse is a monosyllable," the term horse refers to the spoken or written word, and is called "material supposition."[83]

Other topics, or categories, that involved properties of terms on which medieval logicians wrote separate tracts, or parts of treatises, were (1) relatives *(respectivae),* whereby a concept is considered with relation to something else, rather than by itself; (2) copulation *(copulatio),* which concerns the way predicates are related to the subject, or substance, and therefore the way they function in a proposition, as for example, "Man is *running,*" and "Man is *white*"; (3) ampliation *(ampliatio),* which occurs when the personal supposition of a term is extended so that it not only signifies for the object in the present, but also signifies that object in the past and future and even in the realm of possibility, and thus, as Boehner puts it, "the number of individuals signified by the term is enlarged or 'amplified'"; (4) restriction *(restrictio),* the opposite of ampliation, "since it means that the supposition of a common term is limited to a restricted number of individuals," as in the proposition "every blue-eyed man is an animal," where the supposition of the term man is restricted to those men who are blue-eyed; (5) appellation *(apellatio),* which involves the application of the supposition of a term to existing things only, as, for example, the term "Caesar" has no appellation because the Roman man who was Caesar no longer exists – indeed, since the term "Caesar" applies only to a single man, it cannot have amplia-

81. See Spade, "The History and Kinds of Logic," in the *Encyclopedia Britannica* (1995) on CD-ROM disk, the section on "The 'properties of terms' and discussions of fallacies."

82. Another important property of terms was the *signification* of a word, which was its meaning without reference to its context in a proposition. For a very brief, simplified summary of a few aspects of medieval logic, see Rita Guerlac, *Juan Luis Vives Against the Pseudodialecticians: A Humanist Attack on Medieval Logic. The Attack on the Pseudodialecticians* and *On Dialectic,* Book III, v, vi, vii from *The Causes of the Corruption of the Arts with an Appendix of Related Passages by Thomas More,* the Texts, with translation, introduction, and notes by Rita Guerlac (Dordrecht, Holland: D. Reidel Publishing Co., 1979), 3–9.

83. I have drawn this example from Spade, "The History and Kinds of Logic," in the *Encyclopedia Britannica* (1995), the section on "The theory of supposition." See also Wilson, *William of Heytesbury,* 9–10.

tion or restriction, although it can have signification and supposition.[84] With so many thematic subjects, medieval logicians operated with a highly specialized, and even forbidding, language of terms and concepts.

Medieval logicians also wrote works on semantic paradoxes called *insolubilia,* that is, "insolubles." These were

> propositions which, in the very asserting of what they assert, deny their own truth; thus the proposition "I am stating a falsehood," where the term "falsehood" refers precisely to the proposition ("I am stating a falsehood") in which it occurs, is an insoluble.[85]

The term "insoluble" is misleading, as medieval logicians were well aware, since "it does not deal with what cannot be solved but, rather, with what is hard or difficult to solve."[86]

There were also treatises on the theory of consequences *(Tractatus de consequentiis),* which involve "inferences from one simple or compound proposition to another simple or compound proposition,"[87] or "the relationship between the antecedent and the consequent in a conditional proposition."[88] Numerous consequences were distinguished by medieval logicians. John Murdoch believes that

> what is perhaps most impressive in this particular segment of late medieval logic is the stipulation of all manner of laws of valid inference expressible in these consequences. Here one finds considerable resemblance with much to be found among the theorems of the propositional calculus of modern logic, a factor which has undoubtedly done much to direct the attention of historians to this particular aspect of the medieval doctrine of consequences.[89]

84. In this paragraph I draw on Spade, ibid., the section on "The 'properties of terms' and discussions of fallacies." I have also used William Kneale and Martha Kneale, *The Development of Logic* (Oxford: Clarendon Press, 1962), 235. For definitions of the various ways in which terms function in a proposition, I have relied heavily on Boehner, *Medieval Logic,* 10–12.
85. Wilson, *William Heytesbury,* 5. Spade describes it as "primarily certain sorts of self-referential sentences, semantic paradoxes like the 'liar paradox' ('What I am now saying is false')." See Spade, "Insolubilia," in *The Cambridge History of Later Medieval Philosophy,* Chapter 12, 246.
86. Boehner, *Medieval Logic,* 13.
87. Boehner, ibid., 15–16.
88. See Ivan Boh, "Consequences," *The Cambridge History of Later Medieval Philosophy,* Chapter 15, 300.
89. See John Murdoch's section "Logic" in Grant, *A Source Book in Medieval Science,* 77.

Another major category was called "obligations" *(obligationes),* which were derived ultimately from Aristotle's *Topics* VIII 3 (159a. 15–24) and *Prior Analytics* I 13 (32a. 18–20).[90] An obligation involves a situation where if a

> respondent in a disputation adopts a position which is possible and is subsequently compelled to maintain something impossible (something logically incompatible with the position adopted or something impossible in its own right), then he has failed in his job as respondent. Nothing impossible follows from the possible; so, because it is possible, the position adopted by the respondent does not entail anything impossible. If the respondent is then logically compelled to maintain contradictory propositions, it must be because he has made logical mistakes in responding, so that the impossible, which is not entailed by the original position the respondent adopts, is entailed by the respondent's faulty defence of that position. The job of the interlocutor, called the 'opponent' in obligations disputations, is to trap a respondent into maintaining contradictories, and the job of the respondent is to avoid such traps.[91]

Although much about this category of logical treatise is unknown,[92] and "the precise purpose of these exercises is not yet definitely understood"[93] many treatises were written on it. Broadie conjectures that obligations exercises "may well have formed part of the training of students in which they had the opportunity to display, and also extend, their competence at logic." In any event, "such exercises also provided a context within which a wide range of logical and philosophical problems could be investigated. For example the literature of obligations is a major source for medieval discussions of insolubilia, and of counterfactual inferences."

In addition to all the subdivisions of logic that have already been mentioned, two more themes deserve inclusion. Medieval logicians were much concerned with what is today called modal logic, which was a problem set for them by Aristotle. William and Martha Kneale explain that "a modal statement is one containing the word 'necessary' or the word 'possible' or some equivalent of one of these, and a modal syllogism is one in which at least one of the premisses is a modal statement."[94] A modal statement stands in con-

90. See Eleonore Stump, "Obligations: From the Beginning to the Early Fourteenth Century," in *The Cambridge History of Later Medieval Philosophy,* Chapter 16, 318.
91. Ibid., 318–319.
92. Ibid., 315. See also William and Martha Kneale, *The Development of Logic,* 234.
93. This quotation and those that follow in this paragraph are from Broadie, *Introduction to Medieval Logic,* 3.
94. William and Martha Kneale, *The Development of Logic,* 81–82.

trast to an assertoric proposition, which "just states a connection between subject and predicate – or, if it is negative, states a disconnection."[95] The assertoric proposition becomes a modal proposition if the connection or disconnection involves something *necessary, impossible, contingent,* or *possible.* An elaborate medieval literature was based upon Aristotle's modal logic, and it is a theme that continues to engage modern logicians.

A second problem area was future contingents, which concern "singular events or states of affairs that may come to pass, and also may not come to pass, in the future."[96] Although the problem derived from Aristotle, it was one that concerned God's foreknowledge and was, therefore, of great interest to scholastic theologians. When a statement is made about the future, is it the case that "neither the claim nor its denial is necessarily true?"[97] Also, can a future event be contingent and still be foreknown? And "[c]an complete knowledge of the future by an immutable, infallible, impassible God be reconciled with the contingency of some aspects of the future?"[98] Many of the great medieval theologians and natural philosophers grappled with such questions, as did Anselm of Canterbury, Peter Abelard, Peter Lombard, Robert Grosseteste, Thomas Aquinas, John Duns Scotus, Peter Aureoli, William Ockham, Robert Holkot, Thomas Bradwardine, and Pierre d'Ailly.[99]

To convey an idea of how future contingency problems were treated in the late Middle Ages, I cite a summary account of Peter Aureoli's opinions:

> If man is truly free, it follows according to Petrus Aureoli, that a judgment connecting a future free act is neither true nor false. "The opinion of the Philosopher is a conclusion which has been thoroughly demonstrated, namely that no singular proposition can be formed concerning a future contingent event, concerning which proposition it can be conceded that it is true and that its opposite is false, or conversely. No proposition of the kind is either true or false." To deny this is to deny an obvious fact, to destroy the foundation of moral philosophy and to contradict human experience. If it is now true that a certain man will perform a certain free act at a certain future time, the act will necessarily be performed and it will not be a free act, since the man will not be free to act otherwise. If it is to be a free act, then it cannot now be either true or false that it will be performed.

95. Spade, *Thoughts, Words, and Things,* Chapter 10, 318.
96. Simo Knuutila, "Future contingents," in Robert Audi, ed., *The Cambridge Dictionary of Philosophy,* 290.
97. Calvin Normore, "Future Contingents," in *The Cambridge History of Later Medieval Philosophy,* Chapter 18, 359.
98. Normore, ibid.
99. All are discussed in Normore's article.

To say this may appear to involve a denial of the 'law' that a proposition must be either true or false. If we are going to say of a proposition that it is not true, are we not compelled to say that it is false? Petrus Aureoli answers that a proposition receives its determination (that is, becomes true or false) from the being of that to which it refers. In the case of a contingent proposition relating to the future that to which the proposition refers has as yet no being: it cannot, therefore, determine the proposition to be either true or false. We can say of a given man, for example, that on Christmas day he will either drink wine or not drink wine, but we cannot affirm separately either that he will drink wine or that he will not drink wine. If we do, then the statement is neither true nor false: it cannot become true or false until the man actually drinks wine on Christmas day or fails to do so. And Petrus Aureoli appeals to Aristotle in the *De Interpretatione* (9) in support of his view.

As to God's knowledge of future free acts, Petrus Aureoli insists that God's knowledge does not make a proposition concerning the future performance or non-performance of such acts either true or false. For example, God's foreknowledge of Peter's denial of his Master did not mean that the proposition "Peter will deny his Master" was either true or false. Apropos of Christ's prophecy concerning Peter's threefold denial Petrus Aureoli observes: "therefore Christ would not have spoken falsely, even had Peter denied Him thrice." Why not? Because the proposition, "you will deny Me thrice", could not be either true or false. Aureoli does not deny that God knows future free acts; but he insists that, although we cannot help employing the word "foreknowledge" *(praescientia),* there is no foreknowledge, properly speaking, in God. On the other hand, he rejects the view that God knows future free acts as present. According to him, God knows such acts in a manner which abstracts from past, present and future; but we cannot express the mode of God's knowledge in human language.[100]

THE IMPACT OF LOGIC IN MEDIEVAL EUROPE

Judging from the various examples and sophisms that I have cited here, one can hardly avoid the conclusion that medieval logic was an extraordinarily difficult subject, although it undoubtedly appears much stranger to modern eyes than it would have to the medieval undergraduates who regularly grappled with it in their logic courses. Nevertheless, one marvels at the fact that logic courses based on syllogisms, fallacies, sophisms, and numerous other subdivisions of medieval logic were

100. Frederick Copleston, S. J., *A History of Philosophy,* Vol. 3: Ockham to Suárez (Westminster, MD: The Newman Press, 1953), 37–38. Aureoli's discussion, which Copleston summarizes, occurs in Aureoli's commentary on Peter Lombard's *Sentences,* bks. 1 and 2. Copleston provides the references.

taught to all university students in the arts faculties of European universities for more than four centuries. The textbooks and treatises that have been preserved, and from which excerpts have been presented here, were well organized, but enormously complex and difficult. They are a tribute to the masters who wrote them, but even more remarkable is the fact that medieval undergraduates were required to cope with such difficult texts. As Edith Sylla has put it, "Even while thinking of the work as that of masters, we ought also to wonder at the level of logical sophistication that advanced undergraduates in fourteenth century Oxford must have attained."[101] And that sentiment should be extended to the students of John Buridan and the logicians at the University of Paris, as well as to the undergraduates at most of the universities of Europe, where logic was taught in the late Middle Ages. Through their high-powered logic courses, medieval students were made aware of the subtleties of language and the pitfalls of argumentation. Thus were the importance and utility of reason given heavy emphasis in a university education.

By comparison to the Middle Ages, logic as a formal subject of study in the modern university is of little consequence. Students are certainly not required to take it and most shun it as too difficult and demanding. How ironic it is that although we live in an age of triumphant science, a science the very being and existence of which depends on reason and logical thought, there has been a concomitant diminution of the study of logic, the quintessential embodiment of reason.

In the following passage, R. W. Southern has brilliantly captured the power and significance of logic for the Middle Ages. What he says for the period between the end of the tenth and the end of the twelfth centuries is equally applicable to the three or four subsequent centuries that comprise the late Middle Ages:

> The digestion of Aristotle's logic was the greatest intellectual task of the period from the end of the tenth to the end of the twelfth century. Men then debated about logic, as they do now about natural science, whether it was a curse or a blessing. But whichever it was, the process of absorption could not be stopped. Under its influence, the method of theological discussion and the form of the presentation of theological speculation underwent a profound change. It was in theology that the change was felt most keenly and fought most fiercely, but every department of thought was similarly affected. The methods of logical arrangement and analysis, and, still more, the habits of thought associated with the study of logic, penetrated

101. Edith Sylla, "The Oxford Calculators," 563.

the studies of law, politics, grammar and rhetoric, to mention only a few of the fields which were affected. Dante's *De Monarchia* for instance is arranged as a chain of syllogisms: as, for example,

> 6. *Human affairs are best ordered when man is most free*
> 7. *but it is under a monarch that man is most free*
> 8. *therefore the human race is best disposed when under a monarchy.*

and so on through a chain of arguments similarly disposed. No doubt the thing could have been done otherwise, but the parade of logical consistency was the best guarantee which Dante could find of the irrefutability of his arguments. Or, to take a more trifling case, we notice that when a cultivated man like the chronicler Matthew Paris wanted to clinch his objection to a habit which had grown up in the royal chancery of disregarding general phrases in charters unless accompanied by a distinct enumeration of the items included in such phrases, he said that it was against reason and justice, "and even against the rules of logic, the infallible guide to truth". It was as if one would say "All men are free", but deny that the phrase applied to Jones or Brown unless they were specifically mentioned. There was no more to say: logic was the touchstone of truth, and to argue 'by figure and by mode' the foundation of all discussion.[102]

Logic was only one of the ways in which reason was enthroned in the late Middle Ages. But the study of logic in the medieval university showed better than anything else the high regard in which reason was held. Moreover, because logic was thought to be a tool essential for understanding other disciplines, its influence was great. It was nothing less than the foundation on which a university education was built. But not all were taken with the role logic played and the way it was used. Almost from the beginning of its dominance, there were critics of logic who saw it as too narrow and sterile, too much the instrument of futile argumentation and debate, too abstract and, therefore, too remote from ordinary experience and the concerns of the "real world." But the ultimate reaction against scholastic logic is a separate story that will be told in Chapter 7. Now we must move on from logic, the great instrument of medieval learning, to natural philosophy, the most widely studied substantive subject of medieval education.

102. R. W. Southern, *The Making of the Middle Ages,* 181–182. In a note to this passage, Southern gives the reference to Matthew Paris as: Matthew Paris, *Chroncia Majora,* ed. H. R. Luard, Rolls Series, V, 210–11. For this passage see F. M. Powicke, *Henry III and the Lord Edward,* 1948, i.326.

5

REASON IN ACTION

Natural Philosophy in the Faculty of Arts

NATURAL PHILOSOPHY, OR NATURAL SCIENCE AS IT WAS SOMETIMES called, was the most widely taught discipline at the medieval university. For more than four centuries, virtually all students who obtained the master of arts degree had studied natural philosophy, and most undergraduates were exposed to significant aspects of it. What was natural philosophy for university students in the late Middle Ages?

WHAT IS NATURAL PHILOSOPHY?

In the broadest sense, natural philosophy was the study of change and motion in the physical world. In Chapter 3, we saw that it was one of Aristotle's three subdivisions of theoretical knowledge, or knowledge for its own sake. Natural philosophy was concerned with physical bodies that existed independently and were capable of motion, and therefore subject to change. In truth, Aristotle's natural philosophy was also concerned with bodies in motion that were themselves unchanging, as was assumed for all celestial bodies. In general, Aristotle's natural philosophy was concerned with separately existing animate and inanimate bodies that undergo change and possess an innate source of movement and rest.

Because the domain of natural philosophy was the whole of nature, as the name suggests, it did not represent any single science, but could, and did, embrace bits and pieces of all sciences. In this sense, natural philosophy was "The Mother of All Sciences."[1] For example, John Buridan, one of the most important natural philosophers in the Middle Ages, offered cogent explana-

1. This descriptive phrase and a brief discussion appear in Grant, *The Foundations of Modern Science in the Middle Ages: Their Religious, Institutional, and Intellectual Contexts* (Cambridge: Cambridge University Press, 1996), 192–194.

tions of earthquakes and mountain formation in his questions on Aristotle's *On the Heavens* and in his *Questions on the First Three Books of the Meteors.*[2] Anyone writing a history of geology would be obligated to include Buridan's opinions as part of the overall history of the subject. And yet there was no discipline of geology until the eighteenth or nineteenth century. Indeed, Aristotle's *Meteorology* served as the focal point for numerous questions about possible motions of the earth, about the ebb and flow of oceans, about lightning, and about other themes that were discussed in natural philosophy long before any specific sciences had emerged to claim one or another of these subjects. In their commentaries on Aristotle's *Physics* and *On the Heavens,* scholastic natural philosophers presented significant discussions about the causes of motion centuries before the advent of a recognized science of mathematical physics. With so many bits and pieces of sciences, and often enough significant parts of a science, embedded in natural philosophy, it is obvious that natural philosophy forms an important part of the history of many modern sciences, whether or not we choose to designate natural philosophy as science.

But medieval natural philosophy was far more significant than is indicated by the mere fact that embedded within it were bits and pieces of numerous modern sciences. In a culture such as that of the Middle Ages, in which the powerful tools for scientific research and inquiry routinely available to early modern and modern scientists were largely absent, how could nature be interpreted and analyzed in order to arrive at some understanding of a world that would otherwise be unknowable and inexplicable? The most powerful weapon available was human reason, employed in the manner that Aristotle had used it. The idea was to come to know what things seemed to be – and this could be done by empirical means – and then to determine what made them that way, a process that was largely guided by metaphysical considerations. Although "Aristotle was an indefatigable collector of facts – facts zoological, astronomical, meteorological, historical, sociological,"[3] he relied essentially on a priori reasoning to form a picture of the structure and operation of the world. Logic and reason were used to understand the way the world had to be in order to appear and function the way it does. Medieval Latin scholars eagerly embraced Aristotle's

2. For Buridan's question in *On the Heavens,* bk. 2, qu. 7, see Grant, *A Source Book in Medieval Science* (Cambridge, MA: Harvard University Press, 1974), 621–624. Buridan gave a more detailed discussion in his *Questions on the First Three Books of the Meteors,* parts of which have been translated into French by Pierre Duhem, *Le Système du monde,* 10 vols. (Paris: Hermann, 1913–1959), vol. 9, 293–305.
3. See Jonathan Barnes, *Aristotle,* in *Past Masters* (Oxford: Oxford University Press, 1982), 17.

methodology and his approach to the physical world, while adding important ideas about the cosmos from Christian faith and theology.

The conscious and systematic application of logic and reason to the natural world was the first major phase in the process that would eventually produce modern science. That first phase involved the construction of a comprehensive, intelligible system of the world, one that would permit scholars to explain in satisfactory terms a universe that would otherwise be unintelligible. So far as the Latin Middle Ages was concerned, this was brilliantly achieved in Aristotle's natural philosophy. Without the use of controlled experiments, systematic observations, and the regular application of mathematics to physical phenomena, the only powerful tool of analysis available to those who sought to understand the structure and operation of the world was reason, applied in a largely a priori manner, based on a modicum of observation and empirical data. It was in this manner that Aristotle fashioned his natural philosophy. In the ancient and medieval worlds, Aristotle's works represented the apotheosis of reason. Without the application of reason to organize and analyze data, science cannot exist. It is the first indispensable element in the development of science.

The natural philosophy that reached the Latin Middle Ages in the works of Aristotle was a highly structured, comprehensive, rational, discipline. The subsequent institutionalization, systematization, and expansion of it in the medieval universities of Western Europe may quite appropriately be regarded as the first stage in the continuous evolution of modern science.

Once Aristotle's natural philosophy was disseminated in Western Europe in the twelfth century, how did scholars view that discipline? In the twelfth century, Domingo Gundisalvo (fl. ca. 1140),[4] in a treatise on the classification of the sciences titled *On the Division of Philosophy (De divisione philosophiae),* begins his discussion of natural science (it is synonymous with natural philosophy) with a definition: "'Natural Science is the science considering only things unabstracted and with motion.'" He then cites with approval a definition from Avicenna, that "in truth, 'the matter of natural science is body,'"[5] and that, in the most general sense, it is body that is considered "according to what is subjected to motion and rest and change." Gundisalvo regarded the syllogism as the principal instrument of natural philosophy and therefore emphasized its rationality, explaining that "the 'artificer' is the natural philosopher who, pro-

4. Or Dominicus Gundissalinus, his Latin name.

5. The brief passages cited from Gundisalvo's treatise are from the translation by Marshall Clagett and Edward Grant in Grant, *A Source Book in Medieval Science,* 63.

ceeding rationally from the causes to effect and from effect to causes, seeks out principles. This science, moreover, is called 'physical,' that is, 'natural,' because it intends to treat only of natural things which are subject to the motion of nature."[6] In light of what will be discussed later, it should be emphasized that in his description of the meaning and aims of natural science, Gundisalvo makes no mention of God, or the faith, probably because he did not regard such matters as belonging to the province of natural philosophy.

In a chapter on the place of natural philosophy, Richard Kilwardby says that natural philosophy – he also calls it *natural science (naturalis scientia)* – "can be appropriately defined as follows: [It is] that part of speculative science that is perfective of the human view with respect to the cognition of mobile body in so far as it is mobile."[7]

The physicist, or natural philosopher, must consider "the motive principle, which is called nature."[8] A mobile body is one in which there is a motive principle that causes it to be in motion. Kilwardby enumerates two kinds of motive principle: the soul, which causes bodies to move; and the prime mover, which causes the celestial bodies to move. And in a crucial distinction, Kilwardby declares that the natural philosopher is only interested in the soul and in the prime mover insofar as they are principles of motion, and not insofar as they are substances.[9] Natural philosophy is about natural phenomena, about motion and its causes. It is not about the nature of spiritual substances, which is properly the subject matter of theology. This distinction had a great bearing in shaping the medieval attitude toward natural philosophy, because such sentiments helped prevent the theologization of natural philosophy.

For Gundisalvo and Kilwardby, and most other scholastic natural philosophers, natural philosophy was a discipline concerned with bodies that are capable of motion and change. Natural philosophy was the most

6. Grant, ibid., 65.

7. Kilwardby, *De ortu scientiarum,* ed. Albert G. Judy, O. P. (London: The British Academy, 1976), 17, lines 18–20; my translation.

8. "Item quia physicus debet considerare corpus mobile secundum quod huiusmodi, oportet quod consideret principium eius motivum quod dicitur natura." Kilwardby, *De ortu scientiarum,* 22, lines 12–14.

9. "Inde enim naturaliter mobile est quod in se habet huiusmodi principium motivum. Quia igitur principium naturaliter motivum animati corporis est anima, et principium motivum caelestis corporis est motor primus, ideo naturalis de anima et de motore primo considerat non secundum eorum substantias, sed secundum quod sunt principia motus." Kilwardby, *De ortu scientiarum,* p. 22, lines 14–19.

rational of all disciplines to which logic and reason could be applied. Medieval natural philosophers, whether theologians or not, pursued natural philosophy with the same apparent detachment, and with the same rationalistic investigative spirit, as did their most authoritative models: Aristotle, Avicenna (ibn Sina), and Averroës (ibn Rushd), Aristotle's most famous commentator, who was a Muslim natural philosopher, physician, and jurist.

Natural philosophy at the medieval university was essentially Aristotelian natural philosophy. The works of Aristotle that came to constitute the core of natural philosophy were usually referred to collectively as the "natural books," which were: *Physics, On the Heavens (De caelo), On the Soul (De anima), On Generation and Corruption (De generatione et corruptione),* and *Meteorology.* To this we might add Aristotle's biological works, though these treatises were studied and commented upon relatively little by comparison to the "natural books." Indeed, medieval natural philosophy was even broader than the topics covered in Aristotle's extensive works. It also included alchemy, nonmathematical astrology, and perhaps even books on natural magic. But with an occasional exception for astrology, these subjects were not taught at the medieval universities and, hence, play no role in this study.

NATURAL PHILOSOPHY AND THE EXACT SCIENCES

Certain exact sciences – mathematics, astronomy, statics, and optics, to be precise – were taught in the arts faculties of medieval universities. Mathematics was comprised of arithmetic and geometry, and the latter was taught usually as some version of Euclid's *Elements,* which was regarded as theoretical or speculative geometry. There was also practical geometry, usually thought of as the application of theoretical geometry to themes in natural philosophy. Indeed, statics, optics, and astronomy were viewed in that way.[10] During the Middle Ages, sciences that involved the application of mathematics to natural phenomena were described as "middle sciences" *(scientiae mediae),* because, as mentioned in Chapter 3, they were assumed to lie between natural philosophy and pure mathematics.[11] But although they lay between mathematics and natural philosophy, some natural

10. On theoretical and practical geometry and the sciences of statics, optics, and astronomy, see John North, "The Quadrivium," in H. de Ridder-Symoens, ed., *A History of the University in Europe,* Vol. 1: *Universities in the Middle Ages* (Cambridge: Cambridge University Press, 1992), Chapter 10.2, 346–350.
11. Here I follow Grant, *The Foundations of Modern Science in the Middle Ages,* 136.

philosophers, including Thomas Aquinas and Albert of Saxony, regarded the mathematical, or middle, sciences – for example, astronomy, optics, and statics – as lying more in the domain of natural philosophy than in mathematics.[12] The application of mathematics to natural philosophy was not, however, restricted to the middle, or exact, sciences. It could be applied independently, as it was, for example, in problems of motion and in the doctrine of the intension and remission of forms, where arithmetic and geometry were used to compare variations in all kinds of qualities, including velocities, heat, courage, fear, and so on.[13]

Medieval natural philosophers, like logicians, used more than one literary format to present their ideas. They resorted to questions, tractates, compendia, and encyclopedias. In Chapter 3, I observed that the most popular and predominant type of literature was undoubtedly the questions format, which was utilized primarily to explicate Aristotle's natural philosophy as embedded in the natural books. I also described the structure of a typical question. In order to convey the genuine flavor of medieval questions, I now present a translation of a question in natural philosophy by Nicole Oresme, taken from his *Questions on Aristotle's "On the Heavens."*

DOING NATURAL PHILOSOPHY: NICOLE ORESME

In the question by Nicole Oresme on the possible existence of other worlds, I indicate major subdivisions by numbers enclosed in square brack-

12. For Thomas Aquinas, see his Commentary on Aristotle's *Physics* in *S. Thomae Aquinatis In octo libros De physico auditu sive Physicorum Aristotelis commentaria,* ed. P. Fr. Angeli-M. Pirotta O. P. (Naples, 1953), bk. 2, lecture 3, p. 82, pars. 336–339. For an English translation, see *Commentary on Aristotle's "Physics" by St. Thomas Aquinas,* translated by Richard J. Blackwell, Richard J. Spath, and W. Edmund Thirlkel, Introduction by Vernon J. Bourke (New Haven: Yale University Press, 1963), 80–81, pars. 164–165. For Albert of Saxony, see his *Questions on the Physics,* bk. 2, qu. 8, fols. 25r, col. 2–26r, col. 1, in *Questiones et decisiones physicales insignium virorum. Alberti de Saxonia in octo libros Physicorum*... (Paris, 1518).
13. For a brief discussion, see Grant, The *Foundations of Modern Science in the Middle Ages,* 100–104. Scholastic natural philosophers considered the intension and remission of forms in tractates, rather than in Aristotelian commentaries and questions. For the most important of such treatises, see *Nicole Oresme and the Medieval Geometry of Qualities and Motions: A Treatise on the Uniformity and Difformity of Intensities Known as "Tractatus de configurationibus qualitatum et motuum,"* edited with an Introduction, English Translation, and Commentary by Marshall Clagett (Madison: The University of Wisconsin Press, 1968).

ets; further subdivisions within a major subdivision are numbered and enclosed within parentheses.

Nicole Oresme (ca. 1320–1382)

[1] It is sought whether it is possible that other worlds exist.[14]

[2] And it is argued yes [it is possible that other worlds exist].

(1) [First] it is possible by an argument of the ancients: every universal is predicated of many things [that exist] in actuality or in potentiality. [Now] the world is a universal; therefore [world is predicated of many things that exist in actuality or in potentiality]. The major [premise] is obvious by the definition of universal and also because the potentiality would be in vain [if there could not be other worlds]. Indeed, there would be no potentiality if other worlds could not exist; and thus it is also argued that this common and representative concept of many things would be in vain; and the minor [premise] is obvious because "world" is a common noun and therefore one must speak of "the world" or "this world," as is clear in the text.

(2) Second, it would thus be better that what is best and perfect be multiplied than what is not, because, other things being equal, two good things are better than one. Therefore, since nature always produces what is better, and the world is perfect, it follows that there are more [worlds].

(3) Third, because an imperfect species, such as an ass, generates what is similar to itself and is multiplied in individuals, therefore a more perfect species, such as the world [will also be multiplied in individuals], because to generate a thing similar to oneself is a most perfect and natural work, as is obvious in the second [book] of *On the Soul (De anima).*

(4) Fourth, it seems there could be more worlds, at least successively, because the principal parts [of a world], namely the four elements, are successively corrupted and generated, so that perhaps today there is nothing of the element of water that existed a thousand years ago; and the same may be said regarding the other elements.

[3] Aristotle says the opposite.

[4] It must be understood that in one way, "world" is taken as the totality of all beings, or it is the thing aggregated of all beings; and thus the existence of more worlds implies a contradiction.[15] In another way, ["world" is taken] as the aggregate of the heaven, the elements, and mixed bodies [or compounds], that is, it [the world] is taken [to include] everything contained within the convex surface of the

14. Nicole Oresme, *Questions on De celo,* bk. 1, qu. 18 in *The "Questiones super De celo" of Nicole Oresme,* by Claudia Kren (Ph.D. diss., University of Wisconsin, 1965), 266–279. I have freely altered Kren's translation to make it more readable.

15. In a note, Kren explains (ibid., p. 971, 26–28) that "[i]t implies a contradiction because if the world, by definition, contains all possible entities whatsoever, there could not be another world."

last [or outermost] sphere. And in this way, one can imagine that there is one other world similar to this, or many worlds; nor does this imply a contradiction in terms. Further, as was declared, "possible" is taken in one way naturally; in another way, absolutely, which implies a contradiction, either with respect to the terms, or in another way. And it [that is, the term "possible"] can be taken in yet another way, namely as doubtful, just as if someone should propose that [the totality of] all stars is an even number, you would say that it is possible; or [if someone should say] that Aristotle lived one-hundred years, since that implies a contradiction;[16] or that Aristotle lived at the time of Noah; and similarly for other such instances.

Further, the existence of many worlds can be imagined in four ways. In one way, successively, just as Empedocles assumed. In another way, [existing] simultaneously and mutually separate, as Anaximander said. Aristotle disproves these [first] two ways. In another way, they would mutually contain each other and be concentric. In a fourth way, they would contain each other, but not concentrically.

[5] Therefore, let this be the first conclusion: that it is impossible that there be more worlds successively by natural means. This is proved, because then this whole world would be corruptible, including the heaven, the opposite of which [is true], as will be seen later. Indeed, if after corruption, the matter [of any world] would be without form, this would be impossible. If immediately after [the corruption of the form of one world], it had the form of a similar world, this would still be naturally impossible, because generation arises from privation as to species, and therefore from a man, a man is not produced directly, as is obvious from the first [book] of the *Physics;* nor is whiteness produced [immediately] from whiteness.[17] Moreover, this [world] would be corrupted in vain if something similar to it were made immediately [after]. But those [who support this position] say that it [the world] is not totally corrupted, but that a certain disorganization occurs and things are reduced into a confused chaos; after awhile, things are again organized. And this is done by hate and love.[18] But this is not valid, because, although one could prove that this is impossible, yet assuming this, the world would not be corrupted because its organization is an accidental predication, so that the same thing would

16. It is a contradiction if Aristotle lived 100 years, because, as Kren puts it (ibid., p. 971, 39–40), "this length of time exceeds the limit of life afforded to any individual man." This is a poor example, since even in Oresme's day an occasional individual may have lived to 100 years.
17. In the first book of his *Physics,* Aristotle argues (*Physics,* 1.5. 188b.21–25) that "everything that comes to be or passes away comes from, or passes into, its contrary or an intermediate state. But the intermediates are derived from the contraries – colours, for instance, from black and white. Everything, therefore, that comes to be by a natural process is either a contrary or a product of contraries." Oresme applies Aristotle's argument to successive worlds. There is no contrary to a world and, therefore, one world cannot be produced by natural means directly from its immediate predecessor.
18. According to Aristotle, "hate" and "love" were the mechanisms used by Empedocles to produce successive worlds. Hence, Oresme's argument is directed against Empedocles.

sometimes be organized, [and] then it would be called a world, and afterward it would be disorganized [and also be called a world]; and in this way would the noun "world" function. I say that this would be supernaturally possible, [but not naturally possible].

[6] As to the second way, the [second] conclusion is that this is naturally impossible, namely, that many mutually separate worlds should exist [at the same time].

(1) This is proved in the first place: because then the earth of that world would descend to this world, and conversely, which cannot occur naturally or violently, as was proved.[19]

(2) Second: one may argue from metaphysics, because there is only one prime mover, therefore only one prime mobile and one world. The antecedent is evident in Aristotle in the twelfth [book of the *Metaphysics*] where he says that in the universe of beings there is only one prince and the Commentator [i.e., Averroës] says that in things abstracted from matter there is only one individual in one species.[20] And the consequence holds, because if there were many worlds there would be one first heaven and one mover in any one of them; even if there is one prince, his kingdom is one. The Commentator [Averroës] objects because one could say that there might be one universal ruler of all these [worlds]; but in any one world there could be one mover that moves as much because it is universal, as because it operates toward an end. He sets this aside because the Prime Mover moves properly for its own sake, because the conceiver, the conception, and what is conceived are the same [in the Prime Mover].

(3) Third: it is obvious that there would be a vacuum between these worlds, which is impossible, as is obvious in the fourth [book] of [Aristotle's] *Physics*. The consequence is valid because these worlds would be spherical.[21]

19. Aristotle regards the existence of more than one world as naturally impossible. One of his primary arguments against it was the conviction that if the earths of two worlds were of the same identical nature, each earth would rise up in its world and head for the center of the other world (see Aristotle, *On the Heavens*, 1.8.276a. 18–276b.22). In the preceding question (bk. 1, qu. 17), Oresme considered "whether if there were another world, the earth of that world would be moved to the center of this one" (Kren trans., p. 244). As his principal conclusion, he declares that "if there were another world, the earth of that one would be moved to this; this is proved because all simple bodies of the same species naturally tend to the same numerical place ... and distance plays no part in this." And since the earth of our world moves naturally to the center of our world, the two earths will be in the center of our world, which is impossible. Indeed, Oresme argues that the same thing will happen, apparently simultaneously, to the center of the other world. That is, our earth will seek to unite with the other earth in the center of the other world (see Kren trans., pp. 243–264, but especially pp. 261–264). Oresme concludes that other worlds cannot exist naturally because they would produce absurd and impossible conditions.

20. Therefore, in the species of worlds, there is only one world.

21. Because any two spherical worlds could only touch at a point, empty spaces would exist where they did not touch

(4) Fourth: either there would be an infinite number of such worlds, and then the aggregate of all of them would be infinite, which has been disproved; or, if they were finite [in number], it is not apparent how many there would be, but they ought to be sown everywhere in an infinite space.

(5) Fifth: because outside this world nothing exists, since heavy bodies have their places in this world.

(6) Sixth: this is confirmed because all simple motions are inclined toward the same numerical place, as was proved. Therefore, the center of this world is the natural place of a heavy body; therefore no heavy body is naturally outside [of this world]. And the same may be said of the other elements.

Nevertheless, [because] these [six] arguments are based on natural principles, I say that they are probable, so that there are no more worlds. However, God, who is of infinite power, can make another world or [worlds], and could easily solve these arguments.

[7] As to the third way [that other worlds might exist], one can imagine that under this earth there could be another world around the same center, or two or three [other worlds] below each other; and also that there could be one or more [worlds] above these worlds [but below the earth's surface] and they also would move around the same center, [namely, the center of our earth]. And [we may further imagine that] these worlds are absolutely similar, except in quantity, because an inferior world would be smaller. And that all things would appear in one [world] as they do in another, so that if a man of this world were placed in an inferior world, and if he became as small as the men of that world, he would not believe that he was in another world. And it could be said that the earth of any [of these worlds] rests naturally, because its center of gravity is the center of the whole [group of worlds]; and that it is impossible to pierce through this earth to go to another world, just as it is impossible to scale the heaven. I say, therefore, that God could do this, and it does not imply [a contradiction]; nor perchance is it against natural principles, unless in the sense that these worlds ought to have an order with respect to their perfection, and also with respect to their position. I say, secondly, that this [conception] does appear irrational, nor is there any reason why it should be assumed as an actuality, because it would be assumed in vain, although [it could be posited] as possible. Therefore, the opposite [opinion] is more reasonable, although it would not be demonstrable.

[8] As to the fourth way [that other worlds might exist, and this is the fourth conclusion], one could imagine that there are many worlds of which one contains another and they are eccentric, just as if below the body of the moon there was another world composed of spheres, elements, and other things, [and that this world is] smaller than our world. And I say briefly that this is possible for God, nor does it imply [a contradiction]. Nevertheless, it ought not to be assumed, nor is it likely.

[9] [Now, I respond] to [the principal] arguments.

(1) To the first [principal argument], it is denied that every universal [term] can be predicated of many potential things, in the sense that

there can be many individuals [of which the universal term can be predicated]. But I say that it is not inconsistent with the conception, but that it would represent another world, if there could be one. However it does not follow that if a concept is indeterminate, that it is in vain, because with such a concept we can also use it for the imagination. Therefore it has a plural number [that is, a number greater than one] in the imagination.

(2) To the second [principal argument], I say that that which is impossible is not better or good, so that a greater multitude of good things is not always better, but is a commensurate proportion.

(3) To the third [principal argument], I say that to generate an entity similar to itself is a perfect act where the individuals are corruptible, but according to Aristotle, the world is eternal [and therefore cannot generate its own likeness].

(4) To the fourth [principal argument], I say that the principal part of the world is the heaven itself both in magnitude and in nobility. Therefore, according to Plato, these inferior things are not properly part of the world, but is something unnecessary that is placed in some kind of order. Otherwise, it can be said that the sea is always the same by a continuous succession; and so in regard to the earth and the other elements.

And the question is obvious.

It will be helpful to describe the form of Oresme's question. As in all scholastic questions, he first enunciates the question (see Sec. [1]) and then, in Sec. [2] (1)–(4), presents four "principal arguments" *(rationes principales),* which in this question serve as arguments in favor of the thesis that other worlds do exist. Upon seeing this, we can usually infer that the author will defend the opposite thesis, namely, that other worlds do not exist. Following the principal arguments, Oresme proclaims opposition to them by declaring that "Aristotle says the opposite" (Sec. [3]). In the next phase of the question, Sec. [4], Oresme qualifies the question by an analysis of the terms "world" and "possible." He concludes his qualification and analysis of the question by distinguishing four ways that a plurality of worlds can be interpreted, and he then proceeds to the next stage in which he presents four conclusions, each of which is an analysis of one of the four modes for the possible existence of other worlds (Secs. [5] to [8]). As the final step, in Sec. [9] (1)–(4), Oresme responds to each of the four principal arguments (Sec. [2] (1)–(4)). In this question, Oresme included all of the primary structural elements that would ordinarily appear in a scholastic question:

[1], enunciation of the question;
[2] (1)–(4), principal arguments defending the thesis that the author will almost certainly reject;
[3], statement of the opposite opinion, a version of which the author will defend;
[4], qualification of the question and clarification of terms;
[5]–[8], the body of the argument expressing the author's opinions in the form of four conclusions; and, finally,
[9], the response to each of the four principal arguments proclaimed at the very beginning of the question.

Oresme's question includes two other common aspects of medieval questions in natural philosophy. The first involves the use of logical concepts and terms, such as contradiction, predicate, universal, and major and minor premises; the second is invocation of God's absolute power to create any kind of other world that He pleases. Indeed, common aspects are illustrated in section [8], where Oresme mentions that it is possible for God to make other worlds of the kind described in that section and that the making of them does not involve a logical contradiction. The implication is clear: If a contradiction were implied, it would not be possible for God to make such worlds.[22]

In the questions method that was typical of medieval scholasticism, the objective was to present the best arguments for both sides and, ultimately, to explain why one alternative was preferred over the other. To achieve this goal, one had to be perfectly clear about the meaning of the question, for which reason it was not uncommon to explain and qualify terms used in the question itself, as we saw where Oresme sought to explain how he would understand the terms world and possible.[23]

Medieval natural philosophy at the medieval universities was comprised of hundreds of questions like Oresme's, many of which were structured more intricately than his. For centuries, masters and students were conditioned to treat questions in natural philosophy as if they were listening to, or reading

22. On God and contradictions, see Chapter 6.

23. In his *Questions on On the Heavens,* Buridan also devoted a question to the possibility of other worlds and did so before Oresme. Like Oresme, Buridan also discussed the different meanings of the term "world" *(mundus).* For the translated passage from Buridan's question about the possible existence of other worlds, see Grant, *Foundations of Modern Science in the Middle Ages,* 128. In Buridan's Latin text, see bk. 1, qu. 19, in Ernest A. Moody, ed., *Ioannis Buridani Quaestiones super libris quattuor De caelo et mundo* (Cambridge, MA: The Mediaeval Academy of America, 1942), 88.

about, a disputation between two scholars, each defending a position contrary to the other. In formulating a response to a question, an author had to take account of one, or more, opposing opinions. The format of the question method required it. Over the centuries, this reasoned, analytic approach to nature and to a variety of problems became the routine way to learn about the world and to arrive at knowledge and understanding. Even as the questions method was being abandoned in the sixteenth and seventeenth centuries, its legacy of probing was continued and expanded.

REASON AND THE SENSES IN NATURAL PHILOSOPHY: EMPIRICISM WITHOUT OBSERVATION

Few aspects of natural philosophy reveal the role of reason in medieval thought more dramatically than its relationship to empiricism, the other major component of natural philosophy that manifested itself through sense perception, observation, experience, and experiment. With the dissemination of Aristotle's natural philosophy in the thirteenth century, it was not unusual for scholastic natural philosophers to emphasize the vital role of experience and observation, as did Albertus Magnus, for example, when he declares:

> Anything that is taken on the evidence of the senses is superior to that which is opposed to sense observations; a conclusion that is inconsistent with the evidence of the sense is not to be believed; and a principle that does not accord with the experimental knowledge of the senses is not a principle but rather its opposite.[24]

In a similar vein, Roger Bacon asserts that "there are two modes of acquiring knowledge, namely by reasoning and experience." He then goes on to exalt experience over reasoning when he explains:

> Reasoning draws a conclusion and makes us grant the conclusion, but does not make the conclusion certain, nor does it remove doubt so that the mind may rest on the intuition of truth, unless the mind discovers it by the path of experience;

24. Translated by William A. Wallace, *Causality and Scientific Explanation,* 2 vols. (Ann Arbor: The University of Michigan Press, vol. 1, 1972; vol. 2, 1974), vol. 1, 70. Wallace's translation is from the Borgnet edition of the Latin text in Albertus's *Commentary on Aristotle's Physics* (Borgnet edition, *Liber VIII Physicorum,* tr. 2, cap. 2, 564). The Latin text in the modern edition of Albertus's *Opera omnia* is as follows: "Omnis enim acceptio, quae firmatur sensu, melior est quam illa quae sensui contradicit, et conclusio, quae sensui contradicit, est incredibilis, principium autem, quod experimentali cognitioni in sensu non concordat, non est principium, sed potius contrarium principio." See *Alberti Magni Opera Omnia, Physica,* pars II, libri 5–8, ed. Paul Hossfeld (Aschendorff: Monasterii Westfalorum, 1993), liber VIII, tr. 2, cap. 2, 587, col. 2.

> ... For if a man who has never seen fire should prove by adequate reasoning that fire burns and injures things and destroys them, his mind would not be satisfied thereby, nor would he avoid fire, until he placed his hand or some combustible substance in the fire, so that he might prove by experience that which reasoning taught. But when he has had the actual experience of combustion his mind is made certain and rests in the full light of truth. Therefore reasoning does not suffice, but experience does.[25]

In the fourteenth century, John Buridan emphasized the importance of induction when, in a question on the possible existence of a vacuum, he declared that

> every universal proposition in natural science *(in scientia naturali)* ought to be conceded as a principle which can be proved by experimental induction *(per experimentalem inductionem),* just as in many particular [occurrences] of it, it would be manifestly found to be so and in no instances does it fail to appear. For Aristotle puts it very well [when he says] that many principles must be accepted and known by sense, memory, and experience.[26] Indeed, at some time or other, we could not know that every fire is hot [except in this way].[27]

These are powerful statements in favor of experience and observation. Although few medieval natural philosophers matched the enthusiasm for experience exhibited in these passages, we may plausibly assume that most of them were empiricists in the Aristotelian sense and in the sense expressed by Albertus Magnus, Roger Bacon, and John Buridan. That is, ideally, they

25. Roger Bacon, *Opus majus,* part VI ("On Experimental Science"), ch. 1 in *The Opus Majus of Roger Bacon,* a translation by Robert Belle Burke, 2 vols. (New York: Russell and Russell, 1962), vol. 1, 583. Reference is made to this passage in Bacon by N. W. Fisher and Sabetai Unguru, "Experimental Science and Mathematics in Roger Bacon's Thought," *Traditio* 27 (1971), 358.
26. *Posterior Analytics* 2.19. 100a.4–9.
27. "Item omnis propositio universalis in scientia naturali debet concedi tanquam principium que potest probari per experimentalem inductionem sic quod in pluris singularibus ipsius manifeste inveniatur [corrected from *inveniaur*] ita esse et in nullo nunquam apparet instantia, sicut enim bene dicit Aristoteles quod oportet multa principia esse accepta et scita sensu, memoria et experientia; immo aliquando non potuimus scire quod omnis ignis est calidus." *Questions on the Physics,* bk. 4, qu. 7 in *Acutissimi philosophi reverendi Magistri Johannis Buridani subtilissime questiones super octo Phisicorum libros Aristotelis diligenter recognite et revise a Magistro Johanne Dullaert de Gandavo antea nusquam impresse* (Paris, 1509), fol. 73v, col. 1. The translation is mine (slightly altered) from Grant, *A Source Book in Medieval Science,* 326.

thought it desirable to begin with an observation or sense perception as the foundational basis of a generalization or conclusion.

From whence did this emphasis on experience and observation derive? It is another legacy from Aristotle. For in addition to his emphasis on reason, Aristotle also formulated an epistemology that was grounded in sense perception. It is almost a truism that Aristotelian natural philosophy is, in sharp contrast to Platonic philosophy, rooted in sense perception. A superficial look at Plato's thought might suggest that he, too, emphasized observation. Thus in his *Timaeus,* Plato praised the creation of sight by the Demiurge, or creator god, declaring that

> [s]ight ... in my judgment is the cause of the highest benefits to us in that no word of our present discourse about the universe could ever have been spoken, had we never seen stars, Sun, and sky. But as it is, the sight of day and night, of months and the revolving years, of equinox and solstice, has caused the invention of number and bestowed on us the notion of time and the study of the nature of the world; whence we have derived all philosophy, than which no greater boon has ever come or shall come to mortal man as a gift from heaven.[28]

Despite this high praise for sight and the great benefits it produced for humanity, Plato was not in any manner an empiricist. Although he did not repudiate the senses, he had little confidence in them. They were incapable of leading to truth, which lay beyond our senses. Rational abstract thought was vastly superior to observation. Plato's theory of knowledge based truth in eternal, unchanging forms and not in unstable, ever-changing objects in the physical world.[29]

Aristotle differed radically from his teacher. He based the generalizations of scientific knowledge on perception. As one scholar has put it, Aristotle's scientific treatises "are scientific, in the sense that they are based on empirical research, and attempt to organise and explain the observed phenomena."[30] For Aristotle, "knowledge is bred by generalisation out of perception."[31] Thus, it is hardly surprising that medieval natural philoso-

28. *Timaeus* 47A–47B, in *Plato's Cosmology: The "Timaeus" of Plato translated with a running commentary,* by Francis M. Cornford (New York: The Liberal Arts Press, 1957), 157–158.
29. See G. E. R. Lloyd, *Early Greek Science: Thales to Aristotle* (New York: W. W. Norton & Co., 1970), 70–72, 78–79, and David C. Lindberg, *The Beginnings of Western Science: The European Scientific Tradition in Philosophical, Religious, and Institutional Context, 600 B.C. to A.D. 1450* (Chicago: University of Chicago Press, 1992), 37–39.
30. Jonathan Barnes, *Aristotle,* 61.
31. Ibid., 59.

phers emphasized sense perception and observation as the foundation of knowledge and science. Aristotle believed that our senses "give the most authoritative knowledge of particulars. But they do not," he explains, "tell us the 'why' of anything – e.g. why fire is hot; they only say that it is hot."[32] Despite an emphasis on sense perception and observation of the particular, Aristotle's natural books – *Physics, On Generation and Corruption, Meteorology, On the Heavens, On the Soul,* and the *Parva naturalia* – are largely theoretical accounts of their subject matters. They attempt to tell us the "why" of things. The picture of the cosmos that Aristotle constructed in these works seems far removed from its observational foundation. It seems, rather, an account of a world that was made to conform to Aristotle's preconceived ideas of what the universe had to be like in order to function in a manner worthy of a divine cosmos.

Because of Aristotle's own ambiguities and confusions, his legacy to the Latin Middle Ages was a natural philosophy that presented a rather confused relationship between the theoretical and empirical. Just what role did observation play in medieval natural philosophy? Most, if not all, historians of medieval science and natural philosophy regard the Middle Ages as a period in which the habit of observation was not well developed or practiced. Thus, although we might concede that scholars in the Middle Ages were philosophically committed to Aristotelian empiricism, it would be helpful to know the role that observation actually played in the resolution of physical questions.

A. C. Crombie has rightly observed that medieval natural philosophy was based on

> Aristotle's treatment of the subject, which dealt with very general theoretical questions of space, time, motion, causation, the relation of quality to quantity and so on, and in logical and mathematical problems arising out of this, the argument was likewise purely theoretical. Hence there was no experimental science in the context of the natural philosophy of the medieval universities.[33]

The "experimental arguments and practices" that did reach the university environment were introduced there by means of mathematically oriented sciences, such as "optics, acoustics, mechanics and astronomy, through natural magic, and through medicine."[34] Based on available evidence, it appears that

32. Aristotle, *Metaphysics* 1.1.981b.10–11.
33. A. C. Crombie, *Styles of Scientific Thinking in the European Tradition: The history of argument and explanation especially in the mathematical and biomedical sciences and arts,* 3 vols. (London: Gerald Duckworth, 1994), vol. 1, 317.
34. Crombie, ibid.

medieval natural philosophers provided much more by way of "experimental arguments" than by experimental "practices." Although experiments designed to test theories were produced in the Middle Ages, they were rather sporadic occurrences.[35] It was not experiments but, rather, observations and experiences that turn up in medieval commentaries on Aristotle's works. What kind of empiricism was this? That is what we must now investigate.

Medieval natural philosophers described and reported many seeming observations. But they were not expected to have personally witnessed the experiences they reported. Indeed, they frequently derived experiential examples from the works of others – from Greek and Islamic treatises, and from Latin predecessors and contemporaries. Such experiences seem to have been regarded as indistinguishable from genuine, personal observations. Those who produced the numerous commentaries on Aristotle's natural books followed Aristotle's empiricist philosophy. But they were empiricists who rarely observed for the sole purpose of resolving a question under discussion.

The kinds of observations and experiences they relied on may be divided into two major categories. The first includes those in which an author had, at some time in his life, probably observed the experience he reports. Thus experiments, or observations that were personally experienced by a natural philosopher, or that he might plausibly have performed, will be considered in the category of *Personally Observed Experiences*. The second category embraces experiences that were of a hypothetical character – *thought experiments*, as we might now call them. In both categories, many experiences were derived by authors from other treatises they had read or heard about.

Personally Observed Experiences

A major category of personal observations in medieval natural philosophy was of a gross and obvious kind. Nicole Oresme, for example, offers as an experience that fire heats,[36] and John Buridan mentions that the heaven moves.[37] "Often the sense of 'experience' is," as Peter King explains, "no

35. For the most thorough account of experiments in the Middle Ages, see Crombie, ibid., vol. 1, ch. III(a), 313–423.
36. For the passage, see Stefano Caroti, ed., *Nicole Oresme Quaestiones super De generatione et corruptione* (Munich: Verlag der Bayerischen Akademie der Wissenschaften, 1996), bk. 1, qu. 1, 4, lines 45–46.
37. The passage occurs in Buridan's *Questions on the Metaphysics*. See *In Metaphysicen Aristotelis; Questiones argutissime Magistri Ioannis Buridani* ... (Paris, 1518), bk. 2, qu. 1, fols. 8v, col. 2–9r, col. 1.

stronger than the 'experience' of the old farm hand, a simple way of stating 'what everybody knows.' Such appeals are not to be confused with modern experiments, or testing: no question is being put to Nature."[38] But in the same question in which he declares that fire heats, Oresme also explains that "an alteration is when one thing is changed into another, as hotness into coldness, and similarly fire into air." All his readers would have known from experience that a hot thing would eventually become a cold thing. But why should we believe that fire becomes air? It is hardly an obvious or common experience. When a fire is extinguished, it disappears. Does it become air? We may perhaps infer that fire has been altered into air, but we surely do not observe it directly. Why, then, did medieval natural philosophers believe that fire is transformed into air? Probably because it was assumed by all Aristotelian natural philosophers that the four elements were mutually transformable. Just as, for example, water is changeable into air and vice versa, so is air convertible into fire and vice versa. Thus, the conversion of fire into air does not appear to be a proper observation but seems, rather, an "observation" that is required by theory, the theory of the four elements and their mutual conversions. Many "observations" in medieval natural philosophy were of this kind. They were driven by theory, or by what had to be.

Occasionally, it is difficult to determine whether observations are to be taken at face value, or are theory driven. For example, John Buridan, after the passage quoted from his *Questions on the Physics* (bk. 4, qu. 8), asserting the validity of experimental induction, declared, in the next paragraph, that "by such experimental induction it appears to us that no place is a vacuum, because everywhere we find some natural body, namely air, or water, or some other [body]."[39] Is this denial of vacua based on Buridan's limited observations, namely, of the kind wherein he concluded that all the places of which he had any direct knowledge were filled with some kind of matter, and none were empty; or is it really a theoretically derived "observation" where Buridan is simply upholding Aristotle's generalization that vacua cannot exist in nature? It could, of course, be both.

As evidence that a vacuum cannot exist naturally, medieval natural philosophers often cited the siphon experiment. Albert of Saxony, for example, mentions the siphon, or tube argument, which enables us to "see that if some tube

38. Peter King, "Mediaeval Thought-Experiments: The Metamethodology of Mediaeval Science," in Tamara Horowitz and Gerald J. Massey, eds., *Thought Experiments in Science and Philosophy* (Savage, MD: Rowman & Littlefield, 1991), 47.

39. For the passage, see Grant, *Source Book in Medieval Science,* 326.

(fistula) is put in water, and air is drawn from the tube, the water follows by ascending, striving to remain contiguous with the air lest a vacuum be formed."[40] John Buridan presents a similar version with wine, explaining that one takes a hollow reed and places one end in the wine and the other end in the mouth. "By drawing up the air standing in the reed, you [also] draw up the wine by moving it above [even] though it is heavy. This happens because it is necessary that some body always follow immediately after the air which you draw upward in order to prevent the formation of a vacuum."[41]

Did Albert of Saxony or John Buridan actually place a siphon in a cup, or in a glass of water or wine, and draw up the liquid? Did one or both of them see someone else do it? They make no mention of these alternatives. But they did not have to, because both agreed with Aristotle that a naturally occurring vacuum is impossible, so that an explanation of the kind they provided was almost unavoidable. Moreover, direct familiarity with the siphon experience was not essential because this experiment, and a number of others against the existence of vacuum, appear in a treatise titled *On Emptiness and Void (De inani et vacuo)* that was widely known in the Middle Ages, and was probably derived from Greek or Arabic sources. It seems more plausible to assume that Albert, Buridan, and others who cited this experiment were repeating what was commonly known, or they may have read about it in someone's *Questions on Aristotle's Physics,* where it is frequently cited.[42]

Few, if any, medieval natural philosophers were more empirically minded than John Buridan, as we see in his well-known question on impetus theory, where he asks "whether a projectile after leaving the hand of the projector is moved by the air, or by what it is moved."[43] Aristotle had argued that air was the cause of the continued motion of a projectile after it lost contact with its motive cause or projector. To refute Aristotle's theory that air pushes a projectile along after the projectile has lost contact with the agent that originally caused its motion, a theory known as *antiperistasis,* Buridan

40. Albert of Saxony, ibid. Both Buridan and Marsilius of Inghen cite the same experience. Buridan replaces the water with wine (Grant, *Source Book,* 326); for Marsilius's version, see Grant, ibid., 327.

41. Buridan, *Questions on the Physics,* bk. 4, qu. 8. Translated in Grant, *Source Book,* 326.

42. For a discussion of experiments against the vacuum, see Edward Grant, *Much Ado About Nothing: Theories of Space and Vacuum from the Middle Ages to the Scientific Revolution* (Cambridge: Cambridge University Press, 1981), 80–81. For the *Treatise on Emptiness and Void,* see also ibid., 310–311, n. 82.

43. The translation appears in Marshall Clagett, *The Science of Mechanics in the Middle Ages* (Madison: The University of Wisconsin Press, 1959), 532–538. It is reprinted in Grant, *A Source Book in Medieval Science,* 275–280.

presented three experiences *(experientie),* the first and third of which seem to qualify as "personally observed experiences." As Buridan explains:

> The first experience concerns the top *(trocus)* and the smith's mill (i. e. wheel – *mola fabri*) which are moved for a long time and yet do not leave their places. Hence, it is not necessary for the air to follow along to fill up the place of departure of a top of this kind and a smith's mill. So it cannot be said [that the top and smith's mill are moved by the air] in this manner.[44]...
>
> The third experience is this: a ship drawn swiftly in the river even against the flow of the river, after the drawing has ceased, cannot be stopped quickly, but continues to move for a long time. And yet a sailor on deck does not feel any air from behind pushing him. He feels only the air from the front resisting [him].[45]

It is probable that Buridan actually observed the rotary motions of the mill wheel and top. These were experiences that he could readily have recalled, recognizing their utility in his argument against Aristotle. Although it is, of course, possible that Buridan specifically observed these two phenomena in connection with his treatment of the question on impetus, he gives no indication of this, and it is more than likely that he relied on earlier experiences with such phenomena.

The third experiential example, in which a ship continues to move even after the ship haulers cease to pull it, is a phenomenon that Buridan probably observed directly, perhaps many times in his life. It was probably a common experience. Indeed, not only was a similar experience, namely the smith's mill, or wheel, just cited, used to refute the role of air as the mover of the projectile, but it was used to support the existence of impetus, the impressed incorporeal force that Buridan, and others, believed was the real mover of bodies. As evidence that bodies in motion produce their own impetus, Buridan insists that

44. In the final question of his *Questions on the Physics,* bk. 8, qu. 13, Albert of Saxony cites the same experience involving the smith's wheel and the top in a question, "by what is a projectile moved upward after its separation from the projector?" ("Ultimo quaeritur a qua moveatur proiectum sursum post separationem illius a qua proiicit"). Here is what Albert says: "Similiter ista opinio non habet locum in motu mole fabri; similiter in motu troci. Vidimus enim quod trocus post exitum eius a manu proiicientis diu movetur circulariter absque hoc quod aliquis aer ipsum insequatur, movet enim super eodem puncto spatii." See *Questiones et decisiones physicales insignium virorum. Alberti de Saxonia in octo libros Physicorum...* (Paris, 1518), fol. 83v, col. 1.
45. Clagett, *The Science of Mechanics,* 533; Grant, *Source Book,* 275–276. For the Latin text, see *Acutissimi philosophi reverendi Magistri Johannis Buridani subtilissime questiones super octo Phisicorum libros Aristotelis* (Paris, 1509), bk. 8, qu. 12, fols. 120r, col. 2–120v, col. 1.

> you have an experiment *(experimentum)* [to support this position]: if you cause a large and very heavy smith's mill [i.e., a wheel] to rotate and you then cease to move it, it will still move a while longer by this impetus it has acquired. Nay, you cannot immediately bring it to rest, but on account of the resistance from the gravity of the mill, the impetus would be continually diminished until the mill would cease to move.[46]

Thus did Buridan find that the smith's wheel manufactured its own impetus to produce its motion. He also inferred from the fact that the smith's wheel gradually slows and then becomes motionless that the impetus is gradually diminished to zero by the heaviness of the mill functioning as a resistance that dissipates the impetus. This example is one of the few instances in which experience played an essential role in giving credibility to a theoretical argument, namely, the theoretical existence of impetus.

Thought Experiments

Thought experiments range from those that are imagined under conditions that are contrary to nature to those that are fashioned from conditions that are plausible and have an observational basis, but which were nonetheless imagined for the occasion. An example of the latter is the second of Buridan's three experiences mentioned earlier. "The second experience is this: A lance having a conical posterior as sharp as its anterior would be moved after projection just as swiftly as it would be without a sharp conical posterior. But surely the air following could not push a sharp end in this way, because the air would be easily divided by the sharpness."[47] This second experience is a "reasoned" experience. In the absence of any statement to the contrary, we cannot properly assume that Buridan tested this with two lances, one with a conical posterior and one without. It is far more plausible to assume he *reasoned* that a conical – that is, pointed – posterior would readily divide the air that sought to push it. Consequently, a lance with a conical posterior would not be moved by the air, as contrasted with a lance that had a broad, flat, posterior surface against which the air could push. Although this is presented as an experience, it is extremely unlikely that Buridan hurled a conical lance and a nonconical lance to see which was carried farther. Even if he did, it is unlikely that his actions would have resolved the issue. What we have here is a hypothetical "experience" wherein Buridan reasoned about the way a conical posterior would divide the pushing air in contrast to the way a nonconical lance would react.

46. Clagett, *The Science of Mechanics*, 561; Grant, *Source Book*, 282.
47. Clagett, ibid., 533; Grant, ibid., 276.

In Buridan's previous question on the possible existence of vacuum, the siphon experiment, or experience, was one he might have personally observed or experienced. Another "experience" meant to show that vacuum is naturally impossible was one that Buridan could not have performed or directly experienced and is, therefore, categorized as a "thought experiment." In this appeal, Buridan seeks to "show by experience that we cannot separate one body from another unless another body intervenes." As evidence for this claim, Buridan offers the following:

If all the holes of a bellows *(follis)*[48] were perfectly stopped up so that no air could enter, we could never separate their surfaces. Not even twenty horses could do it if ten were to pull on one side and ten on the other; they would never separate the surfaces of the bellows unless something were forced or pierced through and another body could come between the surfaces.[49]

Buridan may have seen how difficult it is to part the sides of a bellows after most of the air has been squeezed out; or, he may have been made aware of it by others. But, in the absence of any claim to the contrary, we ought not to assume that he ever personally tried to separate the sides of a bellows himself. And we can assume with virtual certainty that he never harnessed ten horses on each side of a bellows and witnessed their failure to pull the sides apart. Buridan's imaginary experience bears a striking resemblance to the famous experiment carried out by Otto von Guericke (1602–1686) in Magdeburg in 1657, approximately 300 years after Buridan. Von Guericke built a large copper sphere formed from two half spheres. After evacuating the sphere by means of an air pump, he showed that because of air pressure on the surface of the spheres and the void within, not even two teams of eight horses, one team harnessed to each side, could pull them apart. Thus, where Buridan's imaginary teams of horses sought to show the impossibility of a vacuum, Von Guericke's real horses labored mightily to demonstrate the existence of a vacuum and the force of air pressure.[50]

48. I have translated *follis* as bellows. But Buridan might have had in mind a leather ball or pouch, or something capable of being inflated with air and then deflated by removing the stop, or stops.
49. Translated in Grant, *A Source Book in Medieval Science,* 326. For the Latin text, see Buridan, *Questions on the Physics,* bk. 4, qu. 7, in *Acutissimi philosophi reverendi Magistri Johannis Buridani subtilissime questiones super octo Phisicorum libros Aristotelis,* fol. 73v, col. 1.
50. See Grant, *Source Book,* 563, n. 54. For a brief sketch of Von Guericke's life, see Fritz Krafft's article, "Guericke (Gericke), Otto Von," in *Dictionary of Scientific Biography,* vol. 5 (1972), 574–576.

Significant thought experiments appear in many other questions. Arguments were often based on an imaginary empiricism that was assumed to represent real situations. In the eighth book of his *Physics* (8.8.264a.7–35), Aristotle argues that a continuous motion up and down is not a single motion, but really two separate motions separated by a moment of rest. In their questions on Aristotle's *Physics,* medieval natural philosophers disagreed with Aristotle and denied a moment of rest. Among the prominent authors who discussed this problem in their *Questions on the Physics* were John Buridan,[51] Albert of Saxony,[52] and Marsilius of Inghen. All three invoked a similar experience to deny a moment of rest. In his argument, Marsilius of Inghen denied that "between any motions that turn back *(reflexos)* [over the same path]" there is a moment of rest:

The proof is that if a bean *(fabba)* were projected upward against a millstone *(molarem)* which is descending, it does not appear probable that the bean could rest before descending, for if it did rest through some time it would stop the millstone from descending, which seems impossible.[53]

This argument was often repeated in subsequent centuries, even by Galileo, who describes it in his early treatise *On Motion (De motu)* and refers to the argument as "the well-known one about a large stone falling from a tower" and meeting a pebble thrown up from below.[54] As often happened, scholastic natural philosophers did not rest content with the dramatic example of the millstone and the bean, which they probably inherited from Islamic sources, but invented their own imaginary experiences to show the implausibility, and even impossibility, of a moment of rest between two contrary motions.

51. *Questiones super octo Phisicorum,* bk. 8, qu. 8, fols. 116r, col. 1.
52. *Questiones in octo libros Physicorum,* bk. 8, qu. 12, fol. 82v, cols. 1–2.
53. Translation by Edward Grant in Grant, *A Source Book in Medieval Science,* 286–287. For the Latin text, see *Questiones subtilissime Johannis Marcilii Inguen super octo libros Physicorum secundum nominalium viam ...* (Lyon, 1518), fol. 84r, col. 1. For essentially the same experience, see also Buridan, *Questions on the Physics,* bk. 8, qu. 8, fol. 116r, col. 2, and Albert of Saxony, *Questions on the Physics,* bk. 8, qu. 12, fol. 82v, cols. 1–2. This argument against Aristotle derives from Islamic sources, appearing in Abu'l Barakat al-Baghdadi's (ca. 1080–ca. 1165) counterargument against Avicenna's defense of Aristotle's position. For others in the West who cited this argument, see Grant, *Source Book,* 287, n. 18.
54. See *Galileo Galilei "On Motion" and "On Mechanics" Comprising "De Motu" (ca. 1590),* translated with introduction and notes by I. E. Drabkin, and *"Le Meccaniche" (ca. 1600),* translated with introduction and notes by Stillman Drake (Madison: University of Wisconsin Press, 1960), 97–98.

In addition to the millstone striking the bean, Marsilius also devised a ship experience in which he assumes that

> Socrates *(Sortes)* is moved toward the west in a ship that is at rest. Then it is possible that Socrates might cease moving in any instant. Now let it be assumed that in the [very] same instant in which Socrates should cease to be moved [toward the west], the ship with all its contents, begins to be moved toward the east. Hence, immediately before, Socrates was moved to the west, and immediately after, will be moved toward the east. Therefore, previously he was moved with one motion and afterward with another, and contrary, motion without a moment of rest.[55]

John Buridan achieved the same objective by citing an analogous experience using a fly. He explains that "if a lance is hanging from a tree [and] a fly *(musca)* ascends [in place of descends] on that lance and the cord by which the lance is hanging is broken, and then the lance and the fly fall down, the motion of the fly will be contrary, from up to down, but there will be no moment of rest."[56] Albert of Saxony also found the fly and lance an attractive example, perhaps drawing it from Buridan. Albert abandons the tree and the cord and simply assumes that the fly ascends the lance more quickly than the lance descends. Albert then assumes that the upward speed of the fly diminishes until it becomes less than the speed of descent of the lance. But at the instant in which the speed of ascent of the fly and the speed of descent of the lance are equal, "it is true to say that immediately before [the speeds were equal] this fly was ascending; and it is [also] true to say that immediately after [the speeds were equal] it descends, because immediately after this the descent of the lance will be quicker, from which it again follows that between the ascent and descent of the fly there is no moment of rest."[57]

The experiences of Socrates on the ship and the fly on the lance served a useful purpose. They conjured up situations in which a moment of rest

55. Grant, *Source Book in Medieval Science,* 287.
56. "Similiter si lancea pendente ad trabem musca ascendat [in place of *descendat*] per illam lanceam et rumpatur corda ad quam pendebat lancea, et tunc cadat lancea cum musca deorsum. Motus musce erit reflexus de sursum ad deorsum; et non erit quies media." Buridan, *Questions on the Physics,* bk. 8, qu. 8, fols. 116r, col. 2–116v, col. 1.
57. I give the full text of Albert's example: "Tertio ponatur quod aliqua musca ascendat super aliquam lanceam velocius quam illa lancea descendat. Et remittatur velocitas ascensus illius musce donec fiat minor quam velocitas descensus illius lancee. Tunc in instanti in quo velocitas musce et velocitas lancee sunt equales verum est dicere quod immediate ante hoc musca ascendebat; et verum est dicere quod immediate post hoc descendet quia immediate post hoc descensus lancee erit velocior. Ex quo iterum sequitur quod inter ascendere et descendere ipsius musce non sit quies media." Albert of Saxony, *Questions on the Physics,* bk. 8, qu. 12, fol. 82v, col. 2.

could not occur, thus presenting significant counterinstances, which were then generalized, to subvert Aristotle's argument in favor of a moment of rest. The very fact that such instances could be imagined struck a blow at Aristotle's idea of a moment of rest. Although such experiences against the moment of rest are *secundum imaginationem* (that is, "according to the imagination"), they are not counterfactual. They could conceivably have occurred. Hence, they functioned as if they were genuine observations.

Perhaps the most famous imaginative case in the history of medieval natural philosophy is the derivation of the mean speed theorem from a concern with the variation of qualities. The development of the mean speed theorem occurs in the medieval subject area known as the "configuration of qualities," or the intension and remission of forms or qualities. It was an attempt to compare variations of all kinds of qualities.

The instrument employed for these comparisons was mathematics: arithmetic at Merton College, Oxford, and geometry at the University of Paris, where Nicole Oresme composed the most significant treatise on the subject known thus far. Qualities could be compared in the way they varied and in the effects they had. Although scholars engaged in these pursuits were generally relating and representing variations in virtually all qualities, observable and unobservable, the observational component in their approach is rather strange. Oresme, for example, used geometrical figures to represent the alleged effects of various powers and qualities, such as pains, colors, joys, sounds, and so forth. He shows how two equal qualities might have different effects. One way this can occur is if the qualities are equal but of unequal intensity. For example,

> let *A* and *B* be two pains, with *A* being twice as intensive as *B* and half as extensive. Then they will be equal simply ... although pain *A* is worse than, or more to be shunned than, pain *B*. For it is more tolerable to be in less pain for two days than in great pain for one day. But these two equal and uniform pains when mutually compared are differently figured, ... so that if pain *A* is assimilated to a square, then pain *B* will be assimilated to a rectangle whose longer side will denote extension, and the rectangle and square will be equal.[58]

Thus, the total pain is equal, but its intensity varies in the two cases.

It is on the basis of such comparisons that Oresme took the great step and geometrized the Merton mean speed theorem, which expresses the modern

58. For the passage, see Oresme's *Tractatus de configurationibus qualitatum et motuum,* part II, ch. 39, in Marshall Clagett, ed. and trans., *Nicole Oresme and the Medieval Geometry of Qualities and Motions,* 387.

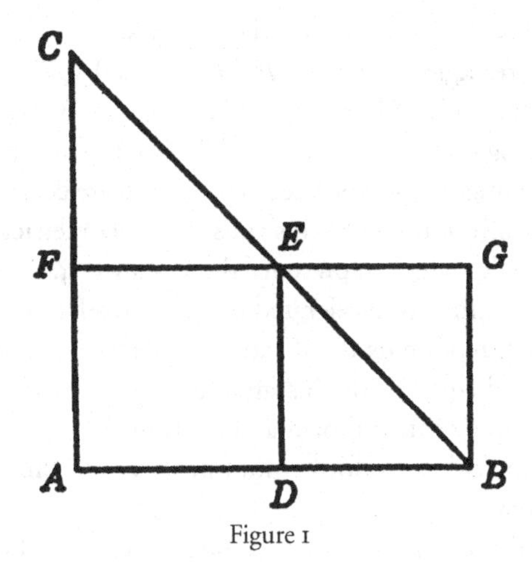

Figure 1

relationship $s = (1/2)\ at^2$, where *s* is distance, *a* is acceleration, and *t* is the time of acceleration.[59] In Figure 1, which accompanies Oresme's proof,[60] let line *AB* represent time and let the perpendiculars erected on *AB* represent the velocity of a body, *Z*, beginning from rest at *B* and increasing uniformly to a certain maximum velocity at *AC*. The totality of velocity intensities contained in triangle *CBA* was conceived as representing the total distance traversed by *Z* in moving from *B* to *C* along line *BC* in the total time *AB*. Let line *DE* represent the instantaneous velocity that *Z* acquires at the middle instant of the time as measured along *AB*. If *Z* were now moved uniformly with whatever velocity it had at *DE*, the total distance it will traverse in moving from *G* to *F* along line *GF* in time *AB* is represented by rectangle *AFGB*.

If it can be shown that the area of triangle *CBA* equals the area of rectangle *AFGB*, it will have been demonstrated that a body accelerated uniformly from rest would traverse the same distance as a body moving during the same time interval at a uniform speed equal to that of the middle instant of the uniformly accelerated motion. That is, $S = (1/2)\ V_f\ t$, the distance traversed by *Z* with uniform motion, equals $(1/2)\ at^2$, the distance traversed by *Z* when it is uniformly accelerated. That the two areas are equal is demonstrated as follows: because $\angle BEG = \angle CEF$ (vertical angles

59. See Oresme, *De configurationibus*, part III, ch. vii, in Clagett, ibid., 409. For William Heytesbury's version of the mean speed theorem, see Marshall Clagett, *The Science of Mechanics*, 270, and also see Chapter 4 of this study where Heytesbury's definition is cited.
60. I draw the proof that follows from Grant, *The Foundations of Modern Science in the Middle Ages*, 102–103.

are equal), $\angle BGE = \angle CFE$ (both are right angles), and $GE = EF$ (line *DE* bisects line *GF*), triangles *EFC* and *EGB* are equal (by Euclid's *Elements*, book I, proposition 26). When to each of these triangles is added area *BEFA* to form triangle *CBA* and rectangle *AFGB*, it is immediately obvious that the areas of triangle *CBA* and rectangle *AFGB* are equal.

Oresme's proof of the theorem was thus done geometrically by equating the area of a triangle, which represents the distance traversed in a certain time interval by a uniformly accelerated quality, or velocity, with the area of a rectangle,[61] which represents the distance traversed by a uniform velocity moving with a speed equal to the instantaneous speed acquired at the middle instant of the time of its uniform acceleration. By equating a uniformly accelerated motion with a uniform motion, it was possible to express the former by the latter.

The mean speed theorem could be applied to velocities because the latter were assumed to be qualities. The rule, or theorem, was applicable to any qualities that changed in the same way, that is, where one quality changed in a manner analogous to a uniform acceleration and the other quality changed uniformly with the mean speed of the quality that was uniformly increasing. But Oresme and his medieval colleagues never sought to determine experimentally or observationally whether qualities and speeds really changed in the manner described in the mean speed theorem, or in accordance with numerous other relationships that were attributed to them. No one is known to have suggested that uniformly accelerated motion might apply to naturally falling bodies until the sixteenth century when Domingo Soto, a Dominican priest, did so. But it was Galileo, in the seventeenth century who not only applied the mean speed theorem to naturally falling bodies but also devised an experiment to determine if bodies really fell with uniform acceleration.

In the *Two New Sciences,* Galileo has Simplicio, the spokesman for Aristotelian natural philosophers, raise the question by declaring that "I am still doubtful whether this is the acceleration employed by nature in the motion of her falling bodies." Simplicio urges his colleagues to devise an experiment that agrees with the conclusions. Salviati, Galileo's spokesman, replies to Simplicio:

> Like a true scientist, you make a very reasonable demand, for this is usual and necessary in those sciences which apply mathematical demonstrations to physical conclusions, as may be seen among writers on optics, astronomers, mechanics,

61. For the relationship of the triangle and rectangle, see Fig. 1, reproduced here from Grant, ibid., 102.

musicians, and others who confirm their principles with sensory experiences that are the foundations of all the resulting structure.[62]

As for the experiment requested by Simplicio, Salviati says that "the Author has not failed to make them, and in order to be assured that the acceleration of heavy bodies falling naturally does follow the ratio expounded above, I have often made the test *[prova]* in the following manner, and in his company." Galileo then presents his famous inclined plane experiment.[63]

Although the configuration of qualities, or intension and remission of forms, was concerned with representing the variation of qualities mathematically, and therefore seems connected to sense perception, it was really an abstract and hypothetical application of mathematics to imaginary qualitative changes that were connected to the real world only in the sense that many, if not most, of the qualities were perceptible in the real world. When we turn to conjectures about the possible movement of bodies in a vacuum, we find ourselves as far removed from the medieval world of experience as one could get. And yet, despite their unanimous view that nature abhorred an extended vacuum, medieval natural philosophers thought it important to answer questions about hypothetical activities of observational entities in hypothetical vacua. Thus, in a question as to whether a body could move in a vacuum, if one existed, Albert of Saxony declares that "we have never experienced the existence of a vacuum, and so we do not readily know what would happen if a vacuum did exist. Nevertheless, we must inquire what might happen if it existed, for we see that natural beings undergo extraordinarily violent actions to prevent a vacuum."[64]

62. Galileo, *Two New Sciences, Including Centers of Gravity & Force of Percussion,* translated, with Introduction and Notes, by Stillman Drake (Madison: The University of Wisconsin Press, 1974), Third Day, 169.
63. See Galileo, ibid., 169–170.
64. Albert of Saxony, *Questions on the Physics,* bk.4, qu. 12 ("Utrum si vacuum esset aliquid posset moveri in ipso velocitate finita seu motu locali seu motu alterationis"), fols. 51r, col. 1–51v, col. 1. The translation is from Grant, *A Source Book in Medieval Science,* 339. The translation of the question is (p. 338): "whether, if a vacuum existed, something could be moved in it with a finite velocity or local motion, or with a motion of alteration." Marsilius of Inghen made a similar statement in his question "Whether a motion could occur in a vacuum, if one existed" ("Utrum in vacuo si esset posset fieri motus," *Questions on the Physics,* bk. 4, qu. 12, fol. 54r, col. 2) when he declares that "because we have never experienced what happens in a vacuum, no one could know what would follow if the existence of a vacuum were assumed. Thus the conclusions stated previously are probable and conjectural." Marsilius of Inghen, ibid., fol. 54v, col. 1. The Latin text of this passage is given by Henri Hugonnard-Roche, "L'hypothétique et la nature dans la physique parisienne du xive siècle," in Stefano Caroti et Pierre Souffrin, *La nouvelle physique du xive siècle* (Florence: Leo S. Olschki, 1997), 175.

The most important questions posed about possible activities in the vacuum involved the motion of bodies. The key to understanding medieval interpretations of motion in hypothetically void space is to realize that medieval natural philosophers analyzed the same bodies in the void that they discussed in the plenum of their ordinary world.[65] They sought to imagine how such bodies would behave in a milieu that was devoid of material resistance. They considered both elemental and mixed, or compound, bodies, the latter consisting of two or more elements. Although some allowed that an elemental body – earth, water, air, or fire – could fall with a finite speed in a void, most rejected such motion because the void lacked a medium that could serve as an external resistance to an elemental body. Without external resistance, an elemental body would fall with an infinitely great speed, which was rejected as impossible. Mixed bodies, however, were quite different. They could fall in a void because they possessed within themselves a motive force (in the form of the heavier element) and an internal resistance (the lighter element), the two essential requirements for finite motion, even in a vacuum.

To render the situation in a void more analogous to that in a plenum, some scholastics assumed a vacuum that was produced by the annihilation of all matter within the concave surface of the lunar sphere, or occasionally all matter below the sphere of fire. Some also assumed that the former natural place of each element, now void, nevertheless retained the properties it had when it functioned as an elemental plenum. Thus, one could speak of the "vacuum of fire" *(vacuum ignis),* where fire once had its natural place, or the "vacuum of air" *(vacuum aeris),* where air was formerly located, or the "vacuum of water" *(vacuum aque),* where water formerly existed. In the examples employed, it became customary to assign degrees of heaviness and lightness to the elements in the compound. One conclusion that was generally accepted was that a mixed body could descend with a successive motion in a vacuum, as Albert of Saxony indicates in the question "whether if a vacuum did exist, a heavy body could move in it":

65. Henri Hugonnard-Roche discusses the role of hypothetical physics in medieval natural philosophy and declares: "Le domaine de cette physique des cas imaginaires, ou physique 'hypothétique', a été construit à l'aide de critères sémantiques touchant les conditions de vérité des propositions du domaine, et d'instruments conceptuels tirés de la physique 'naturelle', ou physique de la *ratio generalis corporum.* Mais dan le même temps qu'elle s'étendait *ad imaginabilia,* cette physique de cas impossibles *secundum quid* s'éloignait de la 'nature' en la dépouillant d'une partie de ses attributs, pour devenir imaginaire." In Caroti and Souffrin, *La nouvelle physique,* 177.

A second conclusion [is this]: By taking "vacuum" in the first way, as it is commonly taken in this question,[66] a heavy mixed body is easily moved in it successively. This is clear, for let there be a heavy mixed body whose heaviness *(gravitas)* is as 2 and lightness *(levitas)* as 1. And let it reach the concave [surface] of air and descend successively until its center of gravity (*medium gravitatis*) is the middle [or center] of the world (*medium mundi*). [This will happen] because it has an internal resistance, for it has one degree of lightness inclining [or tending] upward and two degrees of heaviness inclining downward.[67]

Because a mixed, or compound, body can have varying relationships among its constituent elements, the fall of a mixed body in a void could produce results at variance with the fall of the same body in a plenum. Albert furnishes two significant instances of such differences. In a third conclusion of the question as to whether a heavy body could move in a vacuum, he shows that, under the right circumstances, "a heavy mixed [or compound] body *(mixtum)* could be moved more quickly in a plenum than in a vacuum." Albert then draws another startling consequence in the same third conclusion, arguing that

the natural motion of some heavy body can be quicker in the beginning than in the end. For example, If a mixed [or compound] body of four elements should have one degree of fire, one of air, one of water, and four of earth and if everything were annihilated within the sides of the sky except this mixed body, and if the mixed body were placed where the fire was, then this mixed body would descend more quickly through the vacuum of fire *(vacuum ignis)* than through the vacuum of air *(vacuum aeris),* and so on, as can easily be deduced from this case. But you [now] say, what should be said, therefore, about the common assertion that natural motion is quicker in the end than at the beginning? One can say that is universally true of the motion of heavy and light [elemental] bodies but not of the motion of heavy and light mixed [or compound] bodies.[68]

Although Aristotle had distinguished between elemental and mixed bodies, he had made no use of this distinction in his arguments against the vacuum.[69] Albert shows how scholastic natural philosophers used it to make finite, successive motion (rather than infinite, instantaneous motion) in a

66. That is, as a separate, extended space devoid of body.

67. Grant, *Source Book,* 336. The word "downward" has been changed from "downwards." In the next quotation, the words "more quickly" replace "quicker."

68. Grant, ibid., 337. I have added ["elemental"].

69. On the distinction between elemental and mixed bodies, see Grant, *Much Ado About Nothing,* 44–45.

vacuum seem possible and intelligible. We see this in the two conclusions just described, namely that "a heavy mixed [or compound] body *(mixtum)* could be moved more quickly in a plenum than in a vacuum" and "the natural motion of some heavy body can be quicker in the beginning than in the end." To these two significant deviations from Aristotelian physics we must add one more, the concept that two mixed bodies of homogeneous composition, but of unequal weight, would fall with equal speeds in a vacuum. Thus, Albert declares that

> [m]ixed [or compound] bodies of homogeneous composition *[consimilis compositionis]* are moved with equal velocity in a vacuum but not in a plenum. The first part [concerned with fall in a vacuum] is obvious, because they are of homogeneous mixture. The ratio of motive power to total resistance in one body is the same as in another homogeneous *[consimilis]* body, because they both have only internal resistance.[70]

More than two centuries later, Galileo arrived at the same conclusion in a quite different manner.[71]

Observation did not contribute much to the analysis of motion, other than the fact that natural philosophers could see that a naturally falling body was faster at the end of its motion than at the beginning, a fact that convinced them that falling bodies accelerated. The analysis of motion was, however, largely a rationalistic, rather than an empirical, process. The same conditions that obtained for falling bodies in a material plenum were also applied to falling bodies in a vacuum, with this difference: In the vacuum no material resistance existed. Even though medieval natural philosophers regarded the existence of void space as naturally impossible, they treated it as the limiting case for motion in a plenum. Consequently, motion in a void was treated with as much seriousness as was motion in a plenum.

There is a great anomaly in medieval natural philosophy, which was based overwhelmingly on Aristotle's approach to nature and was, therefore, rooted in empiricism and sense perception. The passages cited at the

70. See Grant, *Source Book in Medieval Science,* 341. This conclusion is the eighth in Albert's twelfth question of the fourth book of his *Questions on the Physics.* Thomas Bradwardine had already asserted this conclusion in 1328 in his *Tractatus de proportionibus,* ch. 3, Theorem XII. See Grant, ibid., 305. Albert goes on to show, in the same conclusion, that the same two homogeneous bodies would fall with different speeds in a plenum. This is so because in a plenum, the plenum itself serves as an external resistance, supplementing the internal resistance.

71. See Grant, *Much Ado About Nothing,* 61–66.

beginning of this section from Albertus Magnus, Roger Bacon, and John Buridan emphasized the importance of empiricism, induction, and observation. In a chapter on physics in the late Middle Ages, A. C. Crombie emphasized theoretical discussions about the nature of induction in the fourteenth century.[72] But theoretical interest in induction by those already mentioned, and by the likes of John Duns Scotus and William of Ockham, did little to encourage and stimulate other scholastic natural philosophers to emphasize observation and experience. We see very little direct observation in the questions literature on Aristotle's natural books. Very few questions were decided by appeals to observation. Despite the emphasis he placed on experience and induction, Crombie recognized that "[f]rom the beginning of the 14th century to the beginning of the 16th there was a tendency for the best minds to become increasingly interested in problems of pure logic divorced from experimental practice, just as in another field they became more interested in making purely theoretical, though also necessary, criticisms of Aristotle's physics without bothering to make observations."[73]

We may properly characterize medieval Aristotelianism as empiricism without observation. It was also empiricism without measurement. What quantification there was in medieval natural philosophy was overwhelmingly of a theoretical and imaginary kind.[74] "The habit of systematic measurement and its instrumentation by appropriate procedures," Crombie rightly explains about medieval natural philosophy, "was characteristically a response not to the theoretical demands of natural philosophy but to the practical demands of the technical arts."[75] Despite the near absence of direct observation and measurement, empiricism served as the foundation of the medieval theory of knowledge. There were numerous empirical elements in medieval questions. But the authors who report them, or use them to support or refute an argument, did not directly observe them and felt no compulsion to do so.

Medieval observations were not introduced for their own sake, namely, to learn more about the world, or to resolve arguments. They were intended rather to uphold an a priori view of the world, or to serve as an example or illustration. The idea of observation was important in the

72. See Crombie, *Medieval and Early Modern Science,* rev. 2nd ed., 2 vols. (Garden City, NY: Doubleday Anchor Books, 1959), vol. 2, 28–35.

73. Crombie, ibid., vol. 2, 22–23.

74. See Crombie, *Styles of Scientific Thinking in the European Tradition,* vol. 1, 411–423.

75. Crombie, ibid., 416.

Middle Ages because it was the basis of Aristotelian epistemology, which was founded on sense perception. But it was clearly not enough, as Aristotle understood when he declared that although the senses "surely give the most authoritative knowledge of particulars … they do not tell us the 'why' of anything – e.g. why fire is hot; they only say that it is hot."[76] Medieval natural philosophers were in agreement with Aristotle on this major point, which explains why, during the late Middle Ages, empiricism was, and remained, the servant of the analytic and a priori, that is, the servant of logic and metaphysics, which provided the "why" of things to explain and interpret the empirical world. John Murdoch has perceptively argued that although it is true that

> empiricist *epistemology* was dominant in the fourteenth century … this did not mean that natural philosophy then proceeded by a dramatic increase in attention being paid to experience and observation (let alone anything like experiment) or was suddenly overwrought with concern about testing or matching its results with nature. On the contrary, its procedures were increasingly *secundum imaginationem* (to use an increasingly frequently occurring phrase) and when some "natural confirmation" of a result is brought forth, more often than not it too was an "imaginative construct."[77]

The most powerful tool medieval natural philosophers possessed was not empiricism as manifested by observation *per se* but, rather, experience as adapted for use in thought experiments *(secundum imaginationem)*. Many, if not most, of the experiences cited in medieval natural philosophy are really thought experiments designed to refute or uphold a theory. But the "experiences" were not actually "experienced" or performed, although they were usually examined and analyzed with great seriousness. They only had to appear plausibly relevant to be accepted and then utilized as part of an overall argument for or against some real or imagined position.

It was one thing to write about induction and observation, and to uphold their importance, as did Roger Bacon, Albertus Magnus, John Buridan, and others; it was quite another to come to the realization that it was essential to make purposeful observations in the real world, and to design experiments to learn things about that world that were not derivable from raw observation and experience, and to make all this a routine

76. Aristotle, *Metaphysics* 1.1.981b.10–11.

77. John E. Murdoch, "The Analytic Character of Late Medieval Learning: Natural Philosophy Without Nature," in Lawrence D. Roberts, ed., *Approaches to Nature in the Middle Ages* (Binghamton, NY: Center for Medieval & Early Renaissance Studies, 1982), 174.

and regular feature of natural inquiry. This stage of development was not reached in the Middle Ages. Indeed, a modern scholar has declared that "we have *no* record of a mediaeval physicist drawing a testable consequence from a theory and then attempting to actually test it, or have it be tested."[78] Such actions had to await the seventeenth century, the century of Newton. But if scholastic natural philosophers developed an empiricism without observation, and focused attention on hypothetical, rather than real and direct observations, they did, at least, recognize that experience and observation, even imaginary experiences and observations, were important ingredients in doing science and natural philosophy. At the very least, they frequently paid lip service to experience and observation, thus recognizing their importance, even if they did not personally engage in these crucial activities.

Because they failed to realize the importance of regular and direct observations and the need for devising experiments to yield nature's patterns of behavior, medieval natural philosophers did the next best thing. They sought to uphold the laws of Aristotle's world as well as they could. Where they found it at variance with reason and observation, they changed those laws and perceptions. But they did this in the way Aristotle had taught them, and also by means of a new tool that they had devised for themselves. That is, they sometimes appealed to observation and sense perception to support positions they opposed or defended, but they relied most heavily on their imaginations, which were guided by reason in the form of analytic techniques and logical analysis. It was in this manner that they concocted thought experiments for the real world, as well as for the world Aristotle had regarded as naturally impossible, the world of imaginary void space. By these methods, they arrived at some rather startling theories and conclusions, such as the mean speed theorem, impetus theory, the possibility of finite motion in a vacuum, the claim that two homogeneous compound bodies would fall with the same speed in a vacuum,[79] and arguments for the existence of extracosmic void space. They achieved all this with a "natural philosophy without nature," to use John Murdoch's perceptive and felicitous phrase, and, perhaps not surprisingly, by employing an "empiricism without observation."

In light of all this, one is inexorably driven to ask: Did medieval scholastic natural philosophers believe that their responses to the multitude of

78. Peter King, "Mediaeval Thought-Experiments," 48.

79. Galileo adopted the same position. For the medieval arguments, see Grant, *Much Ado About Nothing,* 57–60; for Galileo, see ibid., 63–66.

questions they posed about the workings of nature provided them with truths about the structure and operations of the physical world? To this question, we must, I believe, respond in the affirmative, since we have no evidence to the contrary. To reply in the negative is to assume that they knowingly and willingly labored to no purpose, or for the sheer pleasure it provided, both untenable assumptions.

When we realize that the contributions just described, and others, were made without the sophisticated and essential methodologies that would become a routine part of scientific inquiry centuries later, we should recognize that medieval natural philosophers deserve a much greater measure of respect than has hitherto been accorded them. Without those methodologies, and because of an empiricist theory of knowledge that was largely divorced from direct observation and measurement, the requirements and demands of reason dominated medieval natural philosophy. Solutions to problems about the physical world were almost always resolved by appeal to rational and logical arguments. Empiricism served this process only insofar as it provided the ingredients for imaginary thought experiments. But one should never doubt that reason ruled medieval natural philosophy.

REASON AND REVELATION: HOW FAITH AND THEOLOGY AFFECTED NATURAL PHILOSOPHY

By its history and tradition, Aristotle's natural philosophy was a discipline in which it was assumed that reason would guide the students and scholars who studied and wrote about it. The Aristotelian treatises that entered Western Europe in the twelfth century, and reshaped the study of nature, were manifestly secular and rationalistic. But they entered a Western Europe that passionately subscribed to a religion that had been in existence for nearly twelve centuries. By the twelfth century, Western Christianity, in the form of the Catholic Church, was pervasive and ubiquitous in Europe and would soon triumph over its major rival, the heretical, dualistic Cathars, who would be destroyed by the mid–thirteenth century.

Why should Aristotle's books on natural philosophy be of interest or concern to the Catholic Church and the Christian religion? Because it was a large and impressive body of literature about the physical world and its operations, which it described in impersonal, objective, and rational terms. Moreover, there was nothing else remotely comparable to the vast body of literature associated with the name of Aristotle. If God had created the world, as all Christians believed as a matter of faith, and if that world was a rationally structured creation worthy of an omnipotent, omniscient God, then Aristotle's works were an ideal and trustworthy guide for Christians to approach and understand that world. In effect, Aristotelian natural philosophy had applied reason in order to understand the same physical world that Christians had accepted and interpreted on the basis of a Sacred Scripture and, therefore, on the basis of a revealed truth.

Aristotle and the Church

Reason and revelation, as embodied in the works of Aristotle and in Scripture, respectively, were in sharp disagreement on some fundamental issues, however. Where the Church assumed on the basis of Genesis that God had created the world out of nothing and would eventually destroy it, Aristotle had argued that the world had no beginning and would have no end. Christians held that every human being had an immortal soul that would exist eternally after the death of the body, whereas Aristotle argued that only part of the soul – the rational part – is immortal, while the remainder perishes with the body.[80] The doctrine of the Eucharist assumed that God transforms the bread and wine of the Mass into the body and blood of Christ, but that the accidents of the bread and wine continued to exist and remain visible without inhering in any substances. By contrast, Aristotle had argued that all accidents, without exception, must inhere in a substance. One of the most vexing problems with Aristotle's natural philosophy was the simple fact that his natural philosophy seemed to place limits on God's absolute power to do whatever He pleased, short of a logical contradiction. Thus, in Aristotle's philosophy, the existence of a vacuum anywhere at all is impossible, as is the existence of other worlds beyond ours. Did this signify that if God wished to create a vacuum anywhere at all, that He could not do so; or, if He decided to create other worlds, in addition to our world, that He could not do so because it was contrary to Aristotle's natural philosophy?

It was an uneasiness with such problems that prompted Church authorities to place restrictions on the study of Aristotle's natural philosophy at the University of Paris during the thirteenth century and eventually, in 1277, to issue a condemnation of 219 propositions, the advocacy of any one of which

80. On Aristotle's seemingly ambivalent views on the immortality of the soul, see Lloyd, *Aristotle*, 22, 29–33.

could lead to excommunication.[81] Despite these difficulties and the opposition of some theologians, both theologians and natural philosophers warmly received Aristotle's natural philosophy. Theologians welcomed Aristotle's natural works because they saw them as aids to the study of theology. Indeed, they viewed natural philosophy in the traditional sense as handmaiden to theology and Scriptural study. Natural philosophy was thus assigned a vital role to play in explicating matters of faith and doctrine. By contrast, natural philosophers in the arts faculties of medieval universities welcomed Aristotle's works because those works formed the very basis of the arts curriculum. Because it was not the province of arts masters in the arts faculties to interpret Scripture, or matters of faith, they did not view Aristotle's natural philosophy as the means to some other end. Aristotelian natural philosophy, and the works in which it was embodied, was an end in itself. To understand Aristotle's natural philosophy and to expound its meaning and significance was precisely what natural philosophers were expected to do as teaching masters.

Beginning in the thirteenth century, theology and natural philosophy became, along with law and medicine, basic disciplines in the medieval universities. Natural philosophy was a subject required of all students studying for a master of arts degree, which, in turn, was prerequisite for study in the higher faculties: theology, medicine, and law. Students in theological faculties were usually well trained in natural philosophy, whereas students in the arts faculties, as well as teachers who pursued careers in the arts faculties, had no training in theology. It is not surprising that while theologians could use natural philosophy in theology at their pleasure, natural philosophers in the arts faculties, especially at the University of Paris, as we shall see, were expected to refrain from introducing theology and matters of faith into natural philosophy. Nevertheless, the two disciplines inevitably interacted.

Let us now describe and analyze the ways in which each influenced the other, considering first the ways in which theology affected natural philosophy, and leaving the impact of natural philosophy on theology to the chapter on theology (Chapter 6).

The Minimal Impact of Theology and Faith on Natural Philosophy

Earlier in this chapter, I discussed what natural philosophy is, and how medieval scholastics viewed their discipline. It is also important to know

81. I have given a more thorough treatment of the relationship between Aristotelian natural philosophy and Christian theology and faith in Grant, *The Foundations of Modern Science in the Middle Ages*, Chapter 5, 70–85.

what natural philosophy is not. It is not theology in any guise or form, as has been argued.[82] To demonstrate this important point, it is essential to examine the vast body of commentary literature that was produced by natural philosophers and theologians who taught and studied Aristotle's natural books in medieval universities for more than three centuries.[83] Those who wrote these treatises firmly believed that, by His supernatural power, God had created the world from nothing, and was the ultimate cause, or the First Cause *(prima causa),* as He was frequently called, of all events or effects. But these Christian beliefs did not affect the way medieval scholars wrote natural philosophy. Those beliefs did not mean that their objective in doing natural philosophy was essentially theological or religious, or that their aim was to transform natural philosophy into an instrument for the defense and explication of the faith. Indeed, as we shall see, they rarely intruded doctrinal matters into their investigations into natural questions.

During the Middle Ages, theology and natural philosophy were recognized as distinct disciplines, each taught in its own university faculty.[84] Hence, it is wholly appropriate to regard them as distinct disciplines, and simply false to claim that "the distinction between 'science' and theology is a modern day distinction which cannot legitimately be applied to the practice of natural philosophy in the seventeenth and other centuries."[85] We must also keep in mind the fact that those who commented on the natural books of Aristotle were usually teaching masters in arts faculties, although many would subsequently matriculate in a theology faculty and become professional theologians. When they wrote their Aristotelian commentaries, they had every

82. For these arguments, see Andrew Cunningham, "How the Principia Got its Name; or, Taking Natural Philosophy Seriously," in *History of Science* 29 (1991), 377–392. I have presented my opposing views on Dr. Cunningham's article in Grant, "God, Science, and Natural Philosophy in the Late Middle Ages," in Lodi Nauta and Arjo Vanderjagt, eds., *Between Demonstration and Imagination, Essays in the History of Science and Philosophy Presented to John D. North* (Leiden: Brill, 1999), 243–267. For a subsequent presentation of our respective attitudes toward the issue of the relationship of medieval and early modern natural philosophy and theology, see the "Open Forum" in *Early Science and Medicine* 5, no. 3 (2000), 258–300.

83. To obtain a good sense of the extant number and range of these commentaries from 1200 to 1650, see Charles H. Lohr, "Medieval Latin Aristotle Commentaries" (published in *Traditio* 1967–1974) and *Latin Aristotle Commentaries, II: Renaissance Authors* (Florence: Leo S. Olschki Editore, 1988).

84. On theology and natural philosophy as distinct, independent disciplines, see Asztalos, "The Faculty of Theology," 423–424.

85. Cunningham, "How the Principia Got its Name," 389.

incentive to keep their natural philosophy natural, if not by their own inclination, then by command of their own faculty. We saw earlier (Introduction) that the arts faculty at the University of Paris instituted an oath in 1272 that made it mandatory for arts masters to avoid theological discussions in their questions. Where this was unavoidable, they were sworn to resolve the issue in favor of the faith. Even in universities that had no such oath, arts masters would rarely have considered theological issues in their treatises on natural philosophy, largely because they were all too aware that theology was the domain of theologians. Although theology masters who wrote treatises on natural philosophy could have imported theology into their natural philosophy, they rarely did, choosing instead to relegate theological issues to theological treatises, as we shall see.

The most appropriate way to determine the role that theology and faith played in medieval natural philosophy is to examine relevant texts in natural philosophy as written by those who were consciously doing natural philosophy, not theology. That is, we must carefully inspect treatises on natural philosophy per se, not treatises on theology that used natural philosophy in the service of theology (this will be done in the next chapter). To achieve this, I have examined all of the questions in the following five treatises, which constitute the core of Aristotle's natural philosophy: 59 questions in John Buridan's *Questions on On the Heavens (De caelo)*; 35 questions in Albert of Saxony's *Questions on On Generation and Corruption;* 107 questions in Albert of Saxony's *Questions on the Physics;* 65 questions in Themon Judaeus's *Questions on the Meteors;* and 44 questions in Nicole Oresme's *Questions on On the Soul (De anima)*, for a grand total of 310 questions.[86]

An examination of the 310 questions embedded in these five treatises shows clearly that most of the questions had little to do with God, the faith, or theology, but were concerned solely with issues in natural philosophy. Of the 310 questions, 217 are free of any entanglement with theology or faith. Inspection of any of the 217 questions would not reveal whether the author was Christian, Muslim, Jewish, agnostic, or atheist. The remaining 93 questions, approximately 29 percent, mention God and the faith. Of the 93 with at least a trace of theological sentiment, 53 mention God, or something about the faith, in a cursory manner; of the remaining 40 questions, 10 have relatively detailed discussions about God or the faith.

86. The data presented here are drawn from my article, "God, Science, and Natural Philosophy in the Late Middle Ages," 248–263.

Of the 93 instances where God and the faith are mentioned, 80 fall into three significant categories,[87] most of which appear in three of the five treatises: *Questions on the Physics, Questions on De caelo,* and *Questions on De anima* (very few occur in *Questions on Generation and Corruption* and *Questions on the Meteorology*). The three categories are as follows:

Category 1: Twelve of the 93 questions mention arguments, usually those by Aristotle or Averroës, or both, that involve the faith, or are contrary to the faith. For example, in his *Questions on De anima* (bk. 3, qu. 7), Nicole Oresme argues against Averroës's assumption of a single intellect, or Agent Intellect for the whole human race. That is, Averroës argued that each human being has a "passive intellect," or imagination. This passive, or "possible," intellect can acquire knowledge only when it is made receptive for knowledge by the separate Agent Intellect, which "produces intelligible knowledge in individual souls as the sun produces seeing in the eyes through its light."[88] Although the passive intellect remains with a human being for the duration of life, at death "it passed, as a drop into the sea, into the universal intelligence," that is into the Agent Intellect, an eternal substance: "Thus not only freewill, but personal immortality was excluded."[89] Against Averroës's opinion, Oresme declares that "the opposite is obvious from faith and according to truth. Nor is it probable – indeed it is unthinkable – that my intellect is your intellect or [the intellect] of another who is in Rome or elsewhere."[90]

In his *Questions on the Physics,* Albert of Saxony asks "whether there always was motion and always will be motion."[91] He explains that "Aristotle and the Commentator argue the opposite in this eighth [book]. I respond first to this question according to the way that Aristotle and the Commentator respond. Secondly, we must respond to it according to the

87. In my article, "God, Science, and Natural Philosophy in the Late Middle Ages" (see note 86), I include two additional categories. One (the first category in the article; see ibid., 248) concerns reactions to Aristotle's mention of God or gods. The final category, the fifth in the article, includes mentions of God that did not fit any of the other four categories (ibid., 249).

88. Gilson, *History of Christian Philosophy in the Middle Ages,* 224.

89. Knowles, *The Evolution of Medieval Thought* (Baltimore: Helicon Press, 1962), 200–201.

90. Oresme, *Questions on De anima,* bk. 3, qu. 7, in *Nicholas Oresme's 'Questiones super libros Aristotelis De anima': A Critical Edition with Introduction and Commentary* by Peter Marshall (Ph.D diss., Cornell University, 1980), 582.

91. Albert of Saxony, *Questions on the Physics,* bk. 8, qu. 1, in *Questiones et decisiones physicales insignium virorum. Alberti de Saxonia in octo libros Physicorum …* (Paris, 1518), fols. 76v, col. 2–77v, col. 1.

truth."[92] When Albert arrives at the point where he must respond with the truth, he proclaims that "according to our faith and the truth of the matter, sometimes a motion begins which some motion does not precede. And this occurs in this manner: that there was a prime mover eternally, although there was not a prime mobile eternally. But at sometime it [the prime mobile] began and then motion began."[93] Thus did Albert of Saxony opt for the faith against Aristotle and Averroës. Where Aristotle and Averroës insisted that motion had no beginning and is eternal, Albert argues according to "our faith," that motion did have a beginning and that it began when God created the prime mobile, or first movable sphere.

After presenting arguments for both sides of a question that inquired whether an immobile heaven should be assumed beyond the mobile heavens, that is, whether there is an empyrean heaven, John Buridan declares that "you may choose any side you please. But, because of the arguments of the theologians, I choose the first part [that is, the existence of a resting, empyrean heaven]." He then goes on to defend the faith, when he says: "And one can reply to Aristotle's argument that he assumes many things against Catholic truth because he wished to assume nothing that could not be deduced from the senses and experience. Thus it is not necessary to believe Aristotle in many things, namely where he clashes with Sacred Scripture."[94]

Category 2: Thirty-four questions mention God and faith by way of analogy or example. Oresme illustrates this tendency in a supposition in which he asserts that "some power makes this or that operation anew without changing itself, just as is obvious with God who continuously produces new effects without any change in Himself."[95] Similarly, Buridan declares that "[j]ust as all order in the world arises from God, so does order arise in a city from a prince."[96]

Category 3: Another 34 of the 93 questions invoke some aspect of God's absolute power, perhaps the most significant means of producing counterfac-

92. "In oppositum est Aristoteles et eius Commentator in isto octavo. Ad istam questionem primo respondendum est secundum quod responderet Aristoteles et Commentator. Secundo respondendum est ad eam secundum veritatem." Albert of Saxony, ibid., fol. 77r, col. 1.

93. "Quantum ad secundum, dico quod secundum fidem nostram et rei veritatem. Aliquando incepit motus quem non precessit aliquis motus et hoc per istum modum quod etemaliter fuit primum motor, licet non etemaliter fuerit primum mobile; sed aliquando inceperit, et tunc incepit motus." Albert of Saxony, ibid., fol. 77v, col. 1.

94. Buridan, *Questions on De caelo,* bk. 2, qu. 6, 152 (Moody edition).

95. Oresme, *Questions on De anima,* bk. 3, qu. 2, 517–518.

96. Buridan, *Questions on De caelo,* bk. 2, qu. 21, 224. Buridan cites Aristotle's *Metaphysics,* bk. 12, as his source.

tuals in the Middle Ages, some of which raised important questions about motion, other worlds, and the infinite. The numerous appeals to God's absolute power bear witness to the fact that natural philosophy was not just about God and His creation but also about what God had not created, but could create by virtue of His omnipotence. Among numerous invocations of God's absolute power we may mention that God could create as many other worlds as He pleases;[97] that He could move our world with a rectilinear motion;[98] that He could separate a quantity from its extension;[99] that beyond our world, He could create a corporeal space and any corporeal substances He pleases;[100] that, instead of moving the celestial orbs by intelligences, God might have created the heavens and moved them by means of an impressed force that He implanted within them at creation;[101] and that God could create a vacuum by annihilating all matter below the concave surface of the lunar orb.[102] In sum, by His omnipotence, God was always assumed capable of doing anything whatever that was impossible in Aristotle's natural philosophy, provided that it did not involve a logical contradiction.[103]

Although God was deemed capable of doing these naturally impossible acts, it did not follow that He had done them, or would ever do them. In fact, we must assume that most natural philosophers and theologians who invoked God's absolute power did not believe that God had actually performed the act of which He was assumed capable. John Buridan, for example, concedes that God could create a finite space of any size beyond the world, but he reminds his readers that they should seek no other reason for such an action than the simple desire of God to do so. "But, nevertheless," says Buridan, "I think that there is no space [beyond the world], namely [any space] beyond the bodies that appear to us and that we must believe [exist] on the basis of sacred Scripture."[104]

97. Buridan, *Questions on De caelo,* bk. 1, qu. 18, 84, and Albert of Saxony, *Questions on the Physics,* bk. 3, qu. 12, fol. 39v, col. 1.
98. See Buridan, *Questions on De caelo,* bk. 1, qu. 16, 75–76, and Oresme, *Questions on De anima,* bk. 2, qu. 15, 386.
99. Albert of Saxony, *Questions on the Physics,* bk. 1, qu. 6, fol. 5r, col. 2.
100. Buridan, *Questions on De caelo,* bk. 1, qu. 17, 79.
101. Buridan, ibid., bk. 2, qu. 12, 180–181.
102. Buridan, ibid., bk. 1, qu. 20, principal argument 5, 92, 95.
103. For additional examples of God's absolute power, see Grant, "God, Science, and Natural Philosophy in the Late Middle Ages," 260–263.
104. "Deus posset ibi creare spacium finitum quantum placeret sibi de cuius quantitate non esset querenda ratio nisi simplex voluntas Dei. Sed tamen opinor etiam quod ibi non sit aliquod spacium, scilicet ultra corpora nobis apparentia et ea que ex sacra scriptura tenemur credere." Buridan, *Questions on the Physics,* bk. 3, qu. 15, fol. 58r, col. 2. The question is "whether an infinite magnitude exists."

Appeals to God's absolute power had little, if any, religious motivation or content. Wherever we find it used in Aristotelian questions and commentaries, it is rarely intended to make a religious point. It simply became a convenient vehicle for the introduction of subtle and imaginative questions, the responses to which compelled natural philosophers to apply Aristotelian natural philosophy to situations and conditions that were impossible in Aristotle's natural philosophy. In the process, some of Aristotle's fundamental principles were challenged. The invocation of God's absolute power made many aware that things might be quite otherwise than were dreamt of in Aristotle's philosophy.

To underscore the fact that medieval natural philosophy was about the natural, not supernatural, operations of the world, it is important to recognize that in almost any given question *(questio)* in which some element of theology has been introduced, the invocations of religious or theological material usually occupies a small percentage of the total question. Let us recall that of the 310 questions in the five treatises that formed the basis of my investigation of Aristotelian natural philosophy in the fourteenth century, 217 had nothing whatever on God or the faith and only 93 did. Of the 93, however, most had relatively little on theology.

For example, in his *Questions on the Physics,* Albert of Saxony asks whether "from the addition of some whole to some whole another whole is made; similarly, [whether] by the removal of some whole from some whole, another whole is made."[105] Of the 201 lines of text in this question, 10 are devoted to the fourth and fifth (of 10) principal arguments in which Albert rejects the proposition as follows:

Fourthly, it would follow that none of us would be baptized. But this is false and the consequence is proved because many particles are added to us. And thus we are greater than when we were baptized. Therefore by addition of some part to the whole there occurs another whole. Therefore it follows that none of us is the same whole which we were in [our] youth, and, consequently, none of us is that [person] which was baptized.

Fifthly, by similar reasoning, it would follow that none of us is the one who was born of his mother, just as Christ was not the same man who was suspended on the cross and who was born of the purest virgin.

105. Albert of Saxony, *Questions on the Physics,* bk. 1, qu. 8, fols. 7r, col. 2–8r, col. 1. The translation is from Grant, *A Source Book in Medieval Science,* 200.

Not only do these arguments constitute a small portion of the whole question – slightly less than 5 percent – but the discussions about baptism and Christ are examples, and could have been replaced by other examples of a nonreligious character. Moreover, within the structure of a typical question, the principal arguments and the responses to them at the beginning and end of the question, respectively, represent the least important parts. Between them lies the body of the question in which the author presents the main conclusions and qualifications. In the question we are discussing, the religious component occurs only in the principal arguments (indeed, Albert does not even respond to them) and not in the body of the question. Thus, they play no significant role in the question.

Because fewer than one-third of the 310 questions considered here had theologically relevant material, and most include much less than 5 percent that pertains to God, the faith, or Church doctrine (indeed, more than half of the references are little more than passing mentions of God or some aspect of the faith and play insubstantive roles in their respective questions), we may rightly conclude that God and faith played little role in medieval natural philosophy. But why, we must inquire, did medieval natural philosophers virtually ignore these themes in their questions? The answer seems obvious: because they were irrelevant to their objective, which was to provide natural explanations for natural phenomena. Perhaps the most important reason that theology did not significantly penetrate natural philosophy is simply that while theology needed natural philosophy, natural philosophy did not need theology.

It is obvious that religion and theology played a minimal role in treatises on Aristotle's natural philosophy, and, by a process of extrapolation, we may say that they played little role in the works of medieval natural philosophers as a whole. But what about theologians who wrote treatises on natural philosophy, or, more specifically, wrote commentaries on one or more of Aristotle's natural books? Were they more likely to "theologize" their treatises? To answer this question, we can do no better than examine the commentaries on Aristotle's natural philosophy by two of the greatest theologians in the Middle Ages, Albertus Magnus and Thomas Aquinas, who were already masters of theology when they wrote their commentaries on the natural books of Aristotle. As professional theologians, both were free to insert thoughts about God and the faith in their treatises on natural philosophy, wherever such thoughts might be deemed appropriate. It is of importance, therefore, to see how they viewed the relations between natural philosophy and theology, and to determine the extent to which they were prepared to theologize natural philosophy. The evidence shows unequivocally that both chose to keep the theologization of natural philosophy to a minimum.

In the opening words of his commentary on Aristotle's *Physics,* Albertus declares that his Dominican brothers had implored him to "compose a book on physics for them of such a sort that in it they would have a complete science of nature and that from it they might be able to understand in a competent way the books of Aristotle."[106] Perhaps thinking that his fellow friars would expect him to intermingle theological ideas with natural philosophy, Albertus declares that he will not speak about divine inspirations, as do some "extremely profound theologians," because such matters "can in no way be known by means of arguments derived from nature." And he then explains:

Pursuing what we have in mind, we take what must be termed 'physics' more as what accords with the opinion of Peripatetics than as anything we might wish to introduce from our own knowledge ... for if, perchance, we should have any opinion of our own, this would be proffered by us (God willing) in theological works rather than in those on physics.[107]

Albertus thus believed that Aristotle's natural philosophy was to be treated naturally, in the customary manner of Peripatetics. Where theological issues might be involved, they were to be treated in theological treatises. In his *Commentary on De caelo,* Albertus makes it evident that he wishes to uphold his basic conviction that, unless unavoidable, theology should not intrude into natural philosophy. In discussing whether the heaven is ungenerable and incorruptible, Albertus explains:

Another opinion was that of Plato who says that the heaven was derived from the first cause by creation from nothing, and this opinion is also the opinion of the three laws, namely of the Jews, Christians, and Saracens. And thus they say that the heaven is generated, but not from something. But with regard to this opinion, it is not relevant for us to treat it here.[108]

106. Translated in Edward A. Synan, "Introduction: Albertus Magnus and the Sciences," in James A. Weisheipl, O. P., ed., *Albertus Magnus and the Sciences: Commemorative Essays 1980* (Toronto: The Pontifical Institute of Mediaeval Studies, 1980), 9.

107. See Synan, ibid., 10. Synan presents the section of this passage that follows the ellipsis before the lines that precede it. But the order of the passages in Albertus's *Physics* is as they appear here.

108. Albertus Magnus, *Opera Omnia,* vol. 5, pt. 1, *De caelo et mundo,* bk. 1, tr. 1, ch. 8, 19, col. 2–20, col. 1. (Hereafter *Commentary on De caelo.*) Plato did not hold that the world was created from nothing. The translation is mine. Unless indicated otherwise, the translations that follow are mine.

Because he sought to avoid theology, Albertus says that he will, therefore, only inquire about a third opinion,

> which says that the heaven is generated from something preexisting and is corrupted into something that remains after it, just as natural things are generated and corrupted by the actions of qualities acting and being acted on mutually. And because these things alone proceed naturally and from principles of nature, we inquire about this mode, [namely] whether the heaven is generated.[109]

Thus, Albertus will speak not about the generation of the heaven from nothing, which is only possible supernaturally, but about its generation from something preexisting, which is naturally possible, even though it conflicts with a fundamental doctrine of his faith.

It is undoubtedly because of his conviction that a theologian doing natural philosophy should avoid theological discussions to the greatest extent possible that we find relatively little about God and the faith in Albertus's *Commentary on De caelo.*[110] The subject of the third tractate of the first book is "whether there is one world or more" *(Utrum mundus sit unus vel plures),*[111] a theme that often produced mentions of God. Albertus, however, explains:

> If ... someone should say that there can be more worlds but there are not, because God could have made more worlds if He wished and even now could make more worlds, if He wishes, against this, I do not dispute, since here I conclude that it is impossible that there be several worlds, or that more can be made, and that it is necessary that there be only one [world]. Here our understanding is about what is impossible and necessary with respect to the essential and proximate causes of the world. And there is a great difference between what God can do by means of his absolute power and what can be done in nature [or by nature].[112]

With respect "to the nature of the world," Albertus says that "there cannot be more worlds, although God could make more, if He wishes."[113] It is not, however, what God can do that interests Albertus in his *Commentary on De caelo,* but what nature can do. He concludes that nature cannot produce other worlds by its own powers. At the end of the first book, Albertus emphasizes that investigators into nature do not inquire about how God

109. Albertus Magnus, *Commentary on De caelo,* 20, col. 1.

110. As inspection of the index under "deus" (300, col. 1) and "fides" (304, col. 3) reveals.

111. Albertus Magnus, *Commentary on De caelo,* bk. 1, tr. 3, chs. 1–10, 55–77.

112. Albertus Magnus, ibid., bk. 1, tr. 3, ch. 6, 68, col. 2.

113. "Et ideo quantum est de natura mundi, dico non posse fieri plures mundos, licet deus, si vellet, posset facere plures." Albertus Magnus, ibid., bk. 1, tr. 3, ch. 6, 69, col. 1.

uses the things He has created to make a miracle in order to proclaim his power; but, rather, they investigate "what could be done in natural things according to the inherent causes of nature."[114]

Albertus kept theological references in his natural philosophy to a minimum, as is evident in his Aristotelian commentaries. In the 261 chapters that comprise the eight books of his *Commentary on the Physics,* Albertus mentions God (*deus* and its variants) in 24, or in approximately 9 percent of his chapters; and in the 111 chapters that make up the four books of his *Commentary on De caelo,* he mentions God in 9, or in approximately 8 percent of the total. Most of Albertus's uses of the term God in his *Commentary on the Physics* are in direct response to Aristotle's text, especially in the eighth book. Thus, of the 64 occurrences of *primus motor,* that is, first mover, or God, 55 occur in book 8; of the 69 occurrences of *causa prima,* that is, first cause, or God, 37 occur in book 8; and of the 78 occurrences of *deus,* God, 40 occur in the eighth book.

Most of these occurrences are in direct response to Aristotle's own mentions of God, or gods, or something about divinity. They have nothing to do with considerations of faith or theology. But Albertus unhesitatingly defends the faith against those who offer conflicting interpretations. One of the most serious claims that required a defense was Aristotle's arguments for the eternity of the world, which, if ignored, would have denied the creation. A major locus for these arguments was the eighth book of Aristotle's *Physics,* where Aristotle argued more specifically for the eternity of motion. To these kinds of arguments, Albertus replies in a chapter in which he demonstrates that the world began by a creation.[115]

Many mentions of God are minimal, little more than passing references, as when Albertus, in presenting eight ways in which something can be in another, says that "sometimes it is internal, namely when form is a mover with respect to place, just as the soul in a body and God *(deus)* in the world;"[116] or, in a discussion of time, when Albertus says that "they say that, when it is said that God is 'now' *(nunc),* and an intelligence is 'now', and a

114. "Et ideo supra diximus, quod naturalia non sunt a casu nec a voluntate, sed a causa agente et terminante ea, nec nos in naturalibus habemus inquirere, qualiter deus opifex secundum suam liberrimam voluntatem creatis ab ipso utatur ad miraculum, quo declaret potentiam suam, sed potius quid in rebus naturalibus secundum causas naturae insitas naturaliter fieri possit." Albertus Magnus, ibid., bk. 1, tr. 4, ch. 10, 103.

115. Albertus Magnus, *Opera Omnia,* vol. 4, *Physica,* pt. 2, bk. 8, tr. 1, ch. 13, 574–577. (Hereafter cited as *Commentary on the Physics.*)

116. *Commentary on the Physics,* bk. 4, tr. 1, ch. 6, 211. This is the only mention of God in a lengthy chapter that extends over pages 210 to 214. The translations in this paragraph are mine.

motion is 'now', the same 'now' is denoted."[117] In the two instances just cited, Albertus's usage conforms to that common category where theological terms and concepts are used analogically, or to exemplify and illustrate things and processes in the natural world. Of equal interest is the fact that the parts of their respective chapters that these two specific instances comprise are miniscule.

As a theologian, Albertus could easily have inserted passages about God almost anywhere in his physical commentaries. For example, in his lengthy commentary on the infinite, extending over 32 double-columned pages,[118] it might have been tempting to elaborate on God's infinite powers. But Albertus mentions God only twice: once in a context describing the way in which pre-Socratic philosophers used the term infinite,[119] and again, by way of example, in the first of five ways in which the infinite is described, a privative one, where Albertus says that "God *(deus)* is said to be infinite *(infinitus)* and incorporeal *(incorporeus)* and immense *(immensus)*";[120] that is, God is *not finite;* God is *not a body;* and God is *not measurable.* Indeed, Albertus ignores a good opportunity to invoke God when, within the context of the infinite, he launches into a discussion of extracosmic space, place, and vacuum.[121] In theological treatises, God was often mentioned in discussions about space, place, and vacuum. Also surprising is the fact that in his discussion of the celestial orbs in his *Commentary on De caelo,* where he speaks of 10 orbs, Albertus makes no mention of the crystalline orb and the empyrean heaven, the traditional theological spheres.[122]

Thomas Aquinas (ca. 1224–1274) continued the approach that Albertus Magnus developed toward Aristotelian natural philosophy. Like Albertus, Thomas sought to minimize theological intrusions into his commentaries on the natural books of Aristotle. The relatively few occurrences of key terms such as "God," "faith," "creation," "first mover," and "first cause" in Thomas's commentaries on the *Physics* and *On the Heavens,* and their near total absence from his commentaries on *On Generation and Corruption (De generatione et corruptione)* and the *Meteorology* strongly support this interpretation.

117. Ibid., bk. 4, tr. 4, ch. 5, 299, lines 16–18.
118. Ibid., bk. 3, tr. 2 *(De infinito),* 168–200.
119. Ibid., bk. 3, tr. 2, ch. 2, 172, lines 58–62.
120. Ibid., bk. 3, tr. 2, ch. 4, 175, lines 63–65.
121. Ibid., bk. 3, tr. 2, ch. 3, 174–175.
122. See Albertus Magnus, *Commentary on De caelo,* bk. 2, tr. 3, ch. 11, 166–167. Also surprising is the absence of anything of a religious nature in a chapter titled "On the perpetuity of life that exists in the external convexity of the heaven" (ibid., bk. 1, tr. 3, ch. 10, 75–77).

In Thomas's *Commentary on Aristotle's Physics,* we find almost all mentions of God and its medieval scholastic synonyms, as well as all appeals to faith, embedded in the eighth book, a feature that is also true of Albertus Magnus's *Commentary on the Physics,* as we have seen. Only a few isolated citations occur in the rest of his lengthy commentary. This is striking, but not startling, since Aristotle's major demonstration of a first mover in the eighth book caused Thomas, and all who commented on that book, to speak frequently of the first mover and, consequently, to find occasions to mention God. In view of long-held attitudes and opinions about the role of theology and faith in natural philosophy, the relatively few citations that Thomas made involving theology and the faith come as a surprise. A statistical count supports this interpretation when we realize that Thomas found occasion to mention God in only 21 paragraphs out of 2,550;[123] that the 54 occurrences of "Prime Mover" and its variants occur in 43 paragraphs; that the 10 usages of "First Cause" occur in 10 paragraphs; and that matters of faith are mentioned in only 8 paragraphs. If we sum 21, 43, 10, and 8, we arrive at a total of 82 differently numbered paragraphs. Allowing for overlap in two paragraphs, the total number of paragraphs in which some version of God's name or mention of the faith appears is 80, of which 69 are in the eighth book, leaving 11 for the other seven books. The 80 paragraphs represent approximately 3 percent of the 2550 paragraphs.

Like Albertus, Thomas also refrained from introducing theological ideas into natural philosophy. Thus, in his *Commentary on De caelo,*[124] Thomas, like Roger Bacon and Albertus Magnus, makes no mention of the empyrean heaven, although Thomas and Albertus, who both accepted its existence, found occasion to mention it in their theological treatises.[125]

Thomas frequently indicates where Aristotle is in disagreement with the faith. In 1271, however, near the end of his life, he explained why he did not often mix matters of faith with natural philosophy. In considering a question on the rational soul in man, he seemingly dismisses the

123. The data are drawn from *S. Thomae Aquinatis In octo libros De physico auditu sive Physicorum Aristotelis commentaria,* ed. P. Fr. Angeli-M. Pirotta O. P. (Naples, 1953).

124. Thomas's commentary appears in *S. Thomae Aquinatis In Aristotelis libros De caelo et mundo,* ed. Raymundus M. Spiazzi (Turin, 1952).

125. Perhaps Thomas refrained from mentioning it in a treatise on natural philosophy, because, as he explains in his commentary on the *Sentences* (bk. 2, dist. 2, qu. 2, art. 1), "the empyrean heaven cannot be investigated by reason because we know about the heavens either by sight or by motion. The empyrean heaven, however, is subject to neither motion nor sight ... but is held by authority." Cited from Grant, *Planets, Stars & Orbs: The Medieval Cosmos, 1200–1687* (Cambridge: Cambridge University Press, 1994), 377, n. 28.

question by asserting that "I don't see what one's interpretation of the text of Aristotle has to do with the teaching of the faith."[126] In Vernon Bourke's judgment, Aquinas did not think he was "required to make Aristotle speak like a Christian," and he undoubtedly "thought that a scholarly commentary on Aristotle was a job by itself, not to be confused with apologetics or theology."[127]

What are we to make of all this? We may plausibly infer that the overall impact of specific ideas about God and the faith were quite modest and do not alter the conception that the content of late medieval natural philosophy was fundamentally about natural phenomena studied in an essentially rational manner. Natural philosophy was never significantly infiltrated by theology, and natural philosophy was never really about God and His attributes. It was, of course, about God's creation, but it was about that creation as a rational construction that could only be understood by reason.

In the fourteenth century, the natural philosopher's approach to nature is beautifully exemplified by John Buridan, an arts master, and Nicole Oresme, a theologian, both of whom made outstanding contributions to natural philosophy. As a natural philosopher, Buridan recognized that his objective was to describe and explain nature's operations in terms of natural causes and effects, and not to explicate God's supernatural actions and miracles. In speaking about meteorological effects in his *Questions on Aristotle's Meteorology,* Buridan explains:

> There are several ways of understanding the word *natural.* The first [is] when we oppose it to *supernatural* (and the supernatural effect is what we call a miracle).... And it is clear that the meteorological effects are natural effects, as they are produced naturally, and not miraculously.... The philosophers, consequently, explain them by the appropriate natural causes; but common folk, not knowing of causes, believe that these phenomena are produced by a miracle of God, which is usually not true....[128]

126. "Nec video quid pertineat ad doctrinam fidei qualiter Philosophi verba exponatur." The translation and the Latin text are by Vernon J. Bourke in *Commentary on Aristotle's "Physics" by St. Thomas Aquinas,* xxiv. Bourke does not provide a full reference, but the statement occurs in Thomas's *Responsio ad fr. Joannem Vercellensem de articulis* 42 (43), which was printed in Aquinas's *Opera omnia secundum impressionem Petri Fiaccadori Parmae 1852–1873, Photolithographice reimpressa cum nova introductione generali anglice scripta a Vernon J. Bourke* (New York: Musurgia Publishers, 1948–1950), vol. 16, 167.
127. The two quotations are from Vernon J. Bourke's introduction, in *Commentary on Aristotle's "Physics" by St. Thomas Aquinas,* xxiii and xxiv.
128. Cited from *Nicole Oresme and the Marvels of Nature: A Study of his "De causis mirabilium"* with Critical Edition, Translation and Commentary by Bert Hansen (Toronto: Pontifical Institute of Mediaeval Studies, 1985), 59.

Buridan had no problems with his faith. He accepted the truths of revelation as absolute, and acceded to them. But in keeping with the tradition of his fellow natural philosophers, he acknowledged that his task was to explicate problems about natural actions and phenomena, and not to deal with the supernatural. In treating a question as to whether every generable thing will be generated, Buridan immmediately acknowledges that one can treat this problem naturally – "as if the opinion of Aristotle were true concerning the eternity of the world, and that something cannot be made from nothing" – or supernaturally, wherein God could prevent a generable thing from generating naturally by simply annihilating it. "But now," Buridan declares, "with Aristotle, we speak in a natural mode, with miracles excluded."[129] Buridan believed that truth was attainable when "a common course of nature *(communis cursus nature)* is observed in things and in this way it is evident to us that all fire is warm and that the heaven moves, although the contrary is possible by God's power."[130]

Natural philosophers like Buridan were usually careful to allow for God to upset the natural order of things by direct intervention. That is why an expression such as the "common course of nature" was so useful. Natural philosophers were primarily interested in natural, not supernatural, powers, for which reason Buridan insisted that "in natural philosophy, we ought to accept actions and dependencies as if they always proceed in a natural way."[131] Although, by His absolute power, God could move an infinite body, Buridan regards it as obvious that Aristotle's arguments "conclude sufficiently with respect to natural powers."[132] Even if he had to concede that God could use His absolute, unpredictable power to produce any natural impossibilities He

129. See Buridan, *Questions on De caelo,* bk. 1, qu. 25 ("Utrum omne generabile generabitur"), 123.

130. From Buridan, *Questions on the Metaphysics,* bk. 2, qu. 1 ("whether the grasp of truth is possible for us"), in *In Metaphysicen Aristotelis; Questiones argutissime Magistri Ioannis Buridani* (Paris, 1518), fol. 8v, col. 2–9r, col. 1. The translation is by Edith Sylla, "Galileo and Probable Arguments," in Daniel O. Dahlstrom, ed., *Nature and Scientific Method. Studies in Philosophy and the History of Philosophy:* vol. 22 (Washington, DC: Catholic University of America Press, 1991), 216. Cited in Edward Grant, "Jean Buridan and Nicole Oresme on Natural Knowledge," in *Vivarium* 31 (1993), 88.

131. "Modo in naturali philosophia nos debemus actiones et dependentias accipere ac si semper procederent modo naturali, ..." Buridan, *Questions on De caelo,* bk. 2, qu. 9, 164 (Moody edition). Also cited in Grant, "Jean Buridan and Nicole Oresme on Natural Knowledge," in *Vivarium* 31 (1993), 89.

132. "Et sic manifestum est quod rationes Aristotelis sufficienter concludunt quantum ad potentias naturales." From Buridan, *Questions on De caelo,* bk. 1, qu. 17, 77.

wished, Buridan could still save Aristotle and natural philosophy by characterizing Aristotle's arguments as sufficient in the real, natural world, the one he and his fellow natural philosophers sought to understand.

Nicole Oresme also exhibits the rationalistic temperament of a natural philosopher, although his approach is more complex than Buridan's. To capture Oresme's reasoned attitude toward nature and natural philosophy, I draw upon one of his non-Aristotelian works, *On the Causes of Marvels (De causis mirabilium),* also known as the *Quodlibeta,* composed around 1370, by which time Oresme had been a theologian since 1356, or earlier.[133]

In the Prologue to his treatise, Oresme declares:

> In order to set people's minds at rest to some extent, I propose here, although it goes beyond what was intended, to show the causes of some effects which seem to be marvels and to show that the effects occur naturally, as do the others at which we commonly do not marvel. There is no reason to take recourse to the heavens, the last refuge of the weak, or demons, or to our glorious God as if He would produce these effects directly, more so than those effects whose causes we believe are well known to us.
>
> One thing I would note here is that we should properly assign to particular effects, particular causes, but this is very difficult unless a person looks at effects one at a time and their particular circumstances. Consequently, it will suffice for me to show that the things mentioned occur naturally, as I just said, and that no illogicality is involved.[134]

In a brief, concluding recapitulation of the four chapters comprising his treatise, Oresme reiterates his naturalistic approach to phenomena, when he says:

> The above chapters are sufficient to demonstrate to an understanding person that it is not necessary to have recourse, because of the diversity and marvelousness of effects, to the heavens and unknown influence, or to demons, or to our Glorious God as the cause more than for any other things whatsoever, since it has been sufficiently demonstrated in the above chapters that effects just as marvelous (or nearly so) are found here below. And for finding the causes of these, people do not have recourse to the aforesaid [i.e., the heavens, etc.] as causes, but are well satisfied with natural causes.
>
> Second, it has been demonstrated that the natural causes there assigned and the manner of finding [them] are possible and are much more probable than that demons or unknown influence are the causes of the aforementioned effects.[135]

133. See Hansen, ed. and trans., *Nicole Oresme and the Marvels of Nature: A Study of his "De causis mirabilium,"* 26–27 (on the title of the work); 43–48 (on the date of composition).

134. Oresme, ibid., 137.

135. Ibid., 361.

Oresme almost always sought natural casual knowledge of the world and wrote treatises against magic and divination and against predictive astrology. But he also believed that the natural knowledge that we could attain is often uncertain and imprecise. To this end he wrote two treatises in which his aim was to show that the celestial motions are probably incommensurable, and, therefore, our astronomical knowledge of planetary positions was inherently approximate, from which he inferred that precise astrological predictions were impossible. Oresme is perhaps the best illustration that medieval theologians could do outstanding natural philosophy in a completely, rationalistic mode, while also viewing natural knowledge as often imprecise and vague, as vague and imprecise as the articles of faith. We detect this attitude in Oresme's last work, a French commentary on Aristotle's *On the Heavens,* titled *Le Livre du ciel et du monde.* It was a work he wrote at the request of his king, Charles V of France, completing it in 1377, five years before his death.

Among the problems Oresme considered in *Le Livre du ciel et du monde* was one about the possible rotation of the earth, a question that John Buridan had also discussed at considerable length in his *Questions on Aristotle's On the Heavens* (bk. 2, qu. 22), where, after a series of illuminating arguments, he concluded that the earth does not rotate on its axis.[136] Earlier in his career, Oresme had also written *Questions on Aristotle's On the Heavens,* and he agreed with Buridan that the earth did not rotate on its axis, thus accepting the traditional interpretation that the earth lay immobile at the center of the world.[137] But in his French commentary on Aristotle's *On the Heavens,* Oresme's views on this issue altered radically.

After presenting the strongest arguments he could muster for and against the earth's axial rotation, Oresme concludes that

> one cannot demonstrate by any experience whatever that the heavens move with diurnal motion; whatever the fact may be, assuming that the heavens move and the earth does not or that the earth moves and the heavens do not, to an eye in the heavens which could see the earth clearly, it would appear to move; if the eye were on the earth, the heavens would appear to move.[138]

With the evidence equally balanced toward either of the two alternatives, Oresme opts for the traditional opinion, declaring that

136. For a partial translation, see Grant, *A Source Book in Medieval Science,* 500–503.
137. See Oresme, *Questions on De celo,* bk. 2, qu. 13, 667–696, in Kren's edition and translation.
138. *Nicole Oresme: Le Livre du ciel et du monde,* edited by Albert D. Menut and Alexander J. Denomy, translated with an Introduction by Albert D. Menut (Madison: The University of Wisconsin Press, 1968), bk. 2, ch. 25, 537.

[e]veryone maintains, and I think myself that the heavens do move and not the earth: For God hath established the world which shall not be moved,[139] in spite of contrary reasons because they are clearly not conclusive persuasions. However, considering all that has been said, one could then believe that the earth moves and not the heavens, for the opposite is not clearly evident. Nevertheless, at first sight, this seems as much against natural reason as, or more against natural reason than, all or many of the articles of our faith.[140]

Because he could find no good reasons to choose between a rotating earth or a stationary earth, Oresme opted for the traditional, Aristotelian view that the heavens rotate around a stationary earth. He did so, however, because there was Biblical sanction for a stationary earth (he cites Psalms 92:1). But he had departed from Aristotle because unlike the latter, Oresme was convinced that the two opposing theories about the earth's status were equally plausible, and one could not choose between them by evidence or reason. This represented a dramatic departure from medieval interpretations of this old problem. And yet the rotation of the heavy earth seemed counterintuitive to Oresme, as he informs us when he argues that the earth's rotation is as much against natural reason, and perhaps even more against natural reason, than "all or many of the articles of our faith." Thus, natural philosophy could be as difficult and obscure as were the articles of faith.

Although Oresme valued reason, and always used it in his natural philosophy, he was aware, and often emphasized, that reason cannot always decide an issue, just as it could not decide whether or not the earth rotates on its axis. In his *Le Livre du ciel et du monde,* he again considered the possibility of other worlds, a problem he had discussed in his earlier *Questions on Aristotle's On the Heavens.*[141] In the course of his discussion, Oresme replies to two arguments in which certain configurations of other possible worlds were proposed:

To the sixth argument, where it is said that by analogy one could say that there is another world inside the moon, and to the seventh, where it was posited that there are several worlds within our own and several outside or beyond which contain it, etc. I say that the contrary cannot be proved by reason nor by evidence from experience, but also I submit that there is no proof from reason or experience or other-

139. Oresme's reference is to Psalms 92:1.

140. Nicole Oresme, *Le Livre du ciel et du monde,* bk. 2, ch. 25, 537–539.

141. For a comparison of the discussions in the two treatises, see Grant, "Nicole Oresme, Aristotle's *On the Heavens,* and the Court of Charles V," in *Texts and Contexts in Ancient and Medieval Science: Studies on the Occasion of John E. Murdoch's Seventieth Birthday,* edited by Edith Sylla and Michael McVaugh (Leiden: Brill, 1997), 203–205.

wise that such worlds do exist. Therefore, we should not guess nor make a statement that something is thus and so for no reason or cause whatsoever against all appearances; nor should we support an opinion whose contrary is probable; however, it is good to have considered whether such opinion is impossible.[142]

Once again, Oresme disagrees with Aristotle, who had argued that the existence of other worlds is impossible. Because Oresme believed that God had absolute power to create other worlds, he denies Aristotle's claim that other worlds are impossible. But uncertainty guides Oresme's judgment. Neither reason nor experience can determine whether there are other worlds. Nevertheless, he regards the existence of other worlds as improbable and the existence of one world as probable. At the conclusion of his discussion, Oresme gives expression to a deeply felt sentiment: "Now we have finished the chapters in which Aristotle undertook to prove that a plurality of worlds is impossible, and it is good to consider the truth of this matter without considering the authority of any human but only that of pure reason."[143]

In his *Le Livre du ciel et du monde,* Oresme sought to achieve three important goals, the first two of which can be observed in his discussion about the possible axial rotation of the earth. He wished to show that (1) great authorities, especially Aristotle, could be mistaken or misguided or inconclusive about many points in natural philosophy; (2) that conclusions and assumptions about natural philosophy itself were inherently difficult and elusive, sometimes as difficult and elusive as the articles of faith.[144] In another treatise, Oresme twice declares that, with respect to natural knowledge, "I indeed know nothing except that I know that I know nothing."[145] Notwithstanding his Socratic profession of ignorance, Oresme was no skeptic. He always sought natural explanations (as these quotations show) and refused to invoke supernatural or unnatural agents, such as God, or magic, or demons. Although Oresme seems to

142. Oresme, *Le Livre du ciel et du monde,* bk. 1, ch. 24, 171.

143. Oresme, ibid., bk. 1, ch. 24, 167.

144. With a slight emendation, I cite these goals from Grant, "Nicole Oresme, Aristotle's *On the Heavens,* and the Court of Charles V," 207. The third objective was to show "that there is only one truth: what is true in natural philosophy must also be true in religion and theology, and vice versa" (ibid., 207).

145. The translation is by Hansen, ed., *Nicole Oresme and the Marvels of Nature,* 98. See also Clagett, "Some Novel Trends in the Science of the Fourteenth Century," in Charles S. Singleton, ed., *Art, Science, and History in the Renaissance* (Baltimore: The Johns Hopkins Press, 1967), 280 and n. 12, where Clagett cites the two Latin sentences.

have believed that understanding natural philosophy could be as difficult as understanding the articles of faith, he never let his faith intrude into his natural philosophy.

Despite the fact that Oresme was a theologian and Buridan an arts master with no theological credentials, their approach to natural philosophy was quite similar at the most fundamental level. Both always sought natural explanations for natural phenomena, scrupulously avoiding appeals to the supernatural.

Theology did not penetrate natural philosophy because the two were different disciplines. Natural philosophy was taught in the arts faculties of medieval universities, whereas theology was taught in theology faculties. As we saw earlier, natural philosophers at the University of Paris were not to introduce theological ideas into their works, although theologians could use natural philosophy in their theological treatises. Thus, by the very nature of its subject matter, and by the restriction on introducing theology into natural philosophy, the latter discipline found itself relatively free from theology, except in the ways described earlier in this chapter.

An even more fundamental reason prevented the meaningful intrusion of theology into natural philosophy. It is difficult to inject theology into explanations of natural phenomena. Whenever a theological explanation is given in natural philosophy, it converts what should have been a natural explanation to a supernatural explanation and, consequently, defeats the very purpose of a treatise on natural philosophy, which is to explain phenomena by natural causes. If this were done extensively, the treatise in question would be transformed from a work in natural philosophy to one in theology. Conversely, the more that natural philosophy infiltrates a theological treatise, the less will that treatise be concerned with the supernatural, as we shall see in Chapter 6. Everyone seems to have implicitly recognized this in the Middle Ages, so that the boundary between theology and natural philosophy was rarely blurred beyond recognition.

By the seventeenth century, the disciplinary boundaries between theology and natural philosophy no longer existed. One could indeed discourse about God in natural philosophy. But how would discoursing about God advance natural philosophy? It would not and could not. When the great natural philosopher Sir Isaac Newton, a devout individual who was immersed in religious thought, wrote his monumental treatise in mathematical physics, *The Mathematical Principles of Natural Philosophy*, first published in 1687, he found occasion to mention God only once in the entire work, in book 3. Apparently regretting even this action, Newton deleted mention of God from that passage in subsequent editions.[146] As if in replacement of that passage, Newton added his famous General Scholium to the end of the second edition (1713). In a work of 530 pages, Newton saw fit to discourse upon God only in the last four pages, where he praises the deity as the Universal Ruler and Supreme God, and enunciates some of God's attributes. Coming to the end of his encomium on the deity, Newton declares: "And thus much concerning God; to discourse of whom from the appearances of things, does certainly belong to Natural Philosophy."[147]

In the Middle Ages, when theology and natural philosophy were separate disciplines, it was the responsibility of theology, not natural philosophy, to discourse about God. But the Protestant Reformation and much else had destroyed the jurisdictional boundaries between theology and natural philosophy. When Newton wrote, it was regarded as wholly appropriate for a natural philosopher to discourse about God. And yet Newton found few places where he could do so substantively and effectively. Other than singing the praises of the deity, Newton found very little to say about God. Indeed, even the General Scholium was introduced only because of criticisms leveled against Newton's use of attractions and repulsions, which made his system seem mechanical, much like that of Descartes.[148] In the General Scholium, Newton emphasizes that only God could have produced the cosmos: "Blind metaphysical necessity, which is certainly the same always and everywhere, could produce no variety of things."[149] But after the conclusion of his worshipful tribute to God, Newton, in the final two paragraphs of his great work, admits that he has not yet found the cause of gravity. It is enough for us, he says, "that gravity does really exist, and act according to the laws which we have explained, and abundantly serves to account for all the motions of the celestial bodies, and of our sea."[150]

146. On this, see I. Bernard Cohen, *Introduction to Newton's 'Principia'* (Cambridge: Cambridge University Press, 1971), 155.
147. The translation is from *Sir Isaac Newton's Mathematical Principles of Natural Philosophy and His System of the World*, translated into English by Andrew Motte in 1729. The translations revised, and supplied with an historical and explanatory appendix, by Florian Cajori (Berkeley: University of California Press, 1947), 543–547 for the Scholium, and 546 for the lines quoted here.
148. See Richard S. Westfall, *Never at Rest: A Biography of Isaac Newton* (Cambridge: Cambridge University Press, 1980), 744, where Westfall explains that Newton wrote the General Scholium because "Newton and Newtonians were highly aware of the mounting tide of criticism of his natural philosophy and its concepts of attractions and replusions."
149. General Scholium, 546 (Cajori translation).
150. General Scholium, 547 (Cajori translation).

Why did Newton not attribute the cause of gravity to God? It would almost have followed from the immediately preceding three-page discourse on God's power and attributes, as his brilliant biographer, Richard S. Westfall, recognized.[151] But Newton did not do so. Why not? Probably because he recognized that such an explanation would have been to no avail. It would have explained nothing. If you believe that God has created our world and all of its operations, then you cannot invoke God to function as an explanation for the cause of any particular effect. You must assume that God provided a natural cause for that effect, and it is the task of the natural philosopher to discover it. Theologians and natural philosophers, many of whom were both theologian and natural philosopher, recognized this essential feature of natural philosophy. It explains why, from the Middle Ages onward, natural philosophy remained relatively free of theological encroachments. And it also makes it quite plausible to believe that natural philosophy is the real precursor of modern science. Its methods were rational and systematic by the very nature of the discipline. Theology and faith could not enter it in any significant manner because to do so would transform natural philosophy into theology.

In a perceptive analysis of the relations between science and religion, George M. Marsden declared that "[s]cientists and technicians of all sorts, no matter how religious, are expected to check their religious beliefs when they enter the laboratory. Of course, they may pray about their work, and perhaps when they are done ponder how nature reflects God's design; but the activity itself will be, for methodological purposes, essentially secular."[152] Medieval natural philosophers, many of whom were devout Christians, acted similarly. They considered issues in terms of logic and reason and evidence. When they did introduce religion or faith into their deliberations, it was almost always because they could not avoid it. Theology and faith were for theological treatises, not for treatises on natural philosophy.

It is appropriate to view the relations of medieval natural philosophy and theology as bilateral, though very unevenly bilateral. If theologians viewed natural philosophy as the handmaid to theology, natural philosophers would have been justified to regard much, if not most, of the theology that intruded into natural philosophy as the handmaid to natural philosophy, simply because it served the needs of natural philosophy. This is true for

151. Richard S. Westfall, *Never at Rest*, 748.

152. George M. Marsden, *Religion and American Culture* (San Diego: Harcourt, Brace, Jovanovich, 1990), 102.

the theology that was introduced for analogical and comparative purposes, as well as for the appeals to God's absolute power, which enabled natural philosophers and theologians to extend the range of their discussions, while also reserving to themselves the option of denying that God had in fact produced the counterfactuals they were discussing. All agreed, for example, that God could make other worlds, or that He could create a vacuum anywhere within or beyond the world, but no natural philosopher in the Middle Ages believed that God had actually done so.[153]

The theology that was intruded into treatises on natural philosophy was not theology for its own sake, but was solely intended to elucidate this or that question in natural philosophy. Only aspects of natural philosophy that were contrary to faith were affected by theological considerations. But the responses to contrary-to-faith conditions, most of them associated with Aristotle's arguments for an eternal world, became routine and did not affect the substantive character of natural philosophy. Following a perfunctory bow to faith, an author could assume the eternity of the world hypothetically and pursue a variety of arguments. Because natural philosophy remained a highly rational discipline throughout the Middle Ages, theology never transformed natural philosophy. Indeed, it never really tried. In fact, while natural philosophy was largely independent of theology, theology, as we shall now see, was utterly dependent on natural philosophy.

153. For other worlds, see Grant, *Planets, Stars, & Orbs,* 163–166, where John Buridan and Nicole Oresme are discussed; for supernatural creation of vacua, see Grant, *Much Ado About Nothing,* 55, 109.

6

REASON IN ACTION

Theology in the Faculty of Theology

THE NEW THEOLOGY

In the thirteenth century, theology became a professional discipline taught in independent faculties of theology at the universities of Paris and Oxford, the most important schools of theology during the thirteenth and fourteenth centuries.[1] The faculties of theology controlled the content of theological education and the granting of the master's degree in theology. From the early centuries of Christianity, theologians always regarded their discipline as superior to that of any secular subject. Secular learning was viewed traditionally as the handmaid of theology and, therefore, subordinate to it. After all, the objective of theology was to interpret and explicate the mysteries of the faith and the meaning of Sacred Scripture. But what theology lacked until the thirteenth century was knowledge of its place in the scheme of learning. How did it relate to other disciplines, especially logic and natural philosophy?

Is Theology a Science?

In the course of the thirteenth century, many theologians, beginning with Alexander of Hales and continuing on through St. Thomas Aquinas and many others, discussed the question of "whether theology is a science." In posing this question, theologians were inquiring whether theology is a science in the Aristotelian sense of science, namely, science as demonstrative knowledge derived from premises that are "true, necessary, certain, immediate, and appropriate to the phenomenon to be explained."[2] The question "whether theology

1. See Monika Asztalos, "The Faculty of Theology," in H. de Ridder-Symoens, ed. *A History of the University in Europe,* Vol. 1: *Universities in the Middle Ages* (Cambridge: Cambridge University Press, 1992), 414.
2. From Eileen Serene, "Demonstrative Science," in Kretzmann, Kenny, and Pinborg, eds., *The Cambridge History of Later Medieval Philosophy* (Cambridge: Cambridge University Press, 1982), 497–498.

is a science" was regularly discussed by commentators on the *Sentences* at the very beginning of their commentary, in a prologue that was often expanded to include as many as six to eight subquestions. It was also considered in *Summas* of theology, as, for example, in the famous *Summa of Theology (Summa theologiae)* of Thomas Aquinas. Responses to this question varied greatly throughout the thirteenth and fourteenth centuries. Most theologians regarded theology as "the queen of the sciences," but they did not think it was a science in the strict sense. This was not because it was thought inferior to legitimate sciences but, rather, because it was usually assumed superior to all of them. Because theology relied ultimately on revelation, many theologians viewed its knowledge and wisdom as transcending that of the secular sciences.

Thomas Aquinas is usually regarded as the medieval scholar who most emphatically assigned scientific status to theology in his discussion of the question. This is evident in his *Summa theologiae* (Part 1, Question 1, article 2), where, in answer to the question "whether sacred doctrine is a science," Thomas declares that

> [s]acred doctrine is a science. We must bear in mind that there are two kinds of sciences. There are some which proceed from principles known by the natural light of the intellect, such as arithmetic and geometry and the like. There are also some which proceed from principles known by the light of a higher science: thus the science of optics proceeds from principles established by geometry and music from principles established by arithmetic. So it is that sacred doctrine is a science because it proceeds from principles made known by the light of a higher science, namely the science of God and the blessed. Hence, just as music accepts on authority the principles taught by the arithmetician, so sacred science accepts the principles revealed by God.[3]

But whether Thomas really believed that theology was a science in the strict Aristotelian sense is unclear.[4] What is certain, however, is that he and many other theologians in the thirteenth and fourteenth centuries regarded theology as similar to a science, if not actually a science. Science as Aristotle

3. From *Introduction to Saint Thomas Aquinas,* edited with an Introduction by Anton C. Pegis (New York: The Modern Library, 1948), 5–6.

4. B. P. Gaybba believes that Thomas did not regard theology as a true science and observes that this has been disputed from the thirteenth century to the present. On his interpretation of Thomas, see B. P. Gaybba, *Aspects of the Mediaeval History of Theology: 12th to 14th Centuries* (Pretoria: University of South Africa, 1988) 135–141. Gaybba discusses the views of many thirteenth- and fourteenth-century authors on the problem of whether theology is a science. Indeed, this is his book's primary theme. For a detailed discussion of Thomas Aquinas's views on theology as a science, and the concept of theology as a science in general, see Appendix 6: "Theology as Science," in *Summa theologiae,* 60 vols., Latin text with English translation, ed. and trans. T. Gilby et al. (Blackfriars/McGraw-Hill, 1964–1976), vol. 1, 67–87.

described it was the model on which theologians tried to establish their own discipline. Hence, they proceeded as if it were a science and used rigorous argumentation wherever possible.[5] As J. M. M. H. Thijssen explains: "theology still had its origin in divine revelation as communicated in Sacred Scripture and tradition and had as its goal man's salvation. But its method now involved more than ever before intellectual, speculative investigation. Theology employed a scientific discourse not unlike that of other disciplines, and the doctor of theology was its trained expert. He enquired, argued and taught by rational and analytical methods."[6] By the thirteenth century, "the scales had been definitively tipped in favor of a rational conception of theology, as faith seeking understanding, as an investigation of the data of revelation with the help of the sources of reason."[7] Despite a few noteworthy exceptions, the overwhelming mass of rational theology is found in commentaries on the *Sentences* of Peter Lombard.[8]

The "Sentences" of Peter Lombard

Although Peter was not the first to systematize theology, his effort was the most successful, with monumental consequences. Peter Lombard's *Four Books of Sentences,* completed sometime between 1155 and 1158, was an ordered, rationalized collection of patristic opinions on the major topics of theology:

> "Book I treats of God: the Trinity, God's attributes, providence, predestination, evil; Book II of the creation: the work of the six days, angels, demons, the fall, grace, sin; Book III of the Incarnation, Redemption, the virtues, the ten commandments; Book IV of the sacraments, first in general, then the seven in particular, and the four last things death, judgment, hell, heaven."[9]

Peter Lombard's treatise has been characterized as "a systematically organized 'Augustine breviary.' It contains one thousand texts from the works of Augustine," which make up approximately four-fifths of the whole work.[10]

5. See Gaybba, *Aspects of the Mediaeval History of Theology,* 147.
6. J. M. M. H. Thijssen, *Censure and Heresy at the University of Paris 1200–1400* (Philadelphia: University of Pennsylvania Press, 1998), 113.
7. Thijssen, ibid.
8. Two major exceptions are the *Summa theologica* of Alexander of Hales and the *Summa theologiae* of Thomas Aquinas. Their treatment of the questions, and the kinds of questions they included, however, are very much akin to those found in *Sentence Commentaries.*
9. Richard McKeon, ed. and trans., *Selections from Medieval Philosophers,* Vol. 1: *Augustine to Albert the Great* (New York: Charles Scribner's Sons, 1929), 187. The list of contents is by McKeon.
10. See Josef Pieper, *Scholasticism: Personalities and Problems of Medieval Philosophy* (New York: McGraw-Hill Book Co., 1964; first published in German under the title *Scholastik*), 98.

Soon after the appearance of the first great universities around 1200, Alexander Hales adopted the *Sentences* as a textbook in theology at the University of Paris, a status it held for the next four centuries. It was probably also Alexander of Hales who further subdivided each book into "distinctions" *(distinctiones).*[11] In his distinctions, Alexander also introduced the scholastic question, "the organic cell of all the scholastic Commentaries on the Sentences."[12] The format of such questions is much as I described it at the end of Chapter 3. The *Sentences* remained an enduring part of the university curriculum for approximately the same length of time as the works of Aristotle. Along with the Bible, it was one of the two basic textbooks of the faculty of theology. After hearing lectures on the Bible and the *Sentences* for a period of years, a student lectured on various books of the Bible and then became a "biblical bachelor" *(baccalareus biblicus),* after which he became a "Sententiary bachelor" *(baccalaureus sententiarius),* lecturing on the *Sentences* of Peter Lombard.[13]

In the period between 1150 and 1500, it is probable that only the Bible was commented on as much as was the *Sentences* of Peter Lombard. The names of hundreds of commentators have been identified, and in many instances their commentaries are also preserved, some of which have been published in whole, or in part.[14] "So pervasive was the domination of Peter that

Roger Bacon lists among the seven sins of the study of theology the preference of the *Book of Sentences* over the Bible; at Paris, Bacon says, a bachelor who reads the Bible must yield to the reader of the *Sentences.*[15] With the passing of time the dom-

11. See Colish, *Peter Lombard* (Leiden: E. J. Brill, 1994), vol. 1, 78, n. 75.
12. See Gilson, *History of Christian Philosophy in the Middle Ages* (London: Sheed and Ward, 1955), 328.
13. See Monika Asztalos, "The Faculty of Theology," 418; also see David Knowles, *The Evolution of Medieval Thought* (Baltimore: Helicon Press, 1962), 181–182
14. For a list of commentators on the *Sentences,* see Friedrich Stegmüller, *Reportorium Commentatorium in Sententias Petri Lombardi,* 2 vols. (Würzburg: Schöningh, 1947), and Victorinus Doucet, *Commentaire sur les Sentences: Supplément au Répertoire de M. F. Stegmüller* (Quaracchi: Collegii S. Bonaventurae ad Claras Aquas, 1954). Among the many whose commentaries have been preserved and published are St. Bonaventure, St. Thomas Aquinas, Richard of Middleton, John Duns Scotus, Peter Aureoli, William Ockham, Robert Holkot, and Gregory of Rimini.
15. Bacon regarded the overblown respect paid to lecturers on the *Sentences* as "the fourth sin of theology." See A. G. Little, "The Franciscan School at Oxford in the Thirteenth Century," in *Archivum Franciscanum Historicum* 19 (1926), 808; also McKeon, *Selections from Medieval Philosophers,* vol. 1, 72, n. 4.

ination increased rather than diminished; Gerson would have us believe that in the fifteenth century the Bible was almost forgotten in universities.[16]

It is primarily from the large number of extant *Sentence* commentaries that our knowledge of medieval theology derives. From this unusual body of literature, we are surprised to discover that medieval theology was a highly systematized, rationalistic enterprise. Although reason was heavily emphasized in the Middle Ages, as this study seeks to make clear, one does not expect to find reason and reasoned argumentation stressed in a discipline such as Christian theology, which is so heavily dependent on revealed truths derived from a Holy Scripture. But this indeed is what happened.

We saw that theology was already systematized and rationalized in the twelfth century, prior to the entry into Western Europe of the works of Aristotle and numerous other works in science, medicine, and natural philosophy that came along with them. The twelfth-century phase of the rationalization of theology may be viewed as the first stage in a two-stage process. Indeed, the second stage was not only dependent on the impact of natural philosophy and logic on theology, but also dependent on the fact that Peter Lombard's *Sentences* had become the textbook of the theology faculties. For it was in the commentaries on that famous treatise that scholastic theologians exhibited their reasoned arguments in ways that went far beyond anything that could have been envisioned in the twelfth century. Theologians who commented on the *Sentences* were members of the faculty of theology at the University of Paris or the University of Oxford. Until 1347, when another theological faculty was established at the newly founded University of Prague, they were the only two full-fledged theological schools in Europe.[17]

It is not that such an outcome was envisioned. In fact, a work composed in the latter half of the thirteenth century indicates another view for the role of reason and secular learning. The author of a treatise known as the *Summa philosophiae,* falsely attributed to Robert Grosseteste but written

16. Richard McKeon, ed. and trans., *Selections from Medieval Philosophers,* vol. 1: *Augustine to Albert the Great* (New York: Charles Scribner's Sons, 1929), 185. Jean Gerson (1363–1429) was a famous French theologian and student of Pierre d'Ailly. He defended the conciliar theory that for the good of the Church, a council could supersede a pope. On Roger Bacon's attitude toward the *Sentences,* see Peter Raedts, *Richard Rufus of Cornwall and the Tradition of Oxford Theology* (Oxford: Clarendon Press, 1987), 138.

17. See Monika Asztalos, "The Faculty of Theology," 433. There were some smaller schools, but nothing remotely like the schools at Paris and Oxford.

after the latter's death by a contemporary of Thomas Aquinas, declared that the Church did not require reason and philosophy for the faith, which was independent of reason. But, as had been argued in the struggle against the heretical Cathars, the Church should resort to reason and philosophy to defend itself against unbelievers and heretics. It was the theologians who were to make use of philosophy to defend the faith.[18] This approach had been characteristic of many theologians in the late twelfth and early thirteenth centuries, but was rapidly losing favor as the thirteenth century came to an end, and virtually vanished in the fourteenth century, by which time reason was applied to theology to explicate problems in the *Sentences* of Peter Lombard.

Richard Fishacre (d. 1248) and the New Theology

Theology had been changing since the late twelfth century. There was still some resentment against the use of Peter Lombard's *Sentences*. Many theologians thought that the Bible was sufficient as a text. But already at Paris, the theologians had apparently concluded that teaching about the Bible should be confined to straightforward exegesis and moral exhortation. Consequently, they shifted theological problems relevant to the Bible and to the faith to their commentaries on the *Sentences* of Peter Lombard.

Soon after Richard Fishacre became the first Dominican to obtain the doctorate in theology at Oxford University, probably before 1240, he completed the first *Commentary on the Sentences* at Oxford University, during the period 1241 to 1245.[19] Richard adopted the "new theology," dividing theology into two parts, one theoretical, the other practical. Both parts are mixed together indiscriminately in Scripture, but "modern masters" *(magistri moderni)* treat them separately. The practical part is concerned with moral instruction and should be discussed in commenting on the Bible; the theoretical part deals with the more difficult parts of theology and is reserved for discussion in commentaries on the *Sentences*.

Traditionalists were opposed to the use of the *Sentences* as a rival textbook to Scripture. The famous Robert Grosseteste, who was then bishop of

18. See Gilson, *History of Christian Philosophy in the Middle Ages*, 268. See also Charles King McKeon, *A Study of the "Summa philosophiae" of the Pseudo-Grosseteste* (New York: Columbia University Press, 1948), 16–18. This is the theme on which Roger French and Andrew Cunningham focus in their study *Before Science: the Invention of the Friars' Natural Philosophy* (Aldershot Hants, Eng.: Scolar Press, 1996).
19. On Fishacre, I follow R. James Long, "The Science of Theology According to Richard Fishacre," in *Mediaeval Studies* 34 (1972), 71–98.

Lincoln, wrote to the teaching masters at Oxford in 1246, urging them to ignore the *Sentences* and use only the Old and New Testaments as textbooks in theology. Pope Innocent IV intervened and urged the bishop of Lincoln not to impede Richard, of the Order of Preachers, from lecturing on the *Sentences* but, rather, to assist him in every way.[20]

The Problems and Questions of the New Theology

The urge to analyze and dissect theological problems proved irresistible. Even popes supported the new approach, although from time to time, the Church would complain about the overemphasis on philosophy and logic in theological commentaries. Ultimately, theology became thoroughly analytical and philosophical. It was almost as rationalistic as natural philosophy, on which it came to depend so heavily. Theologians had come a long way from the earlier form of theology that was concerned with moral instruction, contemplation of the divine, and what may be called the "theology of the heart." It is almost as if they were determined to understand the mysteries of the faith and to explain them rationally. They felt free to probe and analyze because commentaries on the *Sentences* were not confined to exegesis, but "soon developed into independent systematic statements by the commentators themselves, and quite often took on the character of a *Summa*."[21] Theologians in the late Middle Ages went far beyond anything envisioned by Peter Lombard. His words were but the springboard for the elaboration of theological ideas. We can begin to capture the rational flavor of theological commentary by first examining the kinds of questions that were posed and discussed.

In the descriptions and discussions that follow, we will meet two kinds of questions: those that are fundamentally theological, and those that are not. The former are heavily infiltrated by natural philosophy, and the latter – those that are not really theological – are often questions drawn from natural philosophy. Very few questions are "purely" theological. Most include significant discussions of natural philosophy, a fact that emphasizes the highly rationalistic character of *Sentence* commentaries. The sense of rationality is heightened when we realize the great extent to which the questions are counterfactual, that is, are concerned with what might be or could be, although are probably not. In Chapter 5, I mentioned the Condemnation of 1277. Although it was

20. For all this, see Long, "The Science of Theology According to Richard Fishacre," 72–73; for the Latin text, 96–97. Also see Raedts, *Richard Rufus of Cornwall*, 137–138.

21. Pieper, *Scholasticism*, 99.

by no means the sole reason for the generation of counterfactuals by both natural philosophers and theologians, it did play a significant role.

The Condemnation of 1277, God, and the Theologians

The brief mention of the Condemnation of 1277 in Chapter 5 did not convey the significant impact it had on the ways God's power was viewed. We saw that the Church and some, if not many, of its theologians were disturbed because Aristotle's natural philosophy seemed to place limits on God's power to effect changes in the natural world and to have done things differently than He did. To convey their displeasure, they issued the Condemnation of 1277. Some of the condemned articles quite obviously had placed restrictions on God's power to do things other than the way Aristotle had permitted in his natural philosophy. Most theologians opposed this approach, arguing that by His absolute power, God could indeed do any of the things that Aristotle had said were naturally impossible. Here, then, was a major potential source of counterfactuals.

In the preceding chapter, a few examples were given of the way that God's absolute power was invoked. Most of them resulted from articles condemned in 1277. Here now are the texts of the condemned articles that provoked them.[22]

34. That the first cause [that is, God] could not make several worlds.
48. That God cannot be the cause of a new act [or thing], nor can He produce something anew.
49. That God could not move the heavens [or world] with a rectilinear motion; and the reason is that a vacuum would remain.
141. That God cannot make an accident exist without a subject is an impossible argument that implies a contradiction.

The first article condemns the claim that God cannot make other worlds if He chose to do so, even though almost no one believed that He had made other worlds. The second article was condemned because it probably implied that if God acted to produce a new effect, the effort itself would imply that God is not immutable. Moreover, it also implied that the world was eternal, because if God cannot produce a new effect, and we assume there is a physical

22. The translations are from Grant, *The Foundations of Modern Science in the Middle Ages: Their Religious, Institutional, and Intellectual Contexts* (Cambridge: Cambridge University Press, 1996), 78.

world, it follows that God could not have produced that world, which would have been a new effect.[23] The third article is condemned because it asserts that God could not move our spherically shaped world with a rectilinear motion, because a vacuum would be left in the place formerly occupied by the world. Since Aristotle had argued that a vacuum is impossible, it followed that not even God could move the world and, as a direct consequence, create the impossible vacuum. Here again, God's absolute power to do as He pleases is restricted. After 1277, it had to be conceded by all that God could move the world in a straight line if He wished to do so, notwithstanding the formation of a vacuum. And, of course, it also had to be assumed that He could create other worlds than ours if He wished to do so. The fourth condemned article just cited was directed against two firm principles of Aristotle's physics: namely, that an accident or quality cannot exist without inhering in a subject or substance; and that two bodies – where "body" signifies any three-dimensional entity, whether material, or immaterial, including three-dimensional empty space – cannot exist in the same place simultaneously. This article was condemned because it would have made the Eucharist, or Mass, impossible. It had to be conceded that the accidents of the bread that have been converted to the body of Christ no longer inhered in the bread and did not exist in Christ. Therefore, the accidents of the bread did not inhere in any substance.

If such articles offended theologians, there were others that were apparently designed to antagonize them. The six articles that follow probably originated from natural philosophers in the arts faculty and appear to reflect a degree of hostility between the arts and theology faculties:

37. That nothing should be believed unless it is self-evident or could be asserted from things that are self-evident.
40. There is no higher life than philosophical life.
152. That theological discussions are based on fables.
153. That nothing is known better because of knowing theology.
154. That the only wise men of the world are philosophers.
175. Christian Revelation is an obstacle to learning.[24]

23. For a brief discussion of this article, see Roland Hissette, *Enquête sur les 219 Articles Condamnés à Paris le 7 Mars 1277* (Louvain: Publications Universitaires; Paris: Vander-Oyez, S.A., 1977), 55. Hissette follows another numbering system and numbers this article 22. The number 48 is also given in parentheses.

24. Articles 37, 152, 153, and 154 are taken from Grant, *A Source Book in Medieval Science* (Cambridge, MA: Harvard University Press, 1974), 48–50. Articles 40 and 175 are cited by Etienne Gilson, *Reason and Revelation in the Middle Ages* (New York: Charles Scribner's Sons, 1938), 64. Gilson also cites articles 152–154.

It is doubtful that any natural philosophers actually incorporated such explosive and potentially dangerous articles into their written work. If such assertions were actually made, they were probably communicated orally around the University of Paris. Because of their antitheological character, Etienne Gilson regarded these opinions "as sufficient proof of the fact that pure rationalism was steadily gaining ground around the end of the thirteenth century."[25] He viewed these articles, and the condemnation generally, as strong evidence of Averroism, a rationalistic philosophical current that was based on the Aristotelian commentaries of the Islamic commentator Ibn Rushd, or Averroës, as he was known in the Latin West. Averroism was a potent philosophical influence in sixteenth-century Italy.[26] Gilson declares further that

> [t]he existence of a medieval rationalism should never have been forgotten by those historians who investigate into the origins of the so-called modern rationalism, for indeed the Averroistic tradition forms an uninterrupted chain from the Masters of arts of Paris and Padua, to the "Libertins" of the seventeenth and eighteenth centuries.[27]

Although the Averroistic arts masters who harbored the thoughts Gilson regards as rationalistic are close to the Averroists of Padua and Italy in the sixteenth century, the very theologians who were incensed by the articles just cited, and by similar articles, became as rationalistic in their approach to theology as the arts masters were in their approach to natural philosophy. It was simply less obvious because theology was not supposed to be about reason, but about revelation. Nevertheless, theologians, especially in the fourteenth century, made theology an exercise in reason.

It is obvious why theologians were upset by these condemned articles, and by numerous others. After 1277 it became commonplace to concede that God could do anything whatever, short of a logical contradiction. Both natural philosophers and theologians often argued by appeal to God's absolute power. They had to make certain that they did not restrict that power by defending some important principle of Aristotle's natural philosophy that actually restricted God's power. But this was easily avoided by a pro forma disclaimer, a mere superficial concession that God could do this or that action, but had definitely, or probably, not done so. One could then pursue the Aristotelian argument.

25. Gilson, *Reason and Revelation in the Middle Ages,* 64.
26. For a discussion of Averroës's importance, see Dominick Iorio, *The Aristotelians of Renaissance Italy: A Philosophical Exposition* (Lewiston, NY: The Edwin Mellen Press, 1991), Chapter 2, 25–45.
27. Gilson, *Reason and Revelation,* 65.

Natural philosophers and theologians, however, differed radically in the ways in which they treated and approached God. Natural philosophers could concern themselves with God's absolute power by conceding it, and then usually moving on. For the most part, they did not probe more deeply into God's nature and powers because it was understood by all that such matters were the province of the theologian.

Theologians continued to regard natural philosophy as a servant of theology and understood that its utility was ultimately limited. They were eager to use it and to seek God by "faith-guided reason."[28] But it was always assumed that "philosophy, even as the supreme achievement of the rational mind, could never completely penetrate the mystery of God which lay beyond the powers of mortal comprehension."[29] Although this was the accepted formula among medieval theologians, they nonetheless sought to explicate those mysteries rationally. This is evident from the kinds of questions they raised and the sorts of answers they provided. Medieval theologians sought to explain as much as possible, thus contracting the domain of faith's mysteries.

Throughout the thirteenth century and much of the fourteenth, they sought to explain the Trinity by making various distinctions between God's essence and His various properties. To do this they used Aristotelian logic as much as they could, sometimes in vain, prompting Robert Holkot (or Holcot) (fl. 1330–1334), an English Dominican theologian of the fourteenth century, to call for two systems of logic, which were, as Gelber explains, "a logic appropriate to the natural order, best exemplified in Aristotle's works, and a logic appropriate to the supernatural order, a logic of faith whose rules would be quite different from Aristotle's." Holkot "concluded that Aristotelian logic did not hold universally, but only for the natural order."[30] To my knowledge, Holkot had no followers. Theologians were quite content to apply the logic of the natural order to that of the supernatural order.

Commentaries on the "Sentences" of Peter Lombard

Because commentaries on the *Sentences* were the foundation of medieval theology, it might be useful to describe, briefly, the structural form of a

28. The phrase is one used by Gelber in *Exploring the Boundaries of Reason: Three Questions on the Nature of God by Robert Holcot, OP,* ed. Hester Goodenough Gelber, *Studies and Texts* 62 (Toronto: Pontifical Institute of Mediaeval Studies, 1983), 28.

29. Ibid.

30. Ibid., 26. Gelber provides the Latin text from Holcot's *Commentary on the Sentences,* bk. 1, dist. 4. See ibid., n. 72, 26–27. See also Gilson, *History of Christian Philosophy in the Middle Ages,* 500–501.

medieval commentary on the *Sentences*. Commentaries that have been preserved were presumably first given as lectures by a student or master of theology. If a master reworked his lectures and "published" them by having them officially copied at the university stationers' shop, which served as a bookstore, the finished work was called an "ordinatio." But if the copy we have was made by a student, or someone else who recorded the lectures, that copy would be called a "reportatio." Indeed, an *ordinatio* was often revised from a *reportatio*.[31] For some authors, versions of both types have been preserved.

Although the text of Peter Lombard was constant, the manner of commenting upon it, and the attention paid to its different books and parts, varied greatly. Usually, commentators discussed the distinctions sequentially in each of the four books. But there were enormous variations. A lecturer might choose to lecture on only one, or two, or three, of the four books. If a commentator lectured on all four books, the space he devoted to any particular book might differ radically from the space devoted to that same book by another lecturer. Also, the space allocated to any particular distinction could vary greatly. Occasionally, authors omitted certain distinctions, or combined them with others that immediately preceded or immediately followed.[32] The number of questions an author might include over the whole of a four-book commentary, including a customary prologue on whether theology is a science, varied greatly. Richard of Middleton, for example, considered 1,862 distinct questions, which, although unnumbered, can easily be identified and counted, since virtually all begin with the word "utrum" ("whether"). Thomas Aquinas included approximately 1,700 questions,[33] while Thomas of Strasbourg (fl. 1345) considered 553 questions, Hugolin of Orvieto (d. 1373) 251, and Gabriel Biel (ca. 1425–1495) 542. There are undoubtedly commentaries that include

31. Anneliese Maier briefly characterizes these two forms of *Sentence* commentaries in *Ausgehendes Mittelalter: Gesammelte Aufsätze zur Geistesgeschichte des 14. Jahrhunderts* (Rome: Edizioni di Storia e Letteratura, 1964), 65, n. 60.

32. For example, in his Commentary on the first book of the *Sentences*, Hugolin of Orvieto posed two questions (each with a number of articles, with each article containing subquestions) that covered the fourteenth to eighteenth distinctions. He did this many times (for example, he covered the fourth to seventh distinctions by two questions, and the tenth to thirteenth by a single question; and so on through the other books). See *Hugolini de Urbe Veteri OESA Commentarius in Quattuor Libros Sententiarum*, edited by Willigis Eckermann O. S. A., 4 vols. (Würzburg: Augustinus Verlag, 1980–1988).

33. I counted Thomas's questions from the index of each volume. The majority of his questions as presented in the index begin with the Latin word *an* ("whether").

more than the extraordinary number of questions discussed by Richard of Middleton, and many that have fewer questions than the 251 included by Hugolin of Orvieto.

Many *Sentence* commentaries were of extraordinary length. In the printed versions that began to appear from the late fifteenth century onward, it was not unusual for printers to devote a separate volume to each of the four books. The printed editions of Thomas Aquinas, Richard of Middleton, and Hugolin of Orvieto are all in four substantial volumes, while Biel's is actually five volumes. The edition of Thomas of Strasbourg's *Sentence* commentary is confined to one folio-sized volume of hundreds of pages in double columns of small print. Occasionally, the commentary on a single book reaches truly formidable proportions, as in the modern edition of Peter John Olivi's commentary on the second book of the *Sentences.* His 118 questions appear in a modern edition of three volumes (1922–1926), encompassing more than 1,800 pages of tightly printed text. If Olivi had left lectures on all four books (he seems to have left a few questions on one or more of the other books), instead of only the second, one shudders to think of its length in a printed version.

All *Sentence* commentaries were originally hand-copied. In an age when rag paper and ink were expensive and copying a slow process, it is obvious that significant resources were given over to commentaries on Peter Lombard's *Sentences.* That these precious resources were lavished on *Sentence* commentaries is a reasonable measure of their importance in medieval society. The central role of theology is not surprising when we realize that the Church and many, if not most, of its theologians regarded it as "a body of knowledge which rationally interprets, elaborates and ordains the truths of Revelation."[34]

What did these lengthy commentaries contain? What were the hundreds of questions about? They were about the themes that Peter Lombard had treated in his four books. Peter, however, had not based his treatise on questions, but on marshalling evidence for and against a great variety of topics that he regarded as theologically important. The distinctions into which his four books are divided were, as we saw, a subsequent development, as was the great multiplication of questions within each distinction. To see the commentaries on the *Sentences* in their most dramatic mode, it will be useful to cite questions from the various books and see what they were about.

34. Congar provides this as a provisional definition of theology in general, but it would have been judged appropriate by most medieval theologians. See Yves M-J. Congar, O. P., *A History of Theology,* translated and edited by Hunter Guthrie, S. J. (Garden City, NY: Doubleday, 1968), 25.

Reason in Theology: The Role of Logic and Natural Philosophy

Before citing questions, I wish to emphasize that the objective of this chapter is to show the powerful role that reason played in these commentaries. Indeed, the most striking use of reason in the Middle Ages was in theology, perhaps because we least expect it there. This becomes evident from the massive use of natural philosophy and logic in the four books of the *Sentences,* although some books and questions were more conducive to the introduction of natural philosophy and logic than others. Indeed, sometimes it appears that even the questions that were posed were influenced by, and even shaped by, problems in logic and natural philosophy.

We can see this in the very first problem theological commentators confronted: "whether theology [or sacred doctrine] is a science." In the early part of this chapter, I discussed the view medieval theologians had of their discipline: Was it or was it not a science? The determination of whether or not theology is a science was guided by the criteria for a science which Aristotle had enunciated for mathematics and the exact sciences. The questions about whether theology is a science that appeared in virtually every theological commentator's Prologue to the *Sentences* formed the basis of the medieval understanding of what a science is:

> Although the ultimate purpose of these investigations is naturally that of determining the status of theology as *scientia,* there is little doubt that the fundamental issue at stake is a philosophical one and is accordingly treated as such. What is more, what was done in this, and in other similar, theological contexts in the fourteenth century is so recognizably philosophical that historians have been able without exaggeration to claim that its 'character and direction' were at one with modern philosophy. That is, to say, not only was the problem treated properly philosophical (though the way it was put may frequently have had a theological tinge), but the conceptions and methods utilized in examining and resolving it were also philosophical in the modern analytical, non-speculative, sense of the term.[35]

Over the course of the late Middle Ages, a number of techniques and methodologies were imported from logic and natural philosophy into theology, or were used in common. One that may have been common was use of the concept of the absolute power of God *(potentia Dei absoluta)* to do whatever He pleased short of a logical contradiction, a concept that

35. Murdoch, "From Social into Intellectual Factors: An Aspect of the Unitary Character of Late Medieval Learning," in *The Cultural Context of Medieval Learning,* edited with an Introduction by John E. Murdoch and Edith Dudley Sylla (Dordrecht, Holland: D. Reidel, 1975), 277.

received strong support from the Condemnation of 1277. It produced a kind of approach that became common to both natural philosophy and theology, and which was characterized by the phrase *secundum imaginationem,* "according to the imagination." Theologians were encouraged to introduce situations in which God is imagined to do an act that is naturally impossible in Aristotle's physical world, but which is logically possible. For example, Gabriel Biel asked "whether bare prime matter could be separated from any whatever form and stand [by itself]." In his reply, Biel says that "although matter cannot stand by itself by means of a natural power, it can, by the divine power, be separated from every form, both substantial and accidental, and separately preserved." The fundamental reason for this, says Biel, is "[b]ecause nothing, which does not imply a contradiction, must be denied to God's power; but matter standing by itself does not imply a contradiction; therefore, etc."[36] By contrast, in asking "whether it is possible that the world was created from eternity,"[37] Richard of Middleton arrived at a contradiction and, therefore, denied that God could have created the world from eternity. In the course of the argument, Richard concludes that "if God could have created the world from eternity, He would make contradictory things exist simultaneously, which is false, as was proved in the first book."[38]

36. Secundo dubitatur: Utrum a qualibet forma posset separari et stare simpliciter nuda materia prima." Biel replies briefly "quod licet per naturalem potentiam materia non potest stare nuda, tamen per potentiam divinam potest ab omni forma tam substantiali quam accidentali separari et separatim conservari." Biel follows this statement with the following: "Ratio fundamentalis est: Quia nihil negandum est a potentia Dei, quod non implicat contradictionem; sed materiam stare nudam nullam contradictionem implicat; ergo etc." See Gabriel Biel, *Collectorium circa quattuor libros Sententiarum,* 4 vols. (in five parts) plus index, ed. Wilfredus Werbeck and Udo Hofmann (Tübingen: J. C. B. Mohr [Paul Siebeck], 1973–1992), bk. 2, dist. 12, qu. 1, dubium 2, vol. 2, 304. Peter John Olivi introduced God directly into the same substantial question when he asked "whether God could make matter exist without any form." ("Quaestio 19. Quarto quaeritur an Deus possit facere esse materiam sine omni forma.") See *Fr. Petrus Iohannis Olivi, O. F. M. Quaestiones in Secundum Librum Sententiarum,* edited by Bernard Jansen, S. I. 3 vols. (Ad Claras Aquas [Quaracchi]: Ex Typographia Collegii S. Bonaventurae, 1922–1926), vol. 1, 365–370.
37. See Richard of Middleton, *Clarissimi theologic magistri Ricardi de Media Villa ... Super quatuor libros Sententiarum Petri Lombardi questiones subtilissimae,* 4 vols. (Brixia [Brescia], 1591; facsimile, Frankfurt: Minerva, 1963), bk. 2, dist. 1, qu. 4: "Utrum possibile fuit machinam mundialem ab aeterno creari," vol. 2, 16–19.
38. "Si ergo Deus potuisset machinam mundialem ab aeterno creasse, potuisset facere contradictoria simul esse, quod falsum est, ut in primo libro probatum est." Richard of Middleton, ibid., 18, col. 1.

John Murdoch has shown that theologians imported a number of what he calls "measure" languages into their *Sentence* commentaries. These measure languages were derived from logic, mathematics, and natural philosophy. The languages included intension and remission of forms; Euclidean proportionality theory; and three "limit" languages we have already met in the chapter on logic (Chapter 4), namely, how to deal with the beginning of a process and the ending of it *(de incipit et desinit);* the first and last instant of a process *(de primo et ultimo instanti);* and the setting of boundaries to the range of variable quantities of different types *(de maximo et minimo).* A sixth language identified by Murdoch is the language of continuity and infinity.[39]

Leaving aside the three limit languages discussed briefly in chapter 4, medieval proportionality theory was based on the fifth book of Euclid's *Elements,* while the medieval doctrine of "intension and remission of forms or qualities" was concerned with the mathematical treatment of imaginary variations in all kinds of qualities.[40] Among the numerous variations that were compared, medieval natural philosophers and theologians showed that "a subject that varies uniformly in heat from zero degrees at one extreme to 8 degrees at the other is 'just as hot' as if it were uniformly hot in the degree of 4 throughout."[41] In his *Rules for Solving Sophisms,* William Heytesbury showed that the same kind of comparison could be made between bodies moving with uniform motion and bodies moving with uniformly accelerated motions, a comparison that became known as the mean speed theorem, which was discussed at length in Chapter 5. Arguments that were developed in the treatment of problems of continuity and infinity were also imported into theology from natural philosophy. Theologians learned about the relationship and order of parts in a continuum, as well as about the relationship between things that differ infinitely. Finally, there was the theory of supposition drawn from medieval logic and mentioned earlier in Chapter 4. Theologians applied these powerful tools to many problems in

39. For all of these measure languages, see Murdoch, "From Social into Intellectual Factors."

40. On medieval proportionality theory, see *Nicole Oresme "De proportionibus proportionum" and "Ad pauca respicientes,"* edited with Introductions, English Translations, and Critical Notes by Edward Grant (Madison/Milwaukee: The University of Wisconsin Press, 1966), 14–65. A brief discussion of the intension and remission of forms is given in Grant, *The Foundations of Modern Science in the Middle Ages,* 99–104. For a detailed treatment, see *Nicole Oresme and the Medieval Geometry of Qualities and Motions: A Treatise on the Uniformity and Difformity of Intensities Known as "Tractatus de configurationibus qualitatum et motuum,"* edited with an Introduction, English Translation, and Commentary by Marshall Clagett (Madison: University of Wisconsin Press, 1968), Introduction, 3–121.

41. An example provided by Murdoch, "From Social into Intellectual Factors," 318, n.45.

their *Sentence* commentaries. Even those who are unfamiliar with these peculiarly medieval techniques – and that includes all but a few scholars – can see how unusual, and even extraordinary, it was for such analytical tools to be widely used in theological contexts.

Just as remarkable, however, if not more so, were the kinds of questions that medieval theologians posed and answered. In the process, they regularly employed natural philosophy to formulate their responses, often crowding out theology. They appear to have been driven by a desire to present as coherent and intelligible an explanation as was possible. After all, even if they were not all agreed that theology was a science in the strict Aristotelian sense, they were, nonetheless, determined to make their responses to questions as scientific – and therefore as rational – as possible. It is here that the rationalistic character of medieval theology becomes apparent.

Questions on God's Power: What He Could and Could Not Do

By making God the subject of the first book, Peter probably meant to mark it as the most significant of the four. Peter was concerned not only with how we know God, or how God has revealed Himself to us, but also about God "as the supreme reality in his own right."[42] In the course of the thirteenth and fourteenth centuries, theologians posed a dazzling array of questions about God and His attributes and powers, most of which they confined to their commentaries on the first book. In elaborating themes that Peter Lombard had presented, scholastic theologians posed questions about God's power, His knowledge, what He could or could not do, and what He had intended to do. Reasoning about God's power and omnipotence was essentially rationalistic, because theologians felt obligated to explain the ways a rational God would have acted under given conditions.

Questions about God's powers and capabilities were usually presented in one of two ways. Sometimes God was simply assumed able to do a particular act by virtue of His absolute power; and sometimes a question was posed as to whether God was able to cause, or perform, a particular act. Thus, William Ockham simply assumes in one of his questions on the *Sentences* that "supernaturally speaking, God can make the same body [be] in many places [simultaneously]."[43] Hence, God can easily do what is naturally impossible in

42. See Colish, *Peter Lombard,* 79.

43. "Sed supernaturaliter loquendo potest Deus facere idem corpus in multis locis." See Ockham, *Venerabilis Inceptoris Guillelmi de Ockham, Quaestiones in Librum Secundum Sententiarum (Reportatio),* ed. Gedeon Gal, O. F. M., and Rega Wood (St. Bonaventure, NY: St. Bonaventure University, 1981), bk. 2, qu. 7, 121.

Aristotelian natural philosophy. More often, commentators asked whether God could do this or that act by virtue of His absolute power, as when John Bassolis (d. 1333) asked "whether God, by His absolute power, could either make another world or make this world better,"[44] or when Gregory of Rimini asked "whether God, by His absolute power, could do every possible thing that could be done"[45] and followed this question with another asking "whether God could speak falsely."[46] In these, and in virtually all other questions in *Sentence* commentaries, the form of the arguments is overwhelmingly logical. In formulating their responses, theologians frequently sought to determine whether contradictions were lurking in the arguments. They used the language of logic that had been developed in the universities, often mentioning antecedents, consequents, and contradictories, and frequently forming syllogistic arguments.

Thus, in the question cited from Gregory of Rimini – "whether God, by His absolute power, could do every possible thing that could be done" – we see how logic and contradictions played a significant role. As a subquestion within the larger question, Gregory wants to determine if God could make someone sin. Gregory says that it is possible that some man, say Peter, sins, but God cannot make Peter sin. For

> if God made Peter sin, God would wish Peter to do what He [God] does not wish him to do; or, He would wish him not to do what He does not wish him not to do [that is, God wishes him not to do what He wishes him to do]. Each of these implies a contradiction. And the consequence is obvious, because God can do nothing except what He wishes to do; and so, if He made Peter sin He [obviously] wishes Peter to sin. However, no one can sin except either by doing what God does not wish him to do; or, by not doing what God does not wish him not to do [that is, by not doing what God wishes him to do].[47]

44. "Secundo an de potentia absoluta possit vel potuerit alium mundum facere vel istum mundum meliorare." In John Bassolis, *Opera Joannis Bassolis Doctoris Subtilis Scoti ... In Quatuor Sententiarum Libros (credite) Aurea.* (Venundantur a Francisco Regnault et Ioanne Fellon, Paris [n.d.]), bk. 1, dist. 44, art. 2, fol. 214r, col. 2–214v, col. 2.
45. From *Gregorii Arimensis OESA, Lectura super primum et secundum Sententiarum,* 7 vols. (Berlin: Walter de Gruyter, 1979–1987), vol. 3, edited by A. Damasus Trapp, O. S. A., and Venicio Marcolino (Berlin: Walter de Gruyter, 1984), bk. 1, dists. 42–44, qu. 1, 355–389.
46. From Gregory of Rimini, ibid., vol. 3, bk. 1, dists. 42–44, qu. 2, 389–409.
47. "Praeterea, si deus faceret Petrum peccare, deus vellet Petrum facere illud quod ipse non vellet eum facere, vel vellet eum non facere quod ipse non vellet eum non facere; quorum utrumque implicat contradictionem. Et consequentia patet, quia deus nihil potest facere nisi volens, et ideo, si faceret Petrum peccare, vellet Petrum peccare. Nullus autem potest peccare, nisi vel faciendo quod deus non vult eum facere, vel non faciendo quod deus non vult eum non facere." Gregory of Rimini, ibid., vol. 3, bk. 1, dists. 42–44, qu. 1, 359.

In the next paragraph, Gregory declares further that "if God would make Peter sin, Peter would sin and not sin. That he would sin [if God wanted him to sin] is obvious; but that he would not sin is now proved: because no one sins by doing what God wishes him to do or makes him do.... It is therefore obvious, properly speaking, that God cannot make Peter sin."[48] This is a typical kind of logical argument that medieval theologians introduced over and over again into their theological questions.

Many questions, however, were unrelated to Aristotle's natural philosophy, and many seem to have little, or nothing to do with the faith, although they are certainly about God. Since Peter Lombard had discussed God's foreknowledge of events in the first book of the *Sentences*, distinction 38, commentators regularly asked whether God had foreknowledge of future events. Thus, in book 1, distinction 38, Richard of Middleton, a Franciscan theologian who probably completed his commentary on the *Sentences* by 1294,[49] asks "whether God has foreknowledge" and replies that He does indeed. He follows this up with five more questions on God's foreknowledge. Is God's foreknowledge the cause of future occurrences? And turning this around, Richard asks "whether future events are the cause of foreknowledge." Does foreknowledge apply to opinions or statements? Does God have foreknowledge of all future events? And "whether all future events occur necessarily."[50]

In distinction 42 of the first book, Richard confronts the problem of God's omnipotence. Where Peter Lombard devoted a few modest pages to this problem, his scholastic successors found much on which to elaborate. Richard of Middleton was no exception. Of the eight questions he discusses in this distinction, perhaps the most significant is the fourth, in which he asks "whether God could do contradictory things simultaneously."[51] In his response, Richard manifests one of the most characteristic features of medieval theology: the heavy reliance on natural philosophy. Following tradition, Richard initially

48. "Praeterea, si deus faceret Petrum peccare, Petrus peccaret et non peccaret. Quod enim peccaret, manifestum est; sed quod non peccaret, probatur, quia nullus peccat faciendo id, quod deus vult eum facere aut facere facit.... Patet igitur quod deus non potest facere Petrum peccare proprie loquendo." Gregory of Rimini, ibid., vol. 3, bk. 1, dists. 42–44, qu. 1, 359.

49. See Gilson, *History of Christian Philosophy in the Middle Ages*, 347.

50. Richard of Middleton, *Super quatuor libros Sententiarum Petri Lombardi*, vol. 1, bk. 1, dist. 38, 336, where the questions are announced. See pp. 336–342 for the text of the six questions.

51. "Quarto quaeritur utrum Deus possit simul contradictoria facere." Richard of Middleton, ibid., vol. 1, bk. 1, dist. 42, qu. 4, 374, col. 1–375, col. 2. My translations below are taken from these pages.

gives a few arguments in favor of the position he will eventually reject. In the first of two important arguments in defense of God's ability to produce a contradiction, Richard declares that it seems that God can do so,

because a ball that has been projected against a wall is departing from it [i.e., the wall] in the same instant that it arrives [at the wall], because otherwise it would rest there, which is not seen [to be the case]. But for the same thing to arrive and depart in the same instant involves a contradiction. Therefore [if] a created thing can make contradictories simultaneously, much more so, therefore, can the Creator."[52]

Because a ball projected against a wall cannot remain against the wall for even an instant, it follows that the ball is both arriving at the wall and departing from it at the same instant, which is a contradiction. If a contradiction can occur in the natural world, surely God, the creator of that natural world, can also produce a contradiction.

Richard then offers another example drawn from a widely discussed problem in natural philosophy that had its roots in Aristotle's *Physics.* In the example, Richard assumes that

a bean is projected upward and meets a millstone descending. Then the surfaces of the bean and the millstone are joined in some instant in some indivisible part of the air [through which they were moving] and in the same instant they will be separated from that indivisible [part of space]. Otherwise, the ascending bean would make the descending millstone rest, which seems absurd. Therefore it seems that contradictories can be made simultaneously by created things. Therefore a fortiori contradictories can be made by an uncreated power."[53]

52. "Quarto quaeritur utrum Deus possit simul contradictoria facere et videtur quod sic quia pila proiecta contra parietem in eodem instanti in quo adest parieti abest ei quia alioquin ibi quiesceret quod non videtur. Sed ad esse et ab esse idem eidem in eodem instanti includit contradictionem. Ergo creatura potest facere contradictoria simul, multo fortius ergo creator." Richard of Middleton, ibid., 374, col. 1.

53. "Item ponatur quod faba proijcatur sursum et obviet lapidi molari descendenti, tunc superficies fabae et molae iunguntur in aliquo indivisibili ipsius aeris in aliquo instanti et in eodem instanti ab illo indivisibili separabuntur, aliter faba ascendens faceret lapidem descendentem quiescere, quod absurdum videtur. Ergo videtur quod etiam potestate creata possunt simul fieri contradictoria. Ergo multo fortius potestate increata." Richard of Middleton, ibid. Jean Mirecourt, a fourteenth-century theologian, also used the same example in his commentary on the *Sentences,* bk. 3, qu. 9. See John Murdoch, "From Social into Intellectual Factors," 330, n. 106. In Chapter 5 of this study, we saw that Marsilius of Inghen used the same millstone-bean argument in a strictly natural philosophical context, concluding that there can be no moment of rest for the bean. For a translation of Marsilius's discussion, see Grant *A Source Book in Medieval Science,* 285–289. The millstone-bean example appears on 286–287.

Aristotle had argued that a "moment of rest" *(quies media)* must intervene when a body changes from upward motion to downward motion. Richard shows that even if these arguments are proper, they do not show a contradiction. For example, he concedes that

> [the bean] does not make the millstone rest, nor is there any contradiction from this. Indeed, just as there is no contradiction because something rests per se and is moved accidentally, as is obvious from a man resting in a ship which is moved, so [also] there is no contradiction [just] because a bean meeting a millstone by the motion of that stone is reversed without any intervening time between its ascent and its descent, although it rests per se from necessity before it begins to be reversed.[54]

If the various examples cited are not proper contradictions in the physical world, it does not follow, as the argument would have it, that God can perform a contradiction. Indeed, Richard argues that God cannot perform contradictory actions. "I respond," he declares,

> that God cannot make two contradictories exist simultaneously, not because of any deficiency in His power, but because it does not make any sense to [His] power in any way. And if you should ask why this does not make possible sense, it must be said that with respect to this [problem] no other argument can be given except that such is the nature, or the disposition, of affirmation and negation, just as if we sought why every whole comprehends a part, no other argument would be forthcoming than that such is the nature of whole and part.[55]

Thus did Richard of Middleton arrive at the standard medieval position that not even God can produce simultaneous, contradictory actions.

He repeats this conclusion in the eighth question of the same distinction, when he asks: "whether God should be called omnipotent because he

54. "Ad secundum quod arguebatur de faba obviante lapidi molari dico quod non faceret lapidem molarem quiescere nec ex hoc sequitur aliqua contradictio. Sicut enim non est contradictio quod aliquid quiescat per se et moveatur per accidens, sicut patet de homine quiescente in navi quae movetur ita non est contradictio quod faba obvians lapidi molari per motus illius lapidis revertatur sine tempore intermedio inter suum ascensum et descensum quamvis necessitate per se quiescat antequam incipiat reverti." Richard of Middleton, *Sentences,* vol. 1, bk. 1, dist. 42, qu. 4, 374–375.

55. "Respondeo quod Deus non potest facere duo contradictoria simul esse non propter defectum potentiae ex parte sui, sed quia illud non habet rationem potentie ullo modo. Et si quaeratur quare hoc non habet rationem possibilis, dicendum quod huius non est alia ratio reddenda nisi quod talis est natura, vel habitudo affirmationis et negationis, sicut si quaereretur quare omne totum comprehendit partem et plus non esset reddenda alia ratio nisi quod talis est natura totius et partis." Richard of Middleton, ibid., 374, col. 2.

can make all things that He wishes to make."[56] In his response, Richard declares that

> God is omnipotent and can make all things that He wishes. This power to make all things that He wishes to make is not the precise reason of His omnipotence. But He ought to be called omnipotent because He is able to make everything that is absolutely possible, that is possible as was said in the preceding question, and this applies to everything that does not include a contradiction.[57]

A contradiction is unintelligible; therefore, not even God can make one. Medieval scholars applied reason to this situation.

It might have been otherwise. One can easily imagine scholastic theologians arguing that even if a contradiction is unintelligible and incomprehensible, God's omnipotence allows Him to produce one, as Peter Damian argued in the eleventh century. But after the twelfth century, I know of no one who did, because it would have violated the firm, unqualified belief in the rule of logic.

GOD AND THE INFINITE

In book 1, distinction 43, Peter Lombard discusses what God can or cannot make. Nowhere does he discuss whether God can make anything that is infinite. Most of his later commentators, however, transformed this distinction into a series of questions or articles on whether God could make an infinite entity. They often began by asking whether God's essence is infinite, as Richard of Middleton[58] did, or whether God's power is infinite, as did Durandus de Sancto Porciano.[59]

56. "Octavo quaeritur utrum Deus ideo dicatur omnipotens quia potest omnia quae vult se posse." Richard of Middleton, ibid., bk. 1, dist. 42, qu. 8, 377, col. 2.
57. "Respondeo quod Deus est omnipotens et potest omnia quae vult se posse, sed ipsum posse omnia quae vult se posse non est praecisa ratio omnipotentiae suae, sed ideo esse omnipotens dici debet quia potest omne illud quod est possibile absolute, hoc est possibile quod in praecedenti quaestione dictum est possibile quantum est ex parte sua et hoc est omne illud quod non includit contradictionem." Richard of Middleton, ibid., 378, col. 1.
58. Richard of Middleton, *Sentences,* bk. 1, dist. 43, 380, col. 2, where Richard lists the questions for this distinction.
59. "Quaestio prima: Utrum potentia Dei sit infinita," in *D. Durandi a Sancto Porciano … In Petri Lombardi Sententias Theologicas Commentariorum libri IIII, Nunc demum post omnes omnium editiones accuratissime recogniti et emendati. Auctoris vita, Indexque, Decisionum locupletissimus* (Venetiis [Venice], 1571, Ex Typographia Guerraea), fol. 112r, col. 1.

Can God Make an Infinite Entity?

Soon, however, they asked whether, in His omnipotence, God could make something that is actually infinite. Such a question was posed by Robert Holkot (d. 1349), Johannes Bassolis (d. ca. 1347), Durandus de Sancto Porciano (d. 1334), Gregory of Rimini (d. 1358), and Peter John Olivi, to name only a few. In enunciating his question, Olivi explains specifically what he intends to cover: "Thirdly [i.e., in the third question of the second book], we inquire whether God could make some actual infinite. And I speak about an infinite with respect to dimension, or with respect to any infinite aggregation of parts."[60] Thus, Olivi wants to investigate whether God could make an actually infinite magnitude and whether He could also make an actually infinite multitude of things. As John Bassolis expressed it, we have here an infinite in geometric magnitude (volume, surface, or length) and an infinite with respect to number. Bassolis adds two more kinds of infinite: an infinite in intensity of a quality or form, and an infinite with respect to force.[61] Associated with these questions, though enunciated separately, were questions about the eternity of the world, which usually led into discussions about possible potential and actual infinites.

There were two kinds of basic infinites with which medieval theologians and natural philosophers were concerned. Nicholas Bonetus (d. ca. 1343), a theologian, explained that the actual infinite could be understood in two ways. In one way, an actual infinite is an infinite to which more can be added. For example, an actual infinity of stones might exist, but yet more stones could be added. This way of identifying an infinite was captured in the Latin phrase "quod non sint tot quin plures," namely, "that there are not so many, but that there could be more." This would be regarded as a syncategorematic infinite and is analogous to Aristotle's concept of a potential infinite, although it was often called an actual infinite in the Middle Ages. Richard of Middleton described it as "an actual infinite with a mixture of potentiality, or becoming."[62] As the second kind of actual infinite, Bonetus understands an infinite to which nothing more can be added, because there are no more things that can be added. This approach is

60. Peter John Olivi, *Quaestiones in Secundum Librum Sententiarum,* vol. 1, 30.

61. Duhem cites this from Bassolis's *Sentence* commentary in Duhem, *Medieval Cosmology: Theories of Infinity, Place, Time, Void, and the Plurality of Worlds,* edited and translated by Roger Ariew (Chicago: University of Chicago Press, 1985), 97. Duhem uses the form "Bassols" where I have used Bassolis.

62. "Infinitum in actu permixto potentiae, vel in fieri." Richard of Middleton, *Sentences,* bk. 1, dist. 43, qu. 5, 384, cols. 1–2.

encapsulated in the phrase "quod sint tot in actu quod plura non possint esse, quia omnia sunt actu posita," which says "that there are so many things in actuality that there could not be more, because all actual things have been posited."[63] This definition of an actual infinite corresponds to Aristotle's concept of an actual infinite.

In the discussions that follow, it should be kept in mind that medieval theologians and natural philosophers who considered problems about actual infinites did not argue for the actual existence of such entities. The only actual, existent infinites accepted during the Middle Ages were the infinite attributes and powers of God, including his infinite omnipresent immensity (as we shall see). Those whom we may call infinitists in the late Middle Ages are not called that because they believed in the real existence of an actually infinite dimension or entity. They are called infinitists because they argued that it was *possible* for God to create actual infinites, although He had probably not done so. Their goal was to demonstrate that the existence of an actually infinite body or magnitude, or the existence of an actually infinite multitude of things, was an intelligible concept devoid of contradiction. By logical analysis, they sought to show the conditions under which an actual infinite might exist. With the possible exception of Jean de Ripa (see the section on "God and Infinite Space"), they did not believe that God had created, or would create, an actually infinite entity. This would be especially true for the production of an infinite body. If God had created, or was assumed to create, a real infinite body, the Aristotelian cosmos, with its physics and cosmology, would have to be abandoned.

Few would have rejected the possibility of both kinds of infinites, namely, the infinite to which more things can be added, or a potential infinite, and the actual infinite to which nothing more can be added, because there is nothing outside of it. Peter Aureoli was one who did, rejecting a potential infinite because if the latter were capable of existing, so also would an actual infinite, which he regards as impossible. A potential infinite "is completely impossible," Peter declares,

> and no power can add a magnitude to another magnitude of equal amount, and so forth to infinity, such that any determined and given magnitude can be surpassed.

63. With the exception of the material on Richard of Middleton, this paragraph draws heavily on Grant, *Planets, Stars, & Orbs: The Medieval Cosmos, 1200–1687* (Cambridge: Cambridge University Press, 1994), 106, n. 1. My source was Maier, *Die Vorläufer Galileis im 14. Jahrhundert: Studien zur Naturphilosophie der Spätscholastik* (Rome: Edizioni di Storia e Letteratura, 1949), 214. Maier does not cite the work in which Bonetus made this distinction.

It is in fact impossible, and an impossibility results from it; for if a magnitude can be indefinitely augmented, and progress to infinity by this augmentation, as a result an infinite magnitude would be able to exist in actuality. And this has been declared impossible; therefore it is also impossible that a magnitude can grow to infinity by addition.[64]

Most theological commentators, however, did not equate the potential infinite with the actual infinite, and they believed that God could create a potential infinite. They assumed that God could indefinitely add together things of the same kind, as, for example, numbers, and thus make a potential, or syncategorematic infinite. Richard of Middleton held that "God could make a dimension greater and smaller without end. Nevertheless, the whole that is taken is always finite, which is customarily called an actual infinite with a mixture of potentiality, or becoming."[65] Franciscus de Mayronis (ca. 1285–1328) offers another example of a syncategorematic, or potential, infinite that is treated as if it were an actual infinite. In this approach, God is said to have created an actual infinite magnitude when He makes a magnitude greater than any finite magnitude.

We see this in a few passages from the first book of Mayronis's *Sentence* commentary, where he asks "whether God could produce something actually infinite."[66] Thus, Mayronis declares:

It is not possible to progress to infinity in the domain of finite things; therefore if God cannot create an infinite magnitude in one blow, one could assign a magnitude such that God can create nothing greater in one blow, which is false.... For however large something created by God is, He can still create something larger, which is not possible if He can only create a finite something.[67]

Mayronis follows this with a similar argument, when he declares:

A magnitude surpassing any finite magnitude is infinite; but the ultimate magnitude that God can create in one blow surpasses any finite magnitude. It is then infi-

64. Translated from Peter Aureoli's (Duhem calls him Aureol) commentary on the *Sentences,* bk. 1, dist. 43, art. 1, in Duhem *Medieval Cosmology,* 83.

65. "Deus posset facere dimensionem maiorem et minorem sine fine. Ita tamen quod semper totum acceptum sit finitum quod solet dici infinitum in actu permixto potentiae vel in fieri." Richard of Middleton, *Sentences,* bk. 1, dist. 43, qu. 5, 384, cols. 1–2.

66. "Quaest. X: Utrum Deus potuit producere aliquid actualiter infinitum." Cited by Duhem, *Medieval Cosmology,* 520, n. 41.

67. Translation in Duhem, ibid., 93. Duhem indicates that the translation is made from Mayronis's first book of the *Sentences,* dist. 43, 44.

nite. Proof of the minor [premise]. Given any finite magnitude that God can create, He can create a greater one; that which surpasses all is therefore infinite.[68]

Thus, Mayronis believed that a potential, or syncategorematic, infinite functioned as an actual infinite. He saw a possible objection to his own approach, when he asked "how is it that infinity is something whose magnitude always leaves something outside it capable of being taken up?" Mayronis replies that the definition of infinity he just gave "is a definition relative to our intellect; when it takes up a multitude, no matter how large, there always remains something else to take up; in fact, it takes up a finite multitude, and never an infinite multitude."[69] But the multitude is called an actual infinite because, although it is technically finite, it is greater than any other multitude – until something is added to it.

Mayronis's description agrees with Aristotle's view as we find it in the latter's *Physics* (3.6.206a.27–29), where he declares: "The infinite has this mode of existence: one thing is always being taken after another, and each thing that is taken is always finite, but always different." Richard Sorabji has observed that Aristotle criticized his predecessors because "they thought of infinity as something which is so all-embracing that it has nothing outside it. But the very opposite is the case: infinity is what always has something outside it."[70]

The actual, or categorematic, infinite was quite another matter. Richard of Middleton denied that God could create an actual infinite, specifically rejecting the idea that God could create or produce an actually infinite dimension. "It is impossible," Richard explains, "that God make an actual infinite dimension, which is customarily called an absolutely actual infinite."[71] In the next question, Richard also denies that God can make an actually infinite multitude.[72] Durandus de Sancto Porciano agreed with

68. Duhem, ibid., 93.

69. Duhem, ibid., 95.

70. Sorabji, "Infinity and Creation," in Richard Sorabji, ed., *Philoponus and the Rejection of Aristotelian Science* (Ithaca, NY: Cornell University Press, 1987), 168.

71. Here is Richard's full response to the possibility of a divinely created actual infinite: "Respondeo, quod quamvis Deus possit facere dimensionem maiorem et minorem sine fine, ita tamen quod semper totum acceptum sit finitum quod solet dici infinitum in actu permixto potentiae vel in fieri. Tamen est impossibile quod Deus faciat aliquam dimensionem infinitam in facto esse, quod solet dici infinitum in actu simpliciter." Richard of Middleton, *Sentences*, bk. 1, dist. 43, qu. 5, 384, cols. 1–2.

72. "Respondeo quod Deus non potest facere aliquid infinitum in actu secundum multitudinem." Ibid., bk. 1, dist. 43, qu. 6, 386, col. 1. Also see Duhem, *Medieval Cosmology*, 79, where this statement and the brief argument that follows is translated by Ariew.

Richard. He accepted the possibility of a potential, or syncategorematic infinite, but denied that God could make an actual, or categorematic, infinite, declaring that it "seems more probable, that God cannot produce such an actual infinite, not because of some defect in His power, but because it is repugnant to reality."[73]

One may conjecture that a major reason that an actual infinite was rejected by many was the fact that if God created it, He could not create anything larger, because there is nothing larger than an actual infinite. Thus, God's absolute power would be restricted. That argument was made by the natural philosopher John Buridan, who, in considering the question "whether there is some infinite magnitude,"[74] declares that "it is not necessary to believe that God could create an actually infinite magnitude, because when it has been created he could not create anything that is greater, since it is repugnant [or absurd] that there should be something greater than an actual infinite."[75] In discussing whether God could make an actual infinite, Buridan was painfully aware that he, a master of arts and not a theologian, might be treading into the domain of theology. Wishing to avoid potential difficulties, Buridan declares that "with regard to all of the things that I say in this question, I yield the determination of them to the lord theologians, and I wish to acquiesce in their determination."[76]

In bowing to the theologians, Buridan surely knew that they were divided on the issue of God's ability to make an actual infinite. Numerous theologians in the fourteenth century concluded that God could, if He wished, make an actual infinite and could, therefore, certainly make a potential infinite. Modern scholars have called them "infinitists." Among the members of this group, who believed that God could make an actual infinite, were William of Ockham, Gregory of Rimini, John Bassolis, Franciscus de Mayronis, Jean de Ripa, Gerard of Bologna, Robert Holkot,

73. See Duhem, ibid., 88–89. Duhem also links William of Ockham with the views of Richard of Middleton and Durandus, but Anneliese Maier places him among the infinitists (Maier, "Das Problem des Aktuell Unndlichen" in *Die Vorläufer Galileis,* 206).

74. "Utrum est aliqua magnitudo infinita." Buridan considers this question in his *Questions on the Physics,* bk. 3, qu. 15, in *Acutissimi philosophi reverendi Magistri Johannis Buridani subtilissime questiones super octo Phisicorum libros Aristotelis* (Paris, 1509; facsimile, entitled Johannes Buridanus, *Kommentar zur Aristotelischen Physik.* Frankfurt: Minerva, 1964), fols. 57r, col. 2–58r, col. 2.

75. Cited from Grant, *Planets, Stars, & Orbs,* 111. The translation is mine from Buridan's *Physics,* bk. 3, qu. 15, fol. 57v, col. 1. I give the Latin text on p. 111, n. 24.

76. Buridan, *Physics,* bk. 3, qu. 15, fol. 57v, cols. 1–2. See also Grant, *Much Ado About Nothing: Theories of Space and Vacuum from the Middle Ages to the Scientific Revolution* (Cambridge: Cambridge University Press, 1981), 341, n. 52.

John Baconthorpe, Paul of Perugia, Hugolin of Orvieto, and the sixteenth-century scholastic theologian, John Major.

In assuming that God could make an actual infinite, medieval theologians left no doubt that the actual infinite they believed God capable of creating was not another God equal to Himself. Bassolis addressed this problem when he declared that there can be two kinds of actual infinites: "First one can understand, by these words, absolute *(simpliciter)* infinity, infinity according to all manners of being and all perfections." God cannot create an actual infinite in this sense because a being who met these conditions would be God Himself. But an actual infinite would be possible that is infinite "not in all manners of being and all perfections, but only according to some manner of being or perfection of a special nature … for example, infinite length, or some similar attribute."[77] In the category of possible actual infinites, Bassolis observes that four types of actual infinites do not involve a contradiction, so that God could, if He wished, create them. These are the same four already mentioned, namely, a geometrically infinite magnitude or dimension; an infinity of number or multitude; an infinite in the intensity of a quality; and an infinite of force.

In their commentaries on book 1, distinction 43, of Peter Lombard's *Sentences,* theologians often argued the possibility of God making one or another of these kinds of actual infinites, usually one of the first three. As Duhem has observed, there were standard arguments against the possibility of an actual, or categorematic, infinite. These were based on absurd consequences that were drawn from the possibility of an actual infinite, namely that "one can add something to infinity; there can be something greater than infinity; an infinity can be the multiple of another, etc."[78] Those who opted for the possibility of an actual infinite rejected one or all of these criticisms.

God and Infinite Space

Sometime around 1354 or 1355, Jean de Ripa, a Franciscan theologian, considered problems of the infinite in his commentary on the first book of the *Sentences.*[79] Of all the infinitists in the Middle Ages, de Ripa's interpretations of God and the infinite were perhaps the most extraordinary, because,

77. Translated from Bassolis's *Sentence* commentary, bk. 1, dist. 43 in Duhem, *Medieval Cosmology,* 97.
78. Duhem, ibid., 109.
79. In what follows on de Ripa, I draw upon my more extensive account of his arguments in Grant, *Much Ado about Nothing,* 129–134.

unlike most, if not all, of his colleagues, he may really have believed in the existence of an actual infinite that was not God. De Ripa expressed his opinions in his commentary on distinction 37 of book one of the *Sentences,* which is concerned with the ways God could be in things. In the course of his arguments, de Ripa proclaims that "God is really present in an infinite imaginary vacuum beyond the world."[80] De Ripa seems to argue that God created an actual infinite void space. Indeed, de Ripa describes the infinite void not only as eternal, without beginning or end, but also as something created by God, for it flows from His presence as from a cause and is, thus, totally dependent on Him. Because God is assumed omnipresent in an infinite void of His own creation, should God's infinite, omnipresent immensity be equated with the infinite void space that He created? No, says de Ripa, because "the infinity of a whole possible vacuum or imaginary place is immensely exceeded by the real and present divine immensity."[81]

De Ripa departed from virtually all of his colleagues. He took the unprecedented step of distinguishing two infinites: the ordinary kind exemplified by infinite void space, or by any actual infinite that one might imagine, and a single superinfinite equated with God's immensity. De Ripa was apparently convinced that God could create actual infinites because he believed that God could perfect spiritual substances all the way to infinity and that this did not involve a contradiction. Indeed, in sharp contrast with John Buridan, de Ripa believed that to deny to God the ability to create an actual infinite is tantamount to a rejection of His absolute power. De Ripa did not regard the creation of infinites as a problem, because no divinely created actual infinite is equal to God's infinity. We do justice to God's infinite and incomprehensible power only when we understand not only that He can create all manner of possible infinites, but also that He immeasurably exceeds and circumscribes them. Only God is uncircumscribable.

Did God actually create an infinite void space? Although de Ripa seems at times to say this, he leaves his ultimate position unclear as to whether God had actually created an infinite void space, or could create such a space if He wished to do so but had not actually done so. De Ripa was, however, emphatic and unequivocal in his conviction that God is uncircumscribable and that there is an infinite difference between God's immensity and the mere infinity of any possible or actual void space in which He might be

80. See the Latin text of de Ripa's commentary on bk. 1, dist. 37, in André Combes and Francis Ruello, "Jean de Ripa I Sent. Dist. XXXVII," *Traditio* 23 (1967), 233, lines 1–2.

81. This statement forms the enunciation of de Ripa's second conclusion: "Totius vacui possibilis seu situs ymaginarii infinitas immense exceditur ab immensitate reali et presentiali divina." Combes, "De Ripa," 235, lines 26–28.

omnipresent. It was in defense of this concept that de Ripa launched a criticism against Thomas Bradwardine, who had discussed God and infinite void space in a theological treatise titled *In Defense of God Against the Pelagians (De causa Dei contra Pelagium)*, written some ten years earlier. The gist of de Ripa's criticism is that Bradwardine improperly believes that as an infinite being, God would be no more uncircumscribable than any other infinite that He might make. In de Ripa's view, Bradwardine, whom de Ripa does not actually name (he refers to him as "this doctor" *[iste doctor]*), is seriously mistaken because he assumes that every possible intensively infinite thing or creature would be equal to the divine infinity, rather than being infinitely less than it.

Thomas Bradwardine's explication of the relations between God and the infinite proved more enduring than that of de Ripa. In his theological treatise, *In Defense of God Against the Pelagians,* composed around 1344, Bradwardine, who was also a talented mathematician,[82] used an axiomatic, analytical, and quasi-mathematical approach to theology, following a tradition that included Boethius, Alan of Lille, and Nicholas of Amiens (see Chapter 2). Bradwardine proceeds by enunciating propositions and deriving numerous corollaries.[83] In a chapter titled "That God is not mutable in any way," he presents five corollaries, which declare that God exists everywhere in an imaginary infinite void.[84] But this is not an infinite void that God has created. Indeed, Bradwardine denies that God can create an actual infinite. Such infinites would produce absurdities.[85] The infinite void in which God exists is His own infinite omnipresent immensity. But because God is not an extended being, Bradwardine concludes that "He is infinitely extended without extension."[86] Thus, where de Ripa assumed that God could, or even did, create an actual infinite void space that was independent of Him but which He infinitely exceeded, Bradwardine denied that God could create an actual infinite void, or any other infinite entity. But God did not have to create an infinite void space because such a space

82. For an excellent brief intellectual biography of Bradwardine, which describes his mathematical treatises, see John E. Murdoch, "Bradwardine, Thomas," in *Dictionary of Scientific Biography,* vol. 2 (1970), 390–397.

83. In what follows on Bradwardine, I have relied on my analysis of Bradwardine's discussion of God and infinite space in *Much Ado About Nothing,* 135–144.

84. See Grant, ibid., 135, for the five corollaries. For a translation of the five corollaries and the relevant sections from Bradwardine's *De causa Dei,* see Grant, *A Source Book in Medieval Science,* 556–560.

85. See Grant, *Much Ado About Nothing,* 349, n. 122.

86. For the passage containing this statement, see Grant, ibid., 141.

already existed, namely, the infinite omnipresence of God's own immensity, an immensity that was without extension or magnitude.

In 1618, Bradwardine's *De causa Dei contra Pelagium* was published in London. Because of this, Bradwardine's ideas about infinite space had an impact on seventeenth-century spatial conceptions, especially on Henry More and Isaac Newton. It was Bradwardine and de Ripa, as well as other scholastics, who introduced God into space. Newton was ultimately influenced by Bradwardine because like Bradwardine, he identified space with God's infinite omnipresence and did not assume that God had created the space that He occupied. Newton, however, departed radically from Bradwardine, and all his medieval predecessors, when he assumed that infinite space is three-dimensional, from which it followed that God was also three-dimensional, although Newton did not make this explicit.[87]

Scholastic ideas about space and God that were developed in theological treatises by theologians form an integral part of the history of spatial conceptions between the late sixteenth and eighteenth centuries, the period of the Scientific Revolution. From the assumption that infinite space is God's immensity, scholastics derived most of the same properties for it as nonscholastics did subsequently. As God's immensity, space had to be homogenous, immutable, infinite, and capable of coexisting with bodies, which it received without offering resistance. Except for extension, the divinization of space in scholastic thought produced virtually all the properties that would be conferred on space during the course of the Scientific Revolution.[88]

The Eternity of the World and the Infinite

Acceptance of an actual infinite was advanced considerably by arguments that developed over the possibility of an eternal world. This was a momentous issue in the thirteenth century, especially at the University of Paris.[89] The issue of the eternity of the world was to the relations between science and religion in the Middle Ages what the heliocentric system of Copernicus was in the sixteenth and seventeenth centuries, and what the Darwinian theory of evolution has been since its inception in the nineteenth century.

87. For details about Newton's ideas of God and an extended infinite space, see Grant, ibid., 240–247.
88. See Grant, ibid., 262.
89. For a discussion, see Grant, *Planets, Stars, & Orbs,* Chapter 4 ("Is the World Eternal?"), 63–82; for a brief treatment, see Grant, *The Foundations of Modern Science in the Middle Ages,* 74–76.

This is evident in the Condemnation of 1277 where at least 27 of the 219 condemned articles were directed against some form of Aristotle's arguments in behalf of the eternity of the world, especially his assertion at the very beginning of the second book of *On the Heavens,* where he declares that "the heaven as a whole neither came into being nor admits of destruction, as some assert, but is one and eternal, with no end or beginning of its total duration, containing and embracing in itself the infinity of time."[90]

Aristotle's arguments for an eternal world were at variance with the creation account in Genesis, and with the first canon of the Fourth Lateran Council of the Church in 1215, which declared that "the creator of all visible and invisible things, spiritual and corporeal, who, by His omnipotent power created each creature, spiritual and corporeal, namely angelic and mundane, at the beginning of time simultaneously from nothing; and then made man from spirit and body."[91] Saint Bonaventure was one of the first in the Latin West to argue against Aristotle's powerful concept. Although he died in 1274, Bonaventure's continuing influence was probably a factor in the 1277 condemnation of so many propositions favoring the eternity of the world. For Bonaventure had not only opposed Aristotle's position on the eternity of the world; he was convinced that he could demonstrate its temporal creation. He did this, appropriately, in the second book of his *Sentence* commentary, which was concerned with the creation and where many theologians over the centuries discussed whether the world was eternal.

To subvert the concept of eternity, and the idea of infinity on which it depends, Bonaventure offers six arguments "based on *per se* known propositions of reason and philosophy."[92] Many of those "propositions of reason and philosophy" have a basis in Aristotle's *Physics, Metaphysics,* and *On the Heavens (De caelo),* thus revealing the significant influence of Aristotle's thoughts about the infinite on medieval discussions of that important theme. Four of Bonaventure's six arguments (1, 2, 3, 5) will be summarized here.

90. Aristotle, *On the Heavens,* bk. 2, ch. 1, 283b.26–30. Translation by J. L. Stocks.

91. See Grant, *Planets, Stars, & Orbs,* 83 and n. 2 for the Latin text.

92. Bonaventure discusses the question "whether the world has been produced from eternity, or in time" ("Utrum mundus productus sit ab aeterno, an ex tempore") in his *Opera omnia* (Ad Claras Aquas [Quaracchi]: Collegium S. Bonaventurae, 1882–1901), Vol. 2: *Commentaria in quatuor libros Sententiarum Magistri Petri Lombardi: In secundum librum Sententiarum.* bk. 2, dist. 1, pt. 1, art. 1, qu. 2, 19–24. I use the translation by Paul M. Byrne, *On the Eternity of the World (De aeternitate mundi) St. Thomas Aquinas, Siger of Brabant, St. Bonaventure* (Milwaukee, WI: Marquette University Press, 1964), 105–116. For the brief quotation, see p. 107.

Argument 1: The first denies the eternity of the world because it is impossible to add to an infinite.[93] Here Bonaventure argues that if an actual infinite number of celestial revolutions have occurred to the present, then all additional revolutions will have to be added to that infinite number. However, adding to an infinite cannot make it larger because "nothing is more than an infinite." To further illustrate the absurdity of the infinite, Bonaventure compares the revolutions of Sun and moon, which he assumes to occur in a world that has an infinite past. Since the moon circles the earth 12 times in a year and the Sun only once, the moon will have made 12 times as many revolutions as the Sun. Therefore, the moon's infinite number of revolutions should be larger than the Sun's, which is impossible, because Bonaventure assumes that all infinites are equal. Thus, Bonaventure ignored, or was unaware of, what was already known, namely, that infinite sets contain subsets that are equivalent to the whole, as when Chrysippos the Stoic asserted that "Man does not consist of more parts than his finger, nor the cosmos of more parts than man. For the division of bodies goes on infinitely, and among the infinities there is no greater and smaller nor generally any quantity which exceeds the other, nor cease the parts of the remainder to split up and to supply quantity out of themselves."[94] Or as a medieval author put it in a theological treatise: "There is no objection to the part being equal to its whole, or not being less, because this is found, not ... only intensively but also extensively ... for in the whole universe there are no more parts than in one bean, because in a bean there is an infinite number of parts."[95]

This concept was repeated in the Middle Ages by others, including Roger Bacon. But the ingenious solution to the problem of the relations between the infinite set of a whole and the infinite set of part of that whole would be proclaimed in the fourteenth century by Gregory of Rimini (as we shall see), whose approach was virtually ignored until it was fully articulated in the nineteenth century.

Argument 2: Bonaventure's second argument asserts that "it is impossible that an infinite be ordered."[96] For, as he declares,

93. I have described this argument in Grant, *Planets, Stars, & Orbs,* 67–68. On this proposition, see Aristotle, *On the Heavens,* bk. 1, ch. 12 (283a.9–10).

94. See Samuel Sambursky, *The Physics of the Stoics* (London: Routledge and Paul, 1959), 97.

95. The passage was translated from the *Centiloquium Theologicum,* of unknown authorship. See Crombie, *Medieval and Early Modern Science,* rev. 2nd ed., 2 vols. (Garden City, NY: Doubleday Anchor Books, 1959), vol. 2, 42.

96. "Impossibile est infinita ordinari." Bonaventure, *Commentary on the Sentences,* bk. 2, dist. 1, pt. 1, art. 1, qu. 2, vol. 2, 21, col. 1.

every order [of things] flows from the beginning to the middle. If therefore, there is no first [thing or element], there is no order. But if the duration of the world, or the revolutions of the heaven, are infinite, they do not have a first [thing or element].[97] Therefore, they do not have an order; therefore, one [thing] is not before another [thing]. But this is false, so that it remains that they do [indeed] have a first [thing or element].[98]

Argument 3: In the third argument, Bonaventure insists that "it is impossible that an infinite be traversed."[99] Once again, Bonaventure assumes that if the world had no beginning, then an infinite number of celestial revolutions must have occurred to the present. But an infinite number of revolutions cannot be traversed; therefore, the present revolution could not have been reached. Bonaventure then anticipates objections: "You will not be able to evade this consequence if you say that there was not a first [revolution], or that an infinite number of revolutions can be traversed in an infinite time."[100]

To counter such arguments, Bonaventure shows that if a numerically infinite number of revolutions has not been traversed, because there was no first revolution in a beginningless universe, then either a particular past could have infinitely preceded today's revolution, or none could. If none, then all past revolutions are distant from the present one by a finite number of revolutions, however large the number, and the world must have had a beginning. But if a particular revolution is infinitely distant, Bonaventure asks

[w]hether the revolution immediately following it is infinitely distant. If not, then neither is the former (infinitely) distant, then I ask in a similar way about the third, the fourth, and so on to infinity. Therefore, one is no more distant than another from this present one, one is not before another, and so they are all simultaneous.[101]

All of which Bonaventure regards as absurd consequences of a world that is of infinite duration.

97. See Aristotle, *Physics* 8.5.256a.17–19.
98. Bonaventure, *Commentary on the Sentences,* bk. 2, dist. 1, pt. 1, art. 1, qu. 2, vol. 2, 21, col. 1.
99. "Impossibile est infinita pertransiri." Ibid., 21, cols. 1–2. See Aristotle, *Metaphysics* 11.10. 1066a.30
100. "*Si tu dicas,* quod non sunt pertransita, quia nulla fuit prima, *vel,* quod etiam bene possunt pertransiri in tempore infinito; per hoc non evades." Bonaventure, ibid., 21, col. 1.
101. Translated by Paul Byrne, in *On the Eternity of the World (De aeternitate mundi),* 108.

Argument 5: In the fifth argument, or proposition, Bonaventure declares that "it is impossible that there be simultaneously an infinite number of things."[102] Here Bonaventure draws the obvious consequence that

> if the world is eternal and without a beginning, then there has been an infinite number of men, since it would not be without there being men – for all things are in a certain way for the sake of man[103] and a man lasts only for a limited length of time. But there have been as many rational souls as there have been men, and so an infinite number of souls. But, since they are incorruptible forms, there are as many souls as there have been; therefore an infinite number of souls exist.[104]

Because Bonaventure rejected the existence of an actual infinite number of things, he therefore rejects the idea that an infinite number of souls could exist.

Bonaventure was convinced that he could rationally demonstrate the creation of the world by drawing absurd consequences from the assumption of an eternally existing world. In opposition to Bonaventure, many scholastic theologians believed that neither alternative was demonstrable. One of this group was Thomas Aquinas, who declared: "That the world has not always existed cannot be demonstratively proved but is held by faith alone."[105] Indeed, Thomas believed that it was not possible to demonstrate the creation of the world or its eternity. And yet it may have been Thomas who spurred discussions of the possible eternity of the world when he argued, in his treatise *On the Eternity of the World* that God could have willed the existence of creatures, and therefore the world, without a temporal beginning. That is, the world might have been cre-

102. "Quinta est ista. Impossibile est infinita simul esse." Bonaventure, *Sentences,* ibid., vol. 2, 21, col. 2. Translated by Paul Byrne, *On the Eternity of the World,* 108. See Aristotle, *Physics* 3.5.204a.20–25 and *Metaphysics* 11.10.1066b.11. The fourth argument, omitted here, proposes that "[i]t is impossible for the infinite to be grasped by a finite power." Translated by Paul Byrne, ibid., 108. For the Latin text, we have: "Quarta propositio est ista. Impossibile est infinita a virtute finita comprehendi." Bonaventure, *Sentences,* ibid. In the sixth, and final, proposition, or argument, which is not discussed here, Bonaventure declares that "[i]t is impossible for that which has being after non-being to have eternal being." Translated by Paul Byrne, ibid., 109.
103. As with the other citations of Aristotle in his translation of Bonaventure's discussion of the eternity of the world, Paul Byrne identifies the specific lines in Aristotle's relevant work, this one being *Physics* 2.2.194a.34–35.
104. Translated by Byrne, *On the Eternity of the World,* 108.
105. From Thomas Aquinas's *Summa Theologiae,* pt. 1, qu. 46, art. 2. The translation appears in Thomas Aquinas, *Summa Theologiae,* vol. 8, trans. Thomas Gilby, O. P. (1967), 79.

ated by God and yet have had eternal existence.[106] In an appeal to logic, Thomas insists that "the statement that something was made by God and nevertheless was never without existence ... does not involve any logical contradiction."[107] Because of His absolute power, and as the efficient cause of the world, God "need not precede His effect in duration, if that is what He Himself should wish."[108] God can achieve this because He produces effects instantaneously and, thus, could have created a world that has existed from eternity. Thomas emphasizes, however, that the mutable world is totally dependent on an immutable God, thus guaranteeing that the former cannot be coequal with the latter. Despite the condemnation of the eternity of the world in 1277, the idea that it was not a logical contradiction to hold that the world could have existed from eternity and also have been created was surprisingly popular during the Middle Ages and Renaissance, and it may have encouraged discussion of the possibility that the world may have existed in an eternal past.

From Bonaventure's arguments, we see how questions about the possible eternity of the world generated profound arguments about the possible existence of actual and potential infinites. Richard of Middleton, writing his commentary on the *Sentences* somewhat after Bonaventure, used similar arguments, relying ultimately on the basic idea that because God cannot create an actual infinite, it follows that the world cannot have existed from all eternity.[109] Richard argues:

> If it is possible that the world was created from all eternity, God could have realized an actual infinity, either in number or in magnitude. He could have similarly created men from all eternity; from all eternity these men would have engendered other men, and their successors would have done the same up to today. Since their souls are incorruptible, there would actually exist an infinite multitude of rational souls.
>
> Similarly, God could have moved the heaven continually until today, and for each of these revolutions, He could have created a stone; He could have amassed the stones together. That done, there would be an infinite volume existing actually. But in the first book we proved that God cannot produce an actual

106. See Cyril Vollert, Lottie H. Kendzierski, and Paul M. Byrne, trans., *St. Thomas Aquinas, Siger of Brabant, St. Bonaventure, On the Eternity of the World ("De Aeternitate Mundi")*, translated from the Latin with an Introduction (Milwaukee, WI: Marquette University Press, 1964), 20.
107. Vollert, trans., *Thomas Aquinas, On the Eternity of the World*, 23. Here I follow Grant, *Planets, Stars, & Orbs*, 71–72.
108. Vollert, trans., *Aquinas, On the Eternity of the World*, 20.
109. See Duhem, *Medieval Cosmology*, 80.

infinite multitude or magnitude. God therefore cannot have created the world from all eternity.[110]

Attitudes toward the actual infinite changed dramatically in the fourteenth century. The absurdity that Bonaventure saw in the inequality of two possible coexisting infinites – for example, the revolutions of the moon as compared to revolutions of the Sun – if the world had existed from eternity was rejected by a number of fourteenth-century theologians. Robert Holkot replies to this claim by denying that the infinites are unequal. He first takes cognizance of the charge made by the supporters of Bonaventure's position, who would argue:

It is repugnant for infinity to be surpassed; and if the world had existed from all eternity, there would be an infinite multitude surpassing another infinite multitude. In fact, there would be a greater number of fingers than men, and a greater number of revolutions of the moon than of the sun.[111]

In his reply Holkot declares:

I deny that the infinite cannot be surpassed without contradiction, because one infinite can [indeed] exceed another.... As for the proposition formulated in the proof, that there would be a greater [number of fingers than men and a greater] number of revolutions of the moon than of the sun, one can reply to it by denying it. With a thousand men, there is a greater number *(plures)* of fingers than men; but with an infinity of men, there is no greater number *(plures)* of fingers than men, because there is an infinity of men and an infinity of fingers.[112]

Thus did Holkot and others counter the claim of those who followed St. Bonaventure and argue that the eternity of the world was absurd because one consequence of it is the production of unequal infinites. Holkot insists that the infinites that would derive from an eternal world are really equal, not unequal. Hence there is no absurdity.

110. Richard of Middleton, *Sentences,* bk. 2, dist. 2, art. 3, qu. 4, vol. 2, 17, col. 1. Translated in Duhem *Medieval Cosmology,* 80. Richard had argued in the first book that God could not create an actually infinite dimension or multitude. This has been mentioned earlier in this chapter.

111. Robert Holkot, *In quattuor libros Sententiarum quaestiones* (Lyon, 1518), bk. 2, qu. 2, fol. liiv, col. 2. Each book of Holkot's *Sentences* is divided into questions, but not distinctions. In the edition of 1518, the folios are unnumbered. I have therefore used the signatures as page indicators. I cite Roger Ariew's translation in Duhem, *Medieval Cosmology,* 102.

112. Holkot, *Sentences,* bk. 2, qu. 2, fol. liiir, col. 2. In this paragraph, I have translated a few words that Duhem omitted and made a few other changes.

The most radical of medieval infinitists was Gregory of Rimini (d. 1358), who, in his commentary on the first book of the *Sentences* in 1344, asked "whether God, by His infinite power, could produce an actually infinite effect."[113] In the first conclusion of the first article, Gregory declares that "God can make any actually infinite multitude."[114] As an example, he shows that God could create an actually infinite number of angels in an hour. To illustrate this, Gregory resorts to the use of proportional parts of an hour, a concept that was frequently employed by scholastic authors.

The use of proportional parts depended on the universally accepted assumption that a continuum, such as time, was infinitely divisible. Any number of different proportional parts of an hour could be taken, as, for example, a one-half proportional part, or a one-third proportional part, and so on. A popular choice was the one-half proportional part, where one would take successive halves of an hour, which we can represent as $(1/2)^n$, where $n = 1, 2, 3, 4 \ldots$ This yields the sequence of fractions: $1/2 + 1/4 + 1/8 + 1/16 + 1/32 \ldots$ where, as is evident, each successive fraction is half the value of its predecessor. Thus some activity is performed in the first half hour (1/2), and then repeated in half of the remaining half hour, or in 1/4 of the whole hour. To perform the first two actions, 3/4 of an hour have passed (that is, $1/2 + 1/4$). Half of the remaining part of the hour is 1/8 of an hour, in which time the same activity will be repeated a third time. Thus, we have now passed through 7/8 of an hour – that is, $1/2 + 1/4 + 1/8 = 7/8$. Because there is an infinite number of proportional parts in an hour, the process is capable of being repeated infinitely, causing the original hour to diminish continually toward zero, while the sum of the successive temporal parts approach closer and closer to one hour.

Scholastics usually chose not to specify whether the proportional parts are successive halves or thirds, or fifths, and so on. Rather, they spoke in general terms of the first proportional part of an hour, the second proportional part of an hour, and so on; or they simply mention the first proportional part and then refer generally to all the successive proportional parts, as Gregory of Rimini does in his example illustrating how God could make an actually infinite multitude of angels. God could achieve this, Gregory explains, if, in the first proportional part of an hour, He creates an angel

113. Gregory of Rimini, *Lectura super primum et secundum Sententiarum*, vol. 3, bk. 1, dist. 42–44, qu. 4, 438–481.

114. "Quantum ad primum articulum pono tres conclusiones. Prima est quod deus potest facere aliquam multitudinem actu infinitam." Gregory of Rimini, *Sentences*, vol. 3, 441.

and preserves it, and then creates, and preserves, an angel in every successive proportional part of an hour. At the end of the hour, God would have created an actually infinite number of angels, because there are an infinite number of proportional parts in an hour.[115]

In the second conclusion, Gregory shows that it is possible for God to make "any infinite magnitude or infinite body."[116] As the last of three examples, Gregory once again uses the proportional parts of an hour to derive his actual infinite. Starting with a cubic magnitude, God adds a cube to any side of the initial cube in the first proportional part of an hour; and then adds another cube to another side of the cube; and continues doing this for one hour. In the instant that terminates the hour, God will have made a magnitude that is infinite in every direction.[117]

And in a third conclusion, Gregory shows that "God can make any form intensively infinite in its species, [as], for example, charity, or any whatever other intensible and remissible [form or quality]."[118] To show this, Gregory again divides a temporal continuum into an infinite number of proportional parts. This time, God continually increases charity. Thus, in the first half of the hour, God causes a certain degree of charity. He creates an equal degree of charity in the first half of the remaining half hour; and again, produces another equal degree of charity in the remaining half of the hour, and so on. At the end of the hour, God would have created an infinitely intense degree of charity comprised of an infinity of equal parts of charity.[119]

Thus, God can make three different kinds of actual infinites: infinite multitude, infinite magnitude, and an infinitely intense quality. In the course of discussing terms such as "part," "whole," "greater than," and "less than," and determining that they were also applicable to infinites in a special sense, Gregory arrived at a momentous idea about the relationship between infinites, an idea that lies at the heart of the modern theory of infinite sets. He argues that one infinite can be part of another infinite, but that the infinite that is a part is nevertheless equal to the infinite of which it is a part. Gregory concedes that "some infinite is less than some [other]

115. Gregory of Rimini, ibid., vol. 3, bk. 1, dist. 42–44, 443.

116. "Secunda conclusio est quod possibile est esse aliquam magnitudinem infinitam seu corpus aliquod infinitum." Ibid., vol. 3, 443.

117. Gregory of Rimini, *Sentences,* vol. 3, bk. 1, dist. 42–44, 445.

118. "Tertia conclusio est quod deus potest facere aliquam formam, verbi gratia caritatem, aut aliam quamlibet intensibilem et remissibilem, infinitam intensive in sua specie." Ibid.

119. Gregory of Rimini, *Sentences,* vol. 3, bk. 1, dist. 42–44, 446–447.

1	2	3	4	5	6	.	.	.
↕	↕	↕	↕	↕	↕			
2	4	6	8	10	12	.	.	.

Figure 2

infinite, because the infinite which is the part does not contain all the things which the infinite that is the whole contains."[120] Although Gregory provides no example, the relationship of the natural numbers to the subset of even numbers exemplifies the conditions he wants to describe. Since one can set even numbers in one-to-one correspondence with the infinite set of natural numbers, it follows that there are as many even numbers as natural numbers (see Figure 2).[121] Therefore, despite the fact that the infinite subset of even numbers is a part of the infinite set of natural numbers, the two infinite sets are equal, or, to use modern terminology, they have the same cardinality. Gregory had discovered the counterintuitive idea that in the domain of the infinite, a part can equal the whole.

The idea that one infinite can be part of another infinite is embedded in Gregory's assertion that "some infinite is less than some [other] infinite, because the infinite which is the part does not contain all the things which the infinite that is the whole contains." We may exemplify Gregory's meaning by invoking the two infinite sets in Figure 2, namely, the set of even numbers and the set of natural numbers. The set of even numbers – or "the infinite which is the part" – includes fewer components than does the set of natural numbers, which also includes the infinite set of odd numbers. It is in this sense that Gregory intends his assertion that "the infinite which is the part does not contain all the things which the infinite that is the whole contains." But Gregory insists that the infinite that is the part and the infinite that is the whole are nevertheless equal infinites. They are equal because one thing exceeds another only if it contains more things or units;

120. "Item sumptis toto et parte in eodem modo, ut iam sumebantur, et sumendo minus et maius secundo modo supra posito concedo quod aliquod infinitum est minus aliquo infinito, quoniam infinitum, quod est sic pars, non continet omnia, quae continet infinitum, quod est totum eius ..." Ibid., 459.

121. This is a commonly used figure to represent a one-to-one correspondence between two equal infinites, where one is the subset of the other. It appears on p. 1595 of the article "Infinity" by Hans Hahn in James R. Newman, ed., *The World of Mathematics,* 4 vols. (New York: Simon and Schuster, 1956), vol. 3, 1593–1611.

but this does not apply to infinites, because Gregory believed that all infinites are equal.[122]

In his *Two New Sciences* of 1638, Galileo offered an example of infinites that are equal. His spokesman, Salviati, shows that although it appears there are many more natural numbers than square numbers, there are, in fact, just as many of each. Thus Salviati declares that

> [s]o far as I see, we can only infer that the totality of all numbers is infinite, that the number of squares is infinite, and that the number of their roots is infinite; neither is the number of squares less than the totality of all numbers, nor the latter greater than the former.

Here was an example that Gregory could have put to good use. But Salviati is not done. He goes on to say, finally, that "the attributes 'equal,' 'greater,' and 'less,' are not applicable to infinite, but only to finite, quantities."[123] It is here that Galileo, to his disadvantage, parts company with Gregory of Rimini, who sought to determine how one might apply those terms, along with the terms "part" and "whole," to infinites.

Indeed, Pierre Duhem observes that there is "a clear affinity" between the thoughts of Gregory of Rimini and those of Georg Cantor (1845–1918) in the first few pages of the latter's *Theory of Transfinite Numbers,*

> even though five-and-a-half centuries separate the times during which they were writing. Gregory of Rimini certainly glimpsed the possibility of the system Cantor constructed; he deemed that there was room for a mathematics of infinite magnitudes and multitudes next to the mathematics of finite numbers and magnitudes. He thought that the two doctrines were two divisions of a more general science.[124]

122. "Et, si inferatur 'infinitum est minus infinito, ergo exceditur ab illo,' nego consequentiam sumpto antecedente in sensu concesso, nam nihil proprie dicitur excedi ab alio, nisi quod non continet tanti tot quot aliud; quod de nullo infinito est verum." Gregory of Rimini, *Sentences,* vol. 3, bk. 1, dist. 42–44, qu. 4, 459.

123. For the two passages cited here, see *Dialogues Concerning Two New Sciences by Galileo Galilei,* translated by Henry Crew and Alfonso de Salvio, with an Introduction by Antonio Favaro (New York: Dover Publications, n.d.; copyright by the Macmillan Company, 1914), First Day, 32–33. For relevant references to Galileo and Isaac Newton, see John E. Murdoch, "*Mathesis in Philosophiam Scholasticam Introducta:* The Rise and Development of the Application of Mathematics in Fourteenth Century Philosophy and Theology," in *Arts Libéraux et Philosophie au Moyen Age* (Montreal: Institut d'Études Médiévales, 1969), 222.

124. See Duhem, *Medieval Cosmology,* 112. Immediately following his assessment of Gregory's concepts, Duhem translates the relevant passages illustrating Gregory's ideas (see ibid., 112–113).

Gregory of Rimini anticipated the modern understanding of the relationship between an infinite set and its infinite subset, namely, where an infinite is viewed as equal to one or more of its infinite parts; or, conversely, where an infinite part of an infinite is equal to the whole infinite of which it is a part. Thus did Gregory overcome "the age-old intuition that no part of a thing can be as large as the thing," an idea that Bertrand Russell, the great mathematician and philosopher, regarded as "a model of conceptual analysis."[125]

Gregory of Rimini was ahead of his time. According to John Murdoch:

> Gregory's resolution of the paradox so frequently generated by the assumption of the eternity of the world is by far the most successful and impressive I have discovered in the Middle Ages. Unfortunately, however, it appears hardly ever to have received the understanding and appreciation it clearly deserved. Since the "equality" of an infinite whole with one or more of its parts is one of the most challenging and, as we now realize, most crucial aspects of the infinite, the failure to absorb and refine Gregory's contentions stopped other medieval thinkers short of the hitherto unprecedented comprehension of the mathematics of infinity which easily could have been theirs.[126]

But if scholastic theologians and natural philosophers failed to capitalize on Gregory of Rimini's extraordinary insight:

> Still their deliberations over this particular paradox as involved in problems like the eternity of the world and the continuum again and again led them, as it seldom did their ancient predecessors, to the heart of the mathematics of the infinite. The fact that they seemed to realize that it was the heart, and that in treating it they fared as well as, and at times better than, anyone else before, it appears, the nineteenth century, is unquestionably to their credit.[127]

The medieval fascination with the infinite led another scholastic to anticipate yet another key concept in modern set theory, namely, the idea that one infinite can be greater than another. Although Gregory of Rimini denied this, Henry of Harclay, in the fourteenth century, "firmly believed that infinites can be, and often are, unequal."[128] The idea that one infinite can be greater than

125. I draw the last two brief quotations from the article "infinity" by J. A. B. (José A. Bernadete) in Robert Audi, ed., *The Cambridge Dictionary of Philosophy* (Cambridge: Cambridge University Press, 1995), 372.

126. Murdoch, *"Mathesis in Philosophiam Scholasticam Introducta,"* 224.

127. Murdoch, ibid.

128. Murdoch, ibid., 223.

another was not really developed by Henry of Harclay, but was made a part of set theory in the second half of the nineteenth century by Georg Cantor, who revealed the existence of nondenumerable infinite sets, such as the infinite set of real numbers, which includes all natural numbers, all fractions, and all irrational numbers. That is, Cantor showed that the infinite set of real numbers cannot be paired in one-to-one correspondence with the set of natural numbers and is, therefore, a nondenumerable infinite set that is greater than the infinite set of natural numbers.[129] Thus, in modern set theory, one infinite can be greater than another; and, generally, some infinite parts are equal to their infinite wholes, while others are not.

In dealing with problems and paradoxes of the infinite, scholastic theologians were heavily involved in logico-mathematical techniques, which should immediately strike us as odd. Why did theologians writing theological treatises resort to such techniques? One can scarcely imagine a more rationalistic approach and emphasis. We must assume that they regarded such analyses as, in some sense, furnishing valuable knowledge and insight about God and what He can or cannot do, or about many other aspects of theology. In fact, what God could, or could not, do seems to have fascinated them.

Robert Holkot used the concept of the infinite divisibility of a continuum and the doctrine of first and last instants to determine limits in imaginary theological problems. For example, he imagined a situation in which a man is alternately meritorious and sinful during the final hour of his life. Thus, he is meritorious during the first proportional part of his last hour and sinful in the second proportional part; he is again meritorious in the third proportional part, and again sinful in the fourth proportional part, and so on through the infinite series of decreasing proportional parts up to the last instant, when death occurs. Because the instant of death does not form part of the infinite series of decreasing proportional parts of the man's final hour, it follows that there is no last instant of his life and, therefore, no last instant in which he could be either meritorious or sinful. Since the man was neither meritorious nor sinful in his last instant of life, God cannot judge him.[130] By conjuring up this example, Holkot shows that God could be in ignorance about a person's state of grace or sin in the last moment of

129. For a lucid explanation of these concepts, see the article "Infinity" by Hans Hahn in James R. Newman, ed., *The World of Mathematics,* vol. 3, 1593–1611, especially 1593–1598.

130. Robert Holkot, *In quattuor libros Sententiarum quaestiones* (Lyon, 1518), bk. 1, qu. 3, fol. Biiiiv, col. 2. The Latin text is reproduced by Murdoch, "From Social into Intellectual Factors," 327, n. 101. For a summary of the argument, see Grant, *The Foundations of Modern Science in the Middle Ages,* 154.

life, and he thus, indirectly, sets limits on God's ability to make just rewards and punishments. Holkot follows this example with eight others. In all of them he uses the concept of first and last instants applied to the infinite divisibility of a continuum, as in the article just described, and also uses arguments involving maxima and minima (see Chapter 4).[131]

Medieval theological discussions about the infinite – the actual infinitely large and the infinitely small that followed upon the infinite divisibility of a continuum – were governed by the most fundamental law of logic: the principle of noncontradiction, which asserts that a statement and its negation cannot both be true at the same time. When they applied it to medieval theological discussions about the infinite, scholastic theologians proceeded on the assumption that if a contradiction could be derived from the supposition that God could make this or that actual infinite, it was always concluded that God could not make that actual infinite. Those theologians who were convinced that no logical contradiction was entailed by the assumption that God could make an actual infinite always concluded that God could make an actual infinite. The debate was thus ultimately ruled by logic. To ascertain whether or not one or more contradictions might follow from the assumption of the existence of a particular kind of actual infinite, theologians often employed one or more of the various logico-mathematical techniques that were described earlier.

The scholastic theologian who used the logico-mathematical approach most extensively was Jean Mirecourt, who lived in the fourteenth century. So frequently did Mirecourt introduce logico-mathematical techniques that he ran afoul of university authorities enforcing the statutes of 1366, which forbade the unnecessary use of logic and mathematics in theology.[132] "One might wonder," Murdoch muses,

> if Mirecourt is not rather uncharacteristic in the apparently quite thorough penetration the new analytical languages had made into his theological work. Not so. There are numerous others who exhibit the same pattern (save, of course, for the condemnation). All of the various theologians it has been, and will be, our occasion to cite are cases in point. For they apply the languages not merely in those contexts that are here referred to, but in all manner of other corners of their *Sentence Commentaries*. The proper conclusion is, I believe, that utilization of our languages was in no sense remarkable, but rather quite common.[133]

131. See Murdoch, "From Social into Intellectual Factors," 327, n. 101.

132. For more on the University of Paris statutes of 1366, see the section "The Reaction to Analytic Theology" in this chapter.

133. Murdoch, "From Social into Intellectual Factors," 296–297. The "new languages" Murdoch alludes to were mentioned earlier in this chapter.

By virtue of the logico-mathematical techniques theologians regularly used, medieval theology became highly analytical and quantified. The techniques that have been mentioned thus far were applied to themes and topics throughout the four books of commentaries on the *Sentences*. In the questions they posed, theologians did not simply apply logico-mathematical techniques but also, as we shall see, regularly introduced arguments and evidence from natural philosophy. But another indication of the systematically rational approach found in *Sentence* commentaries are the questions themselves. Many are paradoxical and devised "according to the imagination" *(secundum imaginationem)*. They are usually about whether something or other that seems impossible on the face of it can be done by a supernatural act. To what extent such questions were regarded as serious issues about the faith is unclear. Theologians may have considered them because such provocative questions offered them a chance to use their analytical skills, or to display their command of natural philosophy on important issues. Hugolin of Orvieto illustrates this tendency. Here is a sampling of some of his questions, the first four of which seem to be concerned with future contingents, a problem that was discussed earlier (in Chapter 4):

Hugolin of Orvieto:[134]

VOL. 2: BOOK 1, DISTINCTION 40:

Question 3 (p. 334): "Whether it is possible for someone to make some future event not to be."

Question 3, art. 1 (p. 335): "Whether it is necessary that the future be the future."

Question 3, art. 2 (p. 338): "Whether the foreknowledge of God is the cause of the future being the future."

Question 3, art. 3 (p. 341): "Whether God could make the future not to be."

Question 6 (p. 374): "Whether the number of the elect could be increased."

VOL. 3: BOOK 2, DISTINCTION 2:

Unique question, art. 3 (pp. 97–99): "Whether God could make a creature exist for only an instant."

BOOK 2, DISTINCTION 4:

Unique question (pp. 115–123): "Whether angels could have foreknowledge of their fall in the first instant of their existence."

134. For the title of Hugolin's work, see the bibliography.

VOL. 4: BOOK 4, DISTINCTIONS 8–14:

Question 2, art. 3 (pp. 217–222): "Whether the body of Christ under the sacrament is extended."

Question 3, art. 1 (pp. 225–227): "Whether every accident of bread remains without a subject and what is the mode of their existence."

Question 3, art. 2 (pp. 227–228): "Whether there are equally active accidents lacking a subject."

Question 3, art. 3: (pp. 228–230): "Whether a natural agent could produce an accident lacking a subject."

The questions cited here from Hugolin of Orvieto are typical of fourteenth-century *Sentence* commentaries. In resolving these questions, theologians not only employed analytical logico-mathematical techniques, but also used Aristotle's ideas and concepts as these had been transformed and modified by natural philosophers. The last four questions cited above from Hugolin's *Sentence Commentary* about an extended Christ, and about accidents and their subjects, rely on Aristotle's discussion of quantity and the way qualities inhere in their subjects. They use Aristotelian natural philosophy to show that what is *naturally* impossible in the physical world is possible by *supernatural* means.

NATURAL PHILOSOPHY IN THEOLOGY

It is now time to describe and characterize the more traditional kinds of natural philosophy that theologians imported into their *Sentence* commentaries. In commentaries on the second book, which treats extensively of the creation and angels, natural philosophy played a large role. In treating these themes, theologians found themselves heavily involved in cosmological and physical topics. If we examine the questions that theologians posed in the second book of their *Sentence* commentaries, we can divide them into at least two major categories. One involves questions that were essentially theological, but in which theologians used natural philosophy to a greater or lesser extent; the other concerns questions that were essentially in the domain of natural philosophy, although they were discussed by both theologians and natural philosophers.[135]

135. In distinguishing these categories and identifying the questions, I rely on my "Catalog of Questions on Medieval Cosmology, 1200–1687," in Grant, *Planets, Stars, & Orbs*, Appendix I, 681–741.

How Theologians Used Natural Philosophy in Theological Questions

Cosmology in the creation account is embedded in the first four days, with days one, two, and four being the most significant. The successive days of creation were as follows:

> On the first day, heaven *(caelum),* earth *(terra),* and light *(lux);* on the second, the firmament *(firmamentum)* that divides the waters above from those below, and which God called 'heaven' *(caelum);* on the third day God turned his attention to the earth, where he gathered the seas together in one place, exposing the dry land on which he then placed plants and trees capable of reproducing themselves; and finally on the fourth day, he made the physical light of the heavens by creating all the celestial bodies, assigning the Sun to provide the light of day and the Moon to provide the light of night.[136]

Questions on all of these aspects of creation appear in *Sentence* commentaries because a significant part of the second book of Peter Lombard's *Sentences* was devoted to these themes. Typical of the questions that appear are these:

"Whether light was created on the first day."[137]
"Whether the light made on the first day is corporeal or spiritual."[138]
"How that light [made on the first day] made day and night."[139]
"Whether waters are above the heavens."[140]
"Whether the crystalline heaven is moved."[141]
"Whether the firmament has the nature of fire."[142]
"Whether the firmament has a spherical shape."[143]
"Whether the firmament has the nature of inferior bodies."[144]

136. Grant, *Planets, Stars, & Orbs,* 91–92.
137. In Peter Aureoli, *Commentariorum in primum [-quartum] librum Sententiarum, pars prima [-quarta],* 2 vols. (Rome: Aloysius Zannetti, 1596–1605), vol. 2, bk. 2, dist. 3, qu. 1, art. 1, 180, col. 1–185, col. 2.
138. Bonaventure, *Sentences,* bk. 2, dist. 13, art. 1, qu. 1, in *Opera omnia,* vol. 2, 311–313.
139. Bonaventure, ibid., bk. 2, dist. 13, art. 1, qu. 2, in ibid., vol. 2, 314–316.
140. Thomas Aquinas, *Sentences,* bk. 2, dist. 14, qu. 1, art. 1, in *Scriptum super libros Sententiarum Magistri Petri Lombardi Episcopi Parisiensis,* new ed., 4 vols. (Paris: P. Lethielleux, 1929–1947), vol. 2, 346–349. The same question is discussed by Durandus de Sancto Porciano, *Sentences,* bk. 2, dist. 14, qu. 1, fols. 155v, col. 2–156r, col. 2.
141. Richard of Middleton, *Sentences,* bk. 2, dist. 14, art. 1, qu. 2, vol. 2, 168, col. 2–169, col. 1.
142. Ibid., qu. 3, vol. 2, 169, col. 1–170, col. 2.
143. Ibid., qu. 4, vol. 2, 170, col. 2–171, col. 1.
144. "Utrum firmamentum sit de natura inferiorum corporum." Thomas Aquinas, *Scriptum super libros Sententiarum Magistri Petri Lombardi,* bk. 2, dist. 14, art. 2, vol. 2, 349–351.

"Whether the empyrean heaven is a body."[145]
"Whether the empyrean heaven is moved."[146]
"Whether the empyrean heaven has stars."[147]
"Whether the empyrean heaven is luminous."[148]
"Whether the empyrean heaven exerts an influence on inferior things [that is, on things in the terrestrial region]."[149]

Questions such as these were basically about the biblical creation account, especially about the "theological heavens," that is, about the heavens, or orbs, that were believed to correspond to Biblical references in Genesis. Responses to questions directly relevant to Biblical texts were of crucial importance and had to be answered by trained, professional theologians. It would be quite unusual to see such questions turn up in commentaries on Aristotle's natural books. In fact, the way arts masters coped with the problem of the empyrean heaven is instructive.

The empyrean heaven was a theological construction derived from faith, not by rational argument or empirical appeals. By the thirteenth century, it was conceived as an invisible, immobile orb that enclosed the world of movable orbs. It was often regarded as the "first and highest heaven, the place of angels, the region and dwelling place of blessed men," or as "the dwelling place of God and the elect."[150] Because it was a theological construct, it was not proper for arts masters, who were natural philosophers without theological training, to discuss the empyrean heaven. And yet it was appropriate for natural philosophers to inquire whether there was anything beyond the last moving celestial sphere that Aristotle and Ptolemy, the great Greek astronomer of the second century A.D., had described.

Three fourteenth-century natural philosophers posed such a question without mentioning the empyrean heaven. In his *Questions on De caelo* (bk. 2, qu. 6), John Buridan asked "whether beyond the heavens that are moved, there should be assumed a heaven that is resting or unmoved"; Albert of

145. Alexander of Hales, *Summa theologica,* inq. 3, tr. 2, qu. 2, tit. 1, memb. 1, ch. 1, art. 2, in *Summa Theologica,* Tomus II: Prima Pars secundi libri. (Florence: Collegium S. Bonaventurae, 1928), 328–329.

146. Alexander of Hales, ibid., inq. 3, tr. 2, qu. 2, tit. 1, memb. 1, ch. 1, art. 3, 329–330.

147. Richard of Middleton, *Sentences,* bk. 2, dist. 2, art. 3, vol. 2, 44, cols. 1–2.

148. Thomas of Strasbourg, *Commentaria in IIII libros Sententiarum* (Venice: ex officina Stellae, Iordani Ziletti, 1564; facsimile, Ridgewood, NJ: Gregg, 1965), bk. 2, dist. 2, qu. 2, fols. 133v, col. 1–135v, col. 1.

149. Richard of Middleton, *Sentences,* bk. 2, dist. 2, art. 3, qu. 3, vol. 2, 44, col. 2–45, col. 2.

150. For the sources of these quotations, see Grant, *Planets, Stars, & Orbs,* 371 and notes 3 and 4.

Saxony (in his *Questions on De caelo,* bk. 2, qu. 8) inquired "whether every heaven is mobile, or whether we must assume some heaven that is at rest"; and Themon Judaeus (in his *Treatise on the Sphere,* qu. 8) asked "whether something should be assumed [to exist] beyond the ninth sphere."[151] Buridan and Albert of Saxony had the empyrean heaven in mind because they ask whether there is a heaven that is at rest, a clear reference to the empyrean, and very likely Themon also had it in mind. And yet these three natural philosophers, who had no theological degrees or credentials, may have deliberately omitted the term "empyrean" in order to avoid possible theological entanglements. In most instances, however, questions concerning the creation account were avoided by natural philosophers and left to the theologians.

Angels and Natural Philosophy

Questions about angels were also regarded as the exclusive province of theologians. It was customary to raise questions about angels in *Sentence* commentaries. Most of the questions are about actions and states of angels that are relevant to their moral behavior. They are the kinds of questions we might expect theologians to ask about angels. For example, Thomas Aquinas asked "whether created angels are blessed";[152] "whether angels were created in grace";[153] "whether in angels there could be sin";[154] "whether the sin of the first angel was the occasion of the sins of the other angels";[155] "whether good angels could sin";[156] and "whether angels guard men."[157] Thomas included numerous other questions on angels in his *Sentences,* approaching them all in his usual systematic and rational manner.

Years later, beginning in 1265, Thomas began to compose his *Summa theologiae.* Once again he devoted a large number of questions to the subject of angels, questions 50 to 64 of the first part (there are three parts). These 15 questions are divided into 72 articles, each of which is a question. Thus, there are 72 questions on angels. Many of the questions Thomas treated in his *Sentences* turn up again in the *Summa.* As in his *Sentences,* many of the questions about angels in the *Summa* were

151. For the precise references, see Grant, *Planets, Stars, & Orbs,* 699–700, qu. 106 and n. 97.
152. See Thomas Aquinas, *Sentences,* bk. 2, dist. 4, quaestio unica, art. 1, vol. 2, 132–134.
153. Ibid., art. 3, vol. 2, 136–139.
154. Ibid., bk. 2, dist. 5, qu. 1, art. 1, vol. 2, 143–145.
155. Ibid., bk. 2, dist. 6, quaestio unica, art. 2, vol. 2, 162–164.
156. Ibid., bk. 2, dist. 7, qu. 1, art. 1, vol. 2, 179–182.
157. Ibid., bk. 2, dist. 11, qu. 1, art. 1, vol. 2, 270–272.

prompted by Scripture and Christian tradition. But there is an interesting difference in the questions of the *Summa* that may reflect a changing approach to theology that was taking hold among theologians. Six of the articles, distributed over questions 52 and 53, have little or no connection with theology but seem to have been derived from natural philosophy. They are about the places and motions of angels and are related to the kinds of questions Aristotle asked about the place and motions of bodies in the fourth book of his *Physics*. We can see this at a glance by simply listing the two questions and six articles:

QUESTION 52: ON THE RELATIONSHIP OF ANGELS TO PLACES[158]
Article 1: does an angel exist in a place?
Article 2: can an angel be in several places at once?
Article 3: can several angels be in the same place at once?
QUESTION 53: ON THE LOCAL MOTION OF ANGELS[159]
Article 1: can an angel move from place to place?
Article 2: does an angel, moving locally, pass through an intermediate place?
Article 3: whether an angel's motion occurs in time or in an instant.[160]

During the last decade of the thirteenth century, most of these questions were also asked by Richard of Middleton in the second book of his *Sentences*.[161] Richard probed further with regard to the location of angels when he asked "whether an angel is in space, or in some indivisible part of this space, as in a point."

In the fourteenth century these, and similar, questions were frequently, and even routinely, asked. As intermediate spiritual substances between God and corporeal entities, angels posed special problems for

158. The translations are from St. Thomas Aquinas, *Summa Theologiae*, vol. 9: *Angels;* translated by Kenelm Foster O. P. (Blackfrairs, 1968), prima pars, qu. 52, 44–45. The Latin text of this question is "De comparatione angelorum ad loca." Foster translates this as "Angels and Position in Space," which is not an appropriate rendering, since place and space are not the same in medieval natural philosophy. I therefore offer my own translation.

159. The Latin text is "De motu locali angelorum" (*Summa Theologiae*, vol. 9, 54–55), which Foster translates as "the movement of angels in space." Since Thomas is not discussing space, but place, I present my own translation.

160. The Latin text of Article 3 (ibid., 54) is: "utrum motus angeli sit in tempore vel in instanti." I have replaced Foster's translation, which has "is such movement instantaneous?"

161. See Richard of Middleton, *Sentences*, bk. 1, dist. 37, art. 2, qu. 3, vol. 1, 328, col. 2–329, col. 1; qu. 4, 329, col. 1–330, col. 2; art. 3, qu. 2, 332, cols. 1–2; art. 3, qu. 3, 332 col. 2–334, col. 1.

theologians. Questions about the relationship of angels to God were almost exclusively theological problems, for which natural philosophy was of little help. But, as we just saw, natural philosophy played a significant role with regard to questions about the location of angels: how they filled a space or place, how they moved from one place to another, and other questions about their role and activities. In most instances, when angels were capable of executing an activity that was also done by physical bodies, angels would do it differently. Theologians were expected to explain the differences. To do this, it was necessary to compare each angelic activity with the corresponding manner in which physical bodies or humans performed the same activity. That is how the questions just cited were treated. Descriptions from natural philosophy about the behavior of bodies were used to show that either angels did it the same way, or, as was more likely, did it differently.

Indeed, this was obvious from the way theologians distinguished between the modes in which bodies and spirits, or souls, occupied their respective places. Peter Lombard himself had distinguished between the two. One term, the Latin term *ubi,* or where, was used to signify the location of a body or spirit. For scholastic theologians, a body is a dimension with length, depth, and width, which is coextensive with the place it occupies. It became customary to designate the location of a body by the expression *ubi circumscriptivum.* By contrast, a spiritual being, such as an angel, occupies a place only in the sense that it must be somewhere, since it cannot be everywhere simultaneously, as could God. Thus, a spiritual substance is delimited by the terminus or boundary of its place, or *ubi,* but need not be coextensive with it. Moreover, since an angel is indivisible, it was assumed that the whole of it could be in every part, however small, of the place that delimits it. This kind of location was called *ubi definitivum.*[162]

With this in mind, Thomas Aquinas asked whether an angel can move from place to place. Because an angel does not occupy a place in the same way as a body does, Thomas argues that a body will move from place to place in a different sense than will an angel. He explains that "a body is in a place as contained by and commensurate with it: hence its movement from one place to another must be commensurate to and

162. On these terms and their use, see Grant, "The Concept of *Ubi* in Medieval and Renaissance Discussions of Place," in Nancy G. Siraisi and Luke Demaitre, eds, *Science, Medicine, and the University 1200–1500, Essays in Honor of Pearl Kibre,* Pt. 1, *Manuscripta* 20 (1976), 71–80; and Grant, *Much Ado About Nothing,* 129–130.

conditioned by place; in the sense that its movement through a space continuum must itself be continuous and must occur successively according to the succession of parts of the continuum."[163] An angel moves very differently. In his explanation, Thomas seems to alter radically the definition of an angel's place, arguing that an angel is not in a place, that is, surrounded by it, but actually contains its place. For, as Aquinas explains, "an angel is not in place as commensurate with it, nor as contained by it, but rather as containing it."

Again, a body always moves with continuity, passing successively over all the parts that intervene between two distant places. An angel may also traverse a divisible distance between two places in the same way: continuously and successively. But, Thomas insists, "angelic movement may also take place as an instantaneous transference of power from the whole of one place to the whole of another; and in this case the angel's movement will be discontinuous."[164]

Whatever this may signify, it is clear that Thomas did not mean to say that an angel could move from place to place instantaneously. In the third article of this question, where he asks whether an angel moves instantaneously, Thomas argues that

> whenever an angel moves, there must have been a definite final 'now' in which he was at the place from which he has now moved to a new place. But where there are more than one distinct and successive 'nows,' there, necessarily, is time; since time is only the numbering of before and after in a sequence of motion. So we have to conclude that an angel's movement takes place in time: in continuous time if the motion is continuous; in discontinuous time if it is not. And we have seen that both kinds of motion are possible for an angel; and, as we read in the *Physics*,[165] the continuity of time depends on the continuity of movement.[166]

Thomas's rejection of instantaneous motion for angels was accepted by Richard of Middleton and most other theologians. One of the arguments Richard cites against instantaneous motion is that an instant is the smallest measure, so that if an angel moved through some medium in an instant, the divine power *(divina virtus)*, God, could not move that angel through the same medium in any time less than an instant. This is false because, as Richard expresses it, "however much a force is stronger, so much the more

163. Thomas Aquinas, *Summa theologiae*, qu. 53, art. 1, vol. 9, 57.
164. Ibid.
165. Aristotle, *Physics* 4.2. 219a.13
166. Thomas Aquinas, *Summa theologiae*, qu. 53, art. 3, vol. 9, 69.

ought it to be able to move the same mobile [or angel] through the same distance in a smaller measure."[167] That is, God, the strongest force of all, should be able to move an angel some distance in less time than an instant. But that is impossible, because there is no measure smaller than an instant.

Some theologians in the fourteenth century substantially repeated, with elaborations, the questions about the motion and place of angels, while others ignored them, or included only one or two of them. But they also added questions and applied new techniques. Hugolin of Orvieto asked "whether angels could have foreknowledge of their fall in the first instant of their existence."[168] and then, as the first article of this question, asked "whether they (angels) could in the first instant be freely unmeritorious and also be meritorious"[169] Hugolin thus posed a question about the first instant of an angel's activity, in contrast with Robert Holkot, who, as we saw earlier, sought to determine what would happen to a man in the last instant of his life. Thus was the doctrine of first and last instants brought into play.

Perhaps the most unusual – even extraordinary – discussion about angels was presented by Gregory of Rimini in book 2, distinction 2, of his *Commentary on the Sentences,* where Gregory asks "whether angels were created before time [began], or after time [began]."[170] At the beginning of the question, Gregory explains: "In this question, it is first necessary to see whether time is something created; and if so, what it is. Then we will see about what has been sought."[171] In a question that extends over pages 235 to 277, Gregory does not again mention angels until page 275, devoting pages 235 to 274 to a detailed and technical discussion of time in which he draws heavily on the philosophical works of Aristotle

167. "Contra: si angelus moveret se per medium in instanti, cum instans sit minima mensura, tunc divina virtus non posset movere angelum per idem medium in minori mensura, quod falsum videtur quia quanto virtus est fortior, tanto movet idem mobile per idem spatium in minori mensura." Richard of Middleton, *Sentences,* bk. 1, dist. 37, art. 3, qu. 3, 333, col. 1. The question asks "whether an angel could traverse a medium between two places in an instant." ("Utrum angelus possit pertransire in instanti medium inter duo loca.") Ibid., 332, col. 2.

168. Hugolin of Orvieto, *Commentarius in Quattuor Libros Sententiarum,* bk. 2, dist. 4, quaestio unica, vol. 3, 115–123.

169. Ibid., art. 1, vol. 3, 115–119.

170. "Utrum angeli fuerint creati ante tempus vel post." Gregory of Rimini, *Lectura super primum et secundum Sententiarum,* vol. 4, bk. 2, dist. 2, qu. 1, 235–277.

171. "In hac quaestione primo oportet videre, utrum tempus sit aliqua res creata; et si sic, quae res est. Deinde videbitur de quaesito." Gregory of Rimini, *Sentences,* bk. 2, dist. 2, qu. 1, vol. 4, 236.

and his commentator, Averroës. When he does finally turn to angels, Gregory closes with four conclusions, in which he argues, in the first conclusion, that "no time was created before angels" and, in the fourth conclusion, asserts that "however time is taken, angels were created before any time was time."

In the very next question (question 2), Gregory asks "whether an angel exists in a divisible or indivisible place."[172] As in the previous question, Gregory explains his intent:

> In this question two things must be seen. The first is whether a magnitude is composed of indivisibles, which is considered in the argument; and [the second is] whether in a magnitude there is something indivisible. To arrive at what is sought, it is necessary to understand each of these [two parts of this first part of the question]. The second thing that must be seen in this question is what is sought [namely, the question itself: "whether an angel exists in a divisible or indivisible place."][173]

To expound the two distinct parts of this question, Gregory divides it into two articles. The first is titled "Three conclusions," which he discusses and analyzes on pages 278 to 331 of the printed edition. He describes the three conclusions as follows:

> The first is that no magnitude is composed of indivisibles, from which it follows that any magnitude is composed of magnitudes. The second [conclusion is] that any magnitude is composed of an infinity of magnitudes. The third [conclusion is] that in no magnitude is there something indivisible that is intrinsic to it.[174]

The second article is titled "On the existence of an angel in a place,"[175] which occupies the rest of question 2, ranging from pages 331 to 339.

172. "Utrum angelus sit in loco indivisibili aut divisibili." Ibid., bk. 2, dist. 2, qu. 2, vol. 4, 277–339.

173. "In ista quaestione sunt duo videnda: Primum est, an magnitudo componatur ex indivisibilibus, quod tangitur in argumento; et utrum in magnitudine sit aliquid indivisibile. Utrumque enim horum scire satis est necessarium ad principale quaesitum. Secundum est quod quaeritur." Gregory of Rimini, ibid., bk. 2, dist. 2, qu. 2, vol. 4, 278.

174. "Quantum ad primum pono conclusiones tres: Prima est quod nulla magnitudo componitur ex indivisibilibus, ex qua sequitur quod quaelibet magnitudo componitur ex magnitudinibus. Secunda, quod quaelibet magnitudo componitur ex infinitis magnitudinibus. Tertia, quod in nulla magnitudine est aliquid indivisibile sibi intrinsecum." Ibid., bk. 2, dist. 2, qu. 2, art. 1, vol. 4, 278.

175. "De existentia angeli in loco." Ibid., bk. 2, dist. 2, qu. 2, art. 2, vol. 4, 331–339.

The first lengthy article has nothing to do with angels. Indeed, the word "angel" *(angelus)* occurs only once, on the very last page (p. 331) of the article. In place of angels, Gregory devotes himself wholly to mathematics, physics, and logic as these are applicable to the three conclusions. More specifically, he discusses the mathematics and physics of indivisibles, subjects that immediately involve him in the nature of instants and the mathematical continuum. In Gregory's 53-page discourse on the mathematical themes announced in the three conclusions, he cites Euclid's *Elements* a number of times and includes 14 elaborate geometrical diagrams. All of this is carried on within the usual scholastic format, where Gregory raises numerous doubts against his own conclusions and then answers all the doubts, while also disputing the conclusions of numerous thirteenth- and fourteenth-century scholastic authors.

Let us examine the structure of the first article more closely. At the outset, Gregory says that he will demonstrate the first conclusion of the first article "by mathematical and physical arguments."[176] He presents the mathematical arguments first (on pages 279–285), offering nine specific arguments in which he employs eight geometrical diagrams. At the end of this section, Gregory proclaims that "many other mathematical arguments could be made, but these suffice."[177] Immediately following, he declares that "[n]ext, I prove the same conclusion by physical arguments" and proceeds to cite Aristotle's discussion on indivisible magnitudes in the sixth book of his *Physics*. In demonstrating the second and third conclusions of the first article, Gregory follows the same pattern, using a combination of mathematical and physical arguments, the former presented first, then the latter. Again, in the second conclusion, he introduces three mathematical diagrams (pages 300–301).

It is only in the second article that Gregory discusses angels, inquiring whether an angel exists in a place. Gregory observes that "among theologians, doctors and saints," there is common agreement that "an angel is in a place, not dimensively [or dimensionally], or circumscriptively, according to the term used by moderns, but only definitively."[178] But is the actual substance of an angel in a place, or is it in a place only by its operation

176. "Quod ego ostendo rationibus et mathematicis et physicis." Ibid., vol. 4, 278.

177. "Multae aliae rationes mathematicae possent fieri, sed hae sufficiant." Ibid., vol. 4, 285.

178. "Secundus articulus, qui est de principali quaesito, ut patet, supponit angelum esse in loco. De quo communis concordia est apud theologos, doctores et sanctos. Dicunt enim omnes communiter quod angelus est in loco, non quidem dimensive, seu circumscriptive iuxta moderniorem usum vocabuli, sed tantummodo definitive." Ibid., bk. 2, dist. 2, qu. 2, art. 2, vol. 4, 331.

around that place, as some have held?[179] Gregory rejects this interpretation, citing the Condemnation of 1277, where in article 204, it was decreed an error to assume that a separated substance, like an angel, could not be in a place, or move from one place to another place, without operating there.[180] Gregory leaves no doubt of his position: "I say therefore that an angel is in place not only by [its] operation, but also by its substance," and that it can be in a place even if it does not operate in that place.[181]

Near the conclusion of the second article, Gregory returns to the basic question: Is an angel in a divisible or indivisible place? He explains: "With regard to that which has been primarily sought, I say that, properly speaking about place, namely a corporeal [place], an angel is not in an indivisible place, but it is in a divisible place. The reason for this is that there is no such indivisible in the nature of things, as was proved in the first article."[182]

In closing out his discussion of angels and their places, Gregory concludes with a brief third question, namely, "whether an angel could be in several places [simultaneously]."[183] Indeed, this question also serves as the first article, which is followed by a second article in which Gregory asks "whether several angels could be in the same place."[184]

179. The editor of Gregory's text indicates that Thomas Aquinas and Aegidius Romanus held this opinion. See Gregory of Rimini, *Sentences,* vol. 4, 332 and n. 7.

180. "Contra opinionem istam expresse est articulus ille Parisiensis 203: 'Quod substantiae separatae sunt in loco per operationem; et quod possunt moveri ab extremo in extremum, nec in medium, quia possunt velle operari in extremo aut in extremis, nec in medio: Error, si intelligatur substantiam sine operatione non esse in loco nec transire de loco ad locum.'" Gregory of Rimini, ibid., vol. 4, 333. Although the condemned article is cited here as 203, it is 204 in the *Chartularium* of the University of Paris, where all 219 condemned articles are published. Moreover, the text is incorrect. It is essential that "quod possunt moveri" be changed to "quod non possunt moveri." See Heinrich Denifle and Emil Chatelain, *Chartularium Universitatis Parisiensis,* 4 vols. (Paris: Fratrum Delalain, 1889–1897), vol. 1, 554.

181. "Dico igitur quod non solum per operationem angelus est in loco, sed etiam per substantiam suam, sic intelligendo quod eius etiam substantia praesens est loco et eo definitur et concluditur; quod etiam contingit, etiamsi non ibi operetur." Gregory of Rimini, *Sentences,* vol. 4, 335.

182. "Quantum ad illud quod principaliter quaeritur, dico quod angelus non est in loco indivisibili proprie loquendo de loco scilicet corporali, sed est in loco divisibili. Ratio est, quia nullum tale indivisbile est in rerum natura, ut probatum est in primo articulo." Ibid., vol. 4, 335.

183. Ibid., bk. 2, dist. 2, qu. 3, arts. 1–2, vol. 4, 339–343.

184. "Respondeo. Duo articuli sunt hic, unus de quaesito, alter de implicato in argumento ad oppositum, scilicet utrum plures angeli possint esse in eodem loco." Ibid., bk. 2, dist. 2, qu. 3, ibid., 340. The second article occupies pp. 341–343.

Gregory of Rimini posed a number of other questions about angels, including questions about the motion of angels from place to place. These are analogous to those raised in the thirteenth century by Thomas Aquinas and Richard of Middleton earlier in this section. Thus, in distinction six of his second book, Gregory poses three questions:

> "Whether an angel could be moved locally by itself."
> "Whether an angel could be moved from place to place successively in some time."
> "Whether an angel could be moved from place to place in an instant."

Gregory divides the first of these questions into four articles, of which the third is of interest because it is not concerned with angels, but with bodies. It asks whether simple bodies are moved locally by themselves. Of significance is the fact that of the four articles, the third, on whether bodies can move themselves locally, is by far the longest. Whereas the third article is 17 pages in length, the first is 1 page long (it extends over pages 2 to 3); the second is 9 pages (3–12), and the fourth article is slightly longer than 2 pages (29–31).

In the third article, Gregory specifies that he is concerned only with the natural motion of simple – that is, elemental – bodies. There is no mention of angels. In this article, Gregory presents six conclusions, by means of which he discusses many common problems of motion drawn from Aristotle and Averroës. He shows that simple bodies are not moved directly by the heavens (first conclusion); that they are not moved actively by the places toward which they tend (second conclusion); that they are not moved by the media in which they happen to be (third conclusion); that they are not moved by the things that generated them (fourth conclusion); but that they are moved by some mover that lies within themselves, and not by something external (fifth conclusion); and, finally (sixth conclusion), that these simple bodies move themselves *per se* and not accidentally. In this lengthy article, Gregory includes many basic problems of natural motion that were normally considered in treatises on natural philosophy.[185]

But why did Gregory make the article about the natural motion of simple elemental bodies the longest of the four articles? What has the simple motion of elemental bodies to do with the motion of angels from place to place? The answer lies in a brief statement in the fourth article, where Gregory explains that "since an angel could be moved locally, as is obvious from the first article, it is not impossible that it could move itself locally, as

185. Gregory of Rimini, ibid., bk. 2, dist. 6, qu. 1, art. 3, vol. 5, 12–21 for the six conclusions.

is obvious from the third [article, since] it is possible that God could confer on it the ability to move itself."[186] The lengthy discussion of the natural local motion of bodies was simply intended to suggest that if God conferred the power of self-motion on simple bodies, it was not unreasonable to conjecture that He might also confer it on the more perfect angels.[187] Gregory could easily have made his point without including a lengthy article on the natural motion of bodies. But it was hardly unusual to do so.

The other two questions on the motion of angels draw heavily on natural philosophy. We can see this with a single example. In the second article of the third question – "whether an angel could be moved from place to place in an instant" – Gregory asserts two conclusions. The first[188] is that an angel can change from place to place in an instant, even though it passes through the midpoint of the distance that separates the two places. He proves this conclusion by analogy with the motion of bodies in a vacuum. Aristotle had argued that if a heavy body descended from place to place through a vacuum, it would descend in an instant because there would be no resistance to it in a vacuum. Without resistance, bodies would move instantaneously. Indeed, the absurdity of such a possibility prompted Aristotle to deny the existence of vacua. Gregory applies Aristotle's argument about bodies moving in a resistanceless vacuum to angels moving from one place to another. Here, the crucial move is the assumption that no body or medium offers resistance to an angel; therefore, an angel could move instantaneously from one place to another, even as it moves through the middle point that separates the two places.

The Emergence of Analytic Theology: The Importation of Natural Questions into Theology

We have seen thus far that there was a considerable importation of natural philosophy into theology. Gregory of Rimini is an excellent example

186. "Nam cum angelus possit localiter moveri, ut patet ex articulo primo, nec sit impossibile idem movere se ipsum localiter, ut patet ex tertio, possibile fuit deum sibi conferre, ut ipse se moveret." Ibid., bk. 2, dist. 6, qu. 1, art. 4, vol. 5, 29.

187. Gregory says: "Et satis rationabile videtur aestimare quod sic de facto contulerit, praesertim cum hoc imperfectioribus mobilibus videamus esse concessum." Gregory of Rimini, ibid.

188. I shall not discuss the second, which asserts that an angel "can be changed by God from place to place in an instant, and not cross through the middle." ("Secunda, quod potest a deo mutari de loco ad locum in instanti, non transeundo per medium.") Ibid., bk. 2, dist. 6, qu. 3, art. 2, vol. 5, 47.

of a common tendency. Most scholastic theologians included subquestions or articles that were essentially, and even wholly, drawn from natural philosophy. An even more significant measure of the impact of natural philosophy on theology is represented by the second major category mentioned earlier, namely, questions that traditionally belonged to natural philosophy but appear in both theology and natural philosophy. These are, obviously, questions that many theologians thought desirable for their discipline, and they did not hesitate to appropriate them from natural philosophy. In some instances, the question may not have a direct counterpart in extant works in natural philosophy, but the subject matter is patently from natural philosophy.

Questions that were used by both natural philosophers and theologians were largely concerned with cosmology, since that subject was intimately linked with the creation account in Genesis. In the years before the Condemnation of 1277, when conservative theologians had worried about the relations between theology and natural philosophy, St. Bonaventure provided a revealing insight into the way theologians were approaching natural philosophy in their *Sentence* commentaries in the third quarter of the thirteenth century. In his commentary on the second book of the *Sentences,* Bonaventure asks "whether the heaven is of a spherical shape,"[189] a question that was regularly taken up by natural philosophers in their questions on Aristotle's book *On the Heavens (De caelo).*[190] In his conclusion to this question, Bonaventure declares that

> it is quite easy to respond to this question according to philosophy, although the expositors of Sacred Scripture seem to speak about it dubiously. They do this because of their reverence for Sacred Scripture ...; or they do this to check [or curb] our curiosity, wishing that we would be content to speak about those things that are in the Law and the Prophets, and not to inquire beyond this. Nevertheless, because the persistence of the curious does not cease, it is necessary that the doctors of Sacred Scripture determine many things, which could easily pass by [i.e., be lost] without the weight of preservation. Because of this, they [i.e., the theologians] respond to the aforesaid question and they say, both according to reason and according to the senses, that the heaven has an orbicular shape.[191]

189. "An caelum sit figurae orbicularis." Bonaventure, *Commentaria in quatuor libros Sententiarum* (1885), vol. 2, bk. 2, dist. 14, pt. 1, art. 2, qu. 1, 341–342.
190. For a list of natural philosophers who considered whether the heaven is spherical, see Grant, *Planets, Stars, & Orbs,* 703, qu. 126.
191. Bonaventure, *Sentences,* bk. 2, dist. 14, pt. 1, art. 2, qu. 1, 342, col. 1.

In the atmosphere that prevailed prior to the Condemnation of 1277, there were conservative theologians who resented the invasion of theology by natural philosophy. Although Bonaventure opposed various opinions of Aristotle, he obviously favored the use of natural philosophy in theology and in the *Sentences.* In this question, Bonaventure, as did virtually all scholastic natural philosophers and theologians, accepts Aristotle's opinion that the world is a sphere. Among the counterarguments, Bonaventure mentions two that appeal to the Bible. One of them invokes Psalm 103, where the heaven is said to be stretched like a skin, or is like an arched roof.[192] To this argument, Bonaventure replies that

> Scripture, condescending to poor, simple people, frequently speaks in a common way. And so, when it speaks about the heaven, it speaks in a way that the heaven appears to our senses, and [therefore] says that. With respect to our hemisphere, the heaven is like a skin *(pellis),* or a stretched, arched roof *(camerae extensum).*[193]

Shunning a literal interpretation of the Bible, Bonaventure dismisses counterarguments drawn from Scripture that seem to violate well-attested, scientific opinion – in this case, that the world is a sphere.

In the fourteenth century, Nicole Oresme adopted a similar stance when he mustered arguments for the possible daily axial rotation of the earth. To counter those who cited Biblical passages that clearly indicated that the Sun moved around the earth and that the earth was immobile,[194] Oresme declares that one can reply "by saying that this passage conforms to the customary usage of popular speech just as it does in many other places, for instance, in those where it is written that God repented, and He became angry and became pacified, and other such expressions which are not to be taken literally."[195]

Although he frequently disagreed with his contemporary rival, Thomas Aquinas, Bonaventure shows that he and Thomas were agreed on the relations that should obtain between Scripture and natural philosophy. Their model was St. Augustine, who, in his commentary on *Genesis* explained:

> In matters that are obscure and far beyond our vision, even in such as we may find treated in Holy Scripture, different interpretations are sometimes possible without

192. The phrase in Psalm 103 is "extendens caelum sicut pellem" (Vulgate edition). It is Bonavenure who says that it is also as a *camera,* or arched roof.
193. Bonaventure, *Sentences,* bk. 2, dist. 14, pt. 1, art. 2, qu. 1, 342, col. 2.
194. See *Nicole Oresme: Le Livre du ciel et du monde,* bk. 2, ch. 25, 527.
195. *Ibid.,* bk. 2, ch. 25, 531.

prejudice to the faith we have received. In such a case, we should not rush headlong and so firmly take our stand on one side that, if further progress in the search of truth justly undermines this position, we too fall with it.[196]

Augustine also insisted that Scripture should be taken literally whenever possible and feasible, as, for example, in interpreting the waters above the firmament. Here, Augustine insists that "whatever the nature of that water and whatever the manner of its being there, we must not doubt that it does exist in that place. The authority of Scripture in this matter is greater than all human ingenuity."[197]

In his *Summa theologiae,* Thomas Aquinas embraced the two major opinions proclaimed by St. Augustine when, in a discussion as to whether the firmament was made on the second day, he declares:

Augustine teaches that two points must be kept in mind when resolving such questions. First, the truth of Scripture must be held inviolable. Secondly, when there are different ways of explaining a Scriptural text, no particular explanation should be held so rigidly that, if convincing arguments show it to be false, anyone dare to insist that it is still the definitive sense of the text. Otherwise unbelievers will scorn Sacred Scripture, and the way to faith will be closed to them.[198]

Many theologians subsequently adopted the attitude of Thomas and Bonaventure toward the Bible and its relation to natural philosophy. Richard of Middleton, writing some years after Bonaventure in the thirteenth century, treated the same question ("whether the firmament has a spherical shape") and included the appeal to Psalms 103 "that the heaven is stretched like a skin" and is, therefore, says Richard, like a tent *(tentorium)* and, hence, not a sphere. To this claim, Richard gives essentially the same reply as did Bonaventure: "that Scripture there [i.e., in Psalms 103] really speaks about the shape of the heaven as it appears to our senses, or as it appears in our hemisphere only."[199] When Durandus de Sancto Porciano asked the same question in the fourteenth century, he did not include the

196. St. Augustine. *The Literal Meaning of Genesis: De Genesi ad litteram,* bk. 1, ch. 18, par. 37, ed. and trans. John Hammond Taylor, in Johannes Quasten, Walter, J. Burghardt, and Thomas Comerford Lawler, eds., *Ancient Christian Writers: The Works of the Fachers in Translation* (New York: Newman, 1982), vol. 41, 41.

197. St. Augustine, ibid., vol. 41, 52.

198. Thomas Aquinas, *Summa theologiae,* vol. 10: Cosmogony, pt. 1, qu. 68, art. 1, 71–73.

199. For the question, see Richard of Middleton, *Sentences,* bk. 2, dist. 14, art. 1, qu. 4, 170, col. 2–171, col. 1.

Biblical objections, perhaps because they were no longer regarded as weighty and appropriate counterarguments. It should be emphasized that Bonaventure, Richard of Middleton, and Durandus discuss the sphericity of the world solely in terms of natural philosophy.

Theological commentators differed considerably with regard to the questions they included and those they chose to ignore. Some, perhaps even many, of the questions that were drawn into theology from natural philosophy were also treated in questions treatises on the natural books of Aristotle. As with the strictly theological questions, many of the questions in natural philosophy were not routinely included by all theological commentators on the *Sentences,* although one or more would usually be included.

What were the questions from natural philosophy that found favor with many theologians?[200] To realize fully the degree to which straightforward questions on natural philosophy penetrated *Sentence* commentaries, it will be helpful to cite such questions in the works of a few theologians. In what follows, I shall first cite the enunciations of relevant questions in the second book by Richard of Middleton, a thirteenth-century theologian, and then mention some relevant questions from two fourteenth-century theologians – Peter John Olivi and Gregory of Rimini – all of whom commented on the second book of Peter Lombard's *Sentences.* I should emphasize that although most questions in *Sentence Commentaries* had some natural philosophy intermingled with theology, the questions cited here have nothing substantial about theology or the faith (as will be indicated, a few mention God, or an angel, or even Genesis; but there are no theological discussions). They are solely about problems in natural philosophy.

Richard of Middleton

"Whether matter is of a composite nature" (*Sentences,* bk. 2, dist. 12, art. 1, qu. 3, 145, col. 1–146, col. 1).

"Whether light is an accidental form" (bk. 2, dist. 13, art. 1, qu. 3, 157, col. 2–158, col. 2).

"Whether any light is of the same species with another light" (bk. 2, dist. 13, art. 1, qu. 4, 158, col. 2–159, col. 2).

"Whether light is brought forth from the potentiality of the medium" (bk. 2, dist. 13, art. 2, qu. 1, 160, col. 1–161, col. 1).

"Whether a medium is illuminated in time or in an instant" (bk. 2, dist. 13, art. 2, qu. 2, 161, col. 2–162, col. 2).

200. Most of the questions that I will cite are drawn from my "Catalog of Questions on Medieval Cosmology, 1200–1687," in Grant, *Planets, Stars, & Orbs,* 681–741.

"Whether light *(lux)* continually produces illumination *(lumen)* with respect to the same part of a medium" (bk. 2, dist. 13, art. 2, qu. 3, 162, col. 2–163, col. 1).[201]

"Whether the firmament is of a fiery nature" (bk. 2, dist. 14, art. 1, qu. 3, 169, col. 1–170, col. 2).

"Whether the firmament has a spherical shape" (bk. 2, dist. 14, art. 1, qu. 4, 170, col. 2–171, col. 1).

"Whether the firmament has right and left" (bk. 2, dist. 14, art. 1, qu. 5, 171, col. 1–172, col. 1).

"Whether the heaven is moved by a created intelligence or by its natural form" (bk. 2, dist. 14, art. 1, qu. 6, 172, col. 1–174, col. 2).[202]

"Whether is any element there are several forms" (bk. 2, dist. 14, art. 2, qu. 1, 174, col. 2–176, col. 1).

"Whether an elementary form can receive more and less" (bk. 2, dist. 14, art. 2, qu. 2, 176, col. 1–178, col. 1).

"Whether in the matter of one element there is an active potentiality for the form of another element" (bk. 2, dist. 14, art. 2, qu. 3, 178, col. 1–179, col. 2).[203]

"Whether a part of some element that is assumed outside of its [natural] place could move itself to its place" (bk. 2, dist. 14, art. 2, qu. 4, 179, col. 2–181, col. 1).

"Whether all the luminaries [i.e., planets] that are assumed are in one continuous body" (bk. 2, dist. 14, art. 3, qu. 1, 183, col. 2–184, col. 2).[204]

"Whether any luminaries [i.e., planets] are moved with a proper motion in addition to the motions of the spheres [that carry them] (bk. 2, dist. 14, art. 3, qu. 2, 184, col. 2–186, col. 1).

201. "The term *lux* was associated with light as the luminous quality of a self-luminous body, as, for example, the body of the Sun. The light from a luminous body that emanated into a surrounding medium, such as the Sun's rays, was characterized by the term *lumen.*" See Grant, *Planets, Stars, & Orbs*, 392.

202. In this question, Richard briefly mentions the "first cause" and then says that God could move all orbs without an intervening cause; later he mentions that "an angel under God is the principal mover of the heaven." But these two assertions were often mentioned in treatises on natural philosophy and are not peculiar to a theological treatise. Hence, I have included this question among those that are wholly free of theology.

203. In two lines, Richard mentions that in Genesis 1, semen is said to exist only in living things. That brief citation has no bearing on the arguments.

204. In this question, Richard mentions God once and also says that in Genesis we are told that the luminaries are in the firmament of the heaven. But these statements play no role in the question.

In the sixteen questions just cited, Richard includes nothing substantive about theology or the faith. These questions could easily have fit into relevant treatises on natural philosophy. But Richard also included numerous questions that are essentially concerned with problems in natural philosophy, but in which he adds a few remarks about God or about a Biblical text, or refers to St. Augustine or some other Church Father. For example, in the question "whether the matter of all corruptible bodies is one in number" (bk. 2, dist. 12, art. 1, qu. 9, 151–152, mistakenly paginated 161–162), Richard invokes God's absolute power and his infinite ubiquity; in the question "whether the substantial form of a lucid body, or of any whatever other body, is perceptible by some sense" (bk. 2, dist. 13, art. 1, qu. 2, 157, cols. 1–2), Richard mentions, in a few lines, the transubstantiation of the bread into the body of Christ. In a question titled "whether light *(lumen)* is of the essence of color" (bk. 2, dist. 13, art. 2, qu. 4, 163, col. 2–165, col. 2), Richard discusses, again in a few lines, whether the earth was invisible before the creation of light and, near the end of the question, mentions that the day is taken in different ways in Scripture and cites two ways. In all of these questions, however, and in others that also include a few minimal theological remarks,[205] the theological elements do not affect the main line of argument.

Peter John Olivi included 118 questions in his commentary on the second book of the *Sentences.* Although creation is a major feature of the second book, Olivi omits most of the kinds of questions that many other theologians – such as Richard of Middleton, Bonaventure, and Thomas Aquinas – considered. While he poses two questions on the eternity of the world, and then asks (in a single question) whether there are other worlds or only one, and whether it is possible that there be other worlds, his work is remarkable for what it ignores and for the unusual questions he does include. The most amazing feature of Olivi's *Sentence Commentary* is that he devotes not a single question to the heavens, even though he discusses a total of 118 questions in book 2. He does not discuss Genesis; does not mention the heavens; or the firmament; or the crystalline sphere; or the waters above and below the firmament, and so on. What, then, did he include? It was not theology.

According to Bernard Jansen, the editor of Olivi's treatise, Olivi devoted very few questions to theology proper. Jansen divides the 118 questions into the following categories: questions 1–31 are on philosophical problems; questions 32–48 are on angels; questions 48–89 are on metaphysics; and questions

205. For example, bk. 2, dist. 14, art. 3, qu. 4–6, 186, col. 2–192, col. 2.

90–118 are on various moral problems. Many of the latter, however, seem theological in character, and we may include them among Olivi's theological questions. Thus, even if we add the 17 questions on angels to the 28 on moral problems, and categorize them all as theology, we have a total of 45 questions on theology. Thus, in a work that purports to be on theology, fewer than half are on that subject. Although the remaining questions may not be on theological themes, they do, of course, have references to God and the faith scattered throughout. Nevertheless, they are something other than theological, as, for example, in question 22, which asks "whether a substance is susceptible to more and less." In this lengthy question of some 33 pages, Olivi speaks of divine power (p. 405) and then in a few lines mentions the soul of Christ (p. 406), and in replying to one of the arguments against his position, Olivi mentions God and Antichrist (p. 410). All told, there is perhaps as much as a half page out of 33 pages devoted to what might be appropriately called theology. The lengthy question 22 is essentially on natural philosophy, with theology playing virtually no role in the main thrust of the question, and similarly in question 26, where Olivi asks "whether the first impressions of all agents are made by them in an instant."[206] Not until the final two pages of this lengthy question of some 18 pages does Olivi find occasion to mention God (twice) and Christ (once).

Other questions include only the barest minimum of theological references. A good illustration is question 23, which asks "whether every agent is always present to its patient, or to its first effect."[207] Not until the penultimate sentence of the question does Olivi mention the "divine power" *(de divina virtute).*

There are also questions in which Olivi makes no mention whatever of anything relevant to theology or the faith. In a series of questions on the nature of matter – questions 18 through 21, extending over some 25 pages (363–388) – Olivi found no occasion to mention God, the faith, or anything relevant to theology. The same pertains to questions 24 (pp. 434–438), 29 (pp. 499–504), 87 (pp. 198–203), and 88 (pp. 203–204).

Although questions that have no apparent connection to theology of the kind that have been identified in the *Sentence* commentaries of Richard of Middleton and Peter John Olivi are significant, they are a relatively small percentage of the total number of questions. Most questions in a *Sentence* commentary that are obviously on a theme in natural

206. Olivi, *Sentences,* bk. 2, qu. 26, vol. 1, 446–464.

207. "Primo quaeritur an omne agens sit semper praesens suo patienti seu suo effectui primo." Ibid., bk. 2, qu. 23, vol. 1, 422–433.

philosophy will usually include some reference to God, or to Christ, or to some aspect of religious faith. But these references are analogous to those we find in questions on Aristotle's natural books (see Chapter 5). They rarely affect the substantive arguments and content of the question. To all intents and purposes, they are properly classifiable as belonging to the domain of natural philosophy. It is in this sense that we can legitimately accept Bernard Jansen's assessment that there are many fewer questions on theology than on natural philosophy and metaphysics in the *Sentence Commentary* of Peter John Olivi.

Gregory of Rimini's *Sentence Commentary* on the second book follows the same pattern as Olivi's; that is, Gregory chose not to discuss the cosmological aspects of the first four days of creation by omitting from his commentary distinctions 12, 13, and 14, which were wholly concerned with the creation and therefore with cosmology and physics. Thus, like Olivi, Gregory chose not to discuss the creation and, therefore, missed a great opportunity for the importation of natural philosophy into his theological commentary. Nevertheless, like Olivi, he managed to include much logic and natural philosophy in his commentary, as we already saw earlier in this chapter.

Since Gregory's commentary on the first book of the *Sentences* has also been published along with his commentary on the second book, I will mention one question in the first book before turning to the second. In distinction 3, question 1, of the first book, Gregory considers "whether sensible things are understood by us naturally."[208] In this 68-page question, Gregory mentions God, Christ, and Antichrist briefly on six pages.[209] Most of the citations are by way of examples. Although the first book is about God, the question cited here is devoted to the psychological problem of how our intellect comprehends sensible things. As expected, Gregory draws heavily on Aristotle's *On the Soul (De anima).*

Gregory begins his commentary on the second book with a remarkable question in which he asks "whether Aristotle and his commentator, Averroës, thought that all other beings were made from a first being or, rather, the contrary, [namely,] that they thought several beings did not have an effective beginning."[210] A few lines further, Gregory explains:

208. "Utrum res sensibiles intelligantur a nobis naturaliter." Gregory of Rimini, *Sentences,* bk. 1, dist. 3, qu. 1, vol. 1, 302–370.

209. See ibid., bk. 1, dist. 3, qu. 1, vol. 1, 333, 335, 336, 340, 349, and 360.

210. "Utrum Aristoteles eiusque Commentator Averroes senserint omnia alia entia a primo esse facta, vel potius econtra fuerint opinati plura entia non habere principium effectivum." Ibid., bk. 2, dist. 1, qu. 1, vol. 4, 1.

Because there is a controversy among the theological doctors [as to] what the Philosopher thought about the efficiency of the prime [being] with respect to others, some [theologians] say that he [Aristotle] thought that all other beings were produced mediately or immediately from the first being, which is God. [If] immediately [this signifies that they were made] from eternity; mediately [signifies] indeed that they are generable and corruptible, with their generation flowing by means of the heavenly motion which moves itself actively, as if it had its own mover.

Gregory then divides the question into two articles:

Firstly, we will see whether it is the intention of the Philosopher [Aristotle] and his Commentator [Averroës] that there are several unproduced entities in the universe, or only one and from this [one] others [are derived] mediately or immediately. Secondly, [I ask] whether it is the intention of the same two that the prime [or first] being, which is God, actively causes the first movable sphere to move immediately, as if it had its own mover, and not [to cause it to move] only as an end [or final cause].[211]

Thus, Gregory justifies the inclusion of this question on the grounds that the "theological doctors" were in disagreement about the intention of Aristotle and his Commentator Averroës. Did they or did they not assume a plurality of uncreated, or "unproduced," beings, or gods; or did they only assume one God and then derive other beings from it. And secondly, Gregory asks if God causes the motion of the outermost movable sphere directly and immediately, as if it were self-moved. Such a motion would be in addition to God's ability to move that same sphere as a final cause.

What is extraordinary about these questions is that they are not about God's actions as they would be understood and interpreted within the Christian faith in a treatise on Christian theology. Rather, they are exclusively about the way Aristotle and Averroës interpreted the activities of

211. The parts of the Latin text that I have translated are as follows: "Respondeo. In ista quaestione, quoniam inter theologicos doctores est controversia, quid senserit circa hoc Philosophus de efficientia primi respectu aliorum, quibusdam dicentibus eum sensisse omnia alia entia a primo, quod est deus, fuisse ab illo mediate vel immediate producta, immediate quidem aeterna, mediate vero generabilia et corruptibilia, ad quorum generationem concurrit mediante motu coeli, quod ipse active movet velut motor appropriatus.... Ideo huius quaestionis duo erunt articuli.

Nam primo videbitur, utrum fuerit de intentione Philosophi Commentatorisque eius esse in universo plura entia penitus improducta, vel unum tantum et ab illo cetera mediate vel immediate. Secundo, utrum fuerit de intentione eorundem primum ens, quod est deus, immediate movere active, et non tantum per modum finis, primum mobile velut motor eidem appropriatus." Gregory of Rimini, *Sentences*, bk. 2, dist. 1, qu. 1, vol. 4, 1–2.

God, and not about the Christian faith. The entire, lengthy question, some 41 pages long, is about Aristotle and Averroës and draws most heavily on a variety of their works. There is nothing whatever in Gregory's discussion about God and the Christian faith. This question, with its two articles, falls squarely within the domain of natural philosophy and would be appropriate in a treatise of questions on one of Aristotle's natural books. It seems an alien intrusion into a commentary on the *Sentences,* despite Gregory's attempt to justify its inclusion by citing the disagreement among his colleagues in theology.

In the fourth question of the first distinction, Gregory discusses motion, asking "whether motion is something by itself, distinct from one or more permanent things."[212] At the beginning of the question, Gregory says that the Master, who is Peter Lombard, showed in book 2, distinction 1, that God is the beginning of all creatures, but that the action of secondary created agents, which are accompanied by motion, are better known to us. Therefore, Gregory says, we must first inquire about motion.[213] Again, this forced, weak link to Peter Lombard is the launching pad for a 50-page, highly sophisticated analysis of motion in all its aspects. Gregory includes not a word about theology, God, the faith, or anything remotely connected with Christianity. The entire question is focused on Aristotle and Averroës. Once again, we may rightly claim that this is a question about an important theme in natural philosophy, but it is not about theology.

In the fifth and next question, Gregory is interested in the action of a thing, whether it is distinct from the agent causing the action and from the thing suffering the action. In this lengthy question, Gregory mentions God in four places and Christ in one. In all cases, the usages are introduced as examples, usually to utilize God's absolute power. Thus, in one of these instances, Gregory assumes that the power of God is such that He could cause something to be heated without that object receiving any heat.[214] Apart from these few citations of God and Christ, however, the question is wholly about natural philosophy.

There are other such questions. Even where the question is theological, Gregory frequently employs the logico-mathematical techniques mentioned earlier and which were illustrated in his questions about angels cited in the preceding section ("Angels and Natural Philosophy").

212. Ibid., bk. 2, dist. 1, qu. 4, 124–175.

213. Ibid., 124.

214. For this example, see Gregory of Rimini, ibid., bk. 2, dist. 1, qu. 5, vol. 4, 184.

Richard of Middleton, Peter John Olivi, and Gregory of Rimini are representative of the way *Sentence* commentaries were produced in the late Middle Ages. The questions they posed and the ways they responded are both typical and atypical, since some of the questions by Olivi and Gregory were not commonly discussed. But that itself may have been fairly typical, since the questions included could vary significantly from author to author, although some questions were common to many authors. It is nevertheless true, as we have seen, that many questions, especially in the second book of the *Sentences,* were questions in natural philosophy, rather than theology. Of these, some appear in the *Sentences* only, while others had direct counterparts in the literature of natural philosophy, that is, in the treatises on Aristotle's natural books. Following are some of the questions that were considered in both disciplines:[215]

"Whether the heaven is composed of matter and form."[216]
"On the number of spheres, whether there are eight or nine, or more or less."[217]
"Whether in the heavens there are up and down; in front of and behind; [and] right and left."[218]
"Whether the heaven is spherical in shape."[219]
"Whether the heavens are animated."[220]

215. Those who treated the same question rarely used the same wording. The authors of any particular question that follows usually differed in the way they expressed it, but there is no doubt that they were considering the same substantial question.

216. Peter Aureoli and John Major in their *Sentence* commentaries; and John of Jandun and Johannes de Magistris in their questions on Aristotle's *De caelo.* For the precise citations, see Grant, *Planets, Stars, & Orbs,* 694, qu. 79. For full titles of the works, see the Bibliography of *Planets, Stars, & Orbs.*

217. Thomas Aquinas and John Major in their *Sentence* commentaries; and Albert of Saxony and Themon Judaeus in their questions on *De caelo.* See Grant, *Planets, Stars, & Orbs,* 697, qu. 97.

218. Bonaventure and Richard of Middleton asked only whether there is a right and left in the heaven; Albert of Saxony and John Buridan asked precisely this question in their questions on *De caelo,* and Roger Bacon responded to a similar question in his questions on the *Physics.* See Grant, ibid., *Planets, Stars, & Orbs,* 701–702, qu. 122.

219. St. Bonaventure and Durandus de Sancto Porciano in their *Sentence* commentaries; and Albert of Saxony, Johannes de Magistris, Johannes Versor, and John Major in their *Questions on De caelo;* and Michael Scot, Themon Judaeus, and Pierre d'Ailly in their commentaries and questions on John of Sacrobosco's *Treatise on the Sphere.* See Grant, *Planets, Stars, & Orbs,* 703, qu. 126.

220. Richard of Middleton and Peter Aureoli in their *Sentences;* and John of Jandun and Johannes de Magistris in their *Questions on De caelo;* Benedictus Hesse in his *Questions on the Physics.* See Grant, *Planets, Stars, & Orbs,* 703–704, qu. 128.

"Whether the whole heaven from the convexity of the supreme [or outermost] sphere to the concavity of the lunar orb is continuous or whether the orbs are distinct from each other."[221]

"Whether celestial motion is natural."[222]

"Whether the stars are self-moved or are moved only by the motions of their orbs."[223]

The questions just cited are cosmological and are limited to a select number of authors. The list could be considerably expanded if questions were included about matter, motion, sense perception, and other themes that were derived from questions on the whole range of Aristotle's natural books. There can be little doubt that theologians were heavily into logic and natural philosophy. "One can point to numerous *Sentence Commentaries*," observes Edith Sylla

> in which natural science is used extensively, and there are some *Sentence Commentaries* which in fact seem to be works on logic and natural science in disguise – in response to each theological question raised, the author immediately launches into a logical-mathematical-physical disquisition and then returns only briefly at the end to the theological question at hand.[224]

Why were so many questions in *Sentence Commentaries* squarely in natural philosophy? And why were so many ostensibly theological questions permeated by natural philosophy? Not only did medieval theologians trans-

221. Bonaventure and Richard of Middleton in their *Sentences;* and Albert of Saxony and Paul of Venice in their *Questions on De caelo.* Others who responded to this question in commentaries on the *Sphere* of Sacrobosco and on Aristotle's *Meteorology* are cited in qu. 132, Grant, *Planets, Stars, & Orbs,* 704–705.

222. Durandus de Sancto Porciano in his *Sentences;* and Johannes de Magistris in his *Questions on De caelo.* For others, see qu. 173, Grant, *Planets, Stars, & Orbs,* 709–710.

223. Bonaventure, Richard of Middleton, and Durandus de Sancto Porciano treated this question in their *Sentence* commentaries; Roger Bacon, John of Jandun, John Buridan, Albert of Saxony, Nicole Oresme, and a few others considered this question in their *Questions on De caelo.* See qu. 211 in Grant, *Planets, Stars, & Orbs,* 715.

224. Edith D. Sylla, "Autonomous and Handmaiden Science: St. Thomas Aquinas and William of Ockham on the Physics of the Eucharist," in *The Cultural Context of Medieval Learning,* edited with an Introduction by John E. Murdoch and Edith Dudley Sylla (Dordrecht, Holland: D. Reidel, 1975), 352. For a brief description of the same attitude of Oxford theologians in the fourteenth century, see W. A. Courtenay, "Theology and Theologians from Ockham to Wyclif," in J. I. Catto and Ralph Evans, eds., *The History of the University of Oxford,* vol. 2 (Oxford: Clarendon Press, 1992), 7.

form theology by the importation of massive amounts of natural philosophy, but they also paid considerable attention to those aspects of natural philosophy that had been quantified by the logico-mathematical techniques that we have already described. By applying those techniques to theological problems, medieval theologians quantified certain aspects of theology. Why did this occur?

We have already seen that beginning in the late eleventh century, a strong tendency to systematize and rationalize theology had been under way, despite considerable opposition. But the tendency to expand these beginnings became seemingly irresistible when medieval universities made logic and natural philosophy virtually indispensable prerequisites for study in the higher faculties and, therefore, indispensable for the study of theology. With these powerful tools constantly at their disposal, theologians could not resist the temptation to use them in theological problems of all kinds. This temptation was probably intensified by the fact that most theologians had been arts masters for a few years before matriculating in a faculty of theology for many long years. During that time many also taught natural philosophy to students in the arts faculty. Because their sojourn in the arts faculty was for a relatively short period – only a few years – they did not have much opportunity to develop their philosophical ideas. The opportunity to do so, along with greater maturity, came while they were theological students and masters. Consequently, they often presented their mature thoughts about natural philosophy in their *Sentence Commentaries.*

Indeed, it almost seems as if they devised difficult theological problems in order to challenge themselves to resolve them by means of the logic and natural philosophy they had earlier absorbed.[225] How else explain such questions as "whether God could make the future not to be";[226] "whether an angel is in a divisible or indivisible place";[227] whether God could cause a past thing [or event] to have never occurred";[228] "whether an angel could sin or be meritorious in the first instant of his existence";[229] "whether God could know something that He does not know";[230] "whether [an angel]

225. See Sylla, "Autonomous and Handmaiden Science," 378, n. 9, where she cites as her source for these thoughts van Steenberghen, *The Philosophical Movement in the Thirteenth Century,* 378.
226. Hugolin of Orvieto, *Sentences,* bk. 1, dist. 40, qu. 3, art. 3, vol. 2, 341.
227. Gregory of Rimini, *Sentences,* bk. 2, dist. 2, qu. 2, vol. 4, 277.
228. Richard of Middleton, *Sentences,* bk. 1, dist. 42, art. 1, qu. 5, vol. 1, 375, col. 2.
229. Gregory of Rimini, *Sentences,* bk. 2, dist. 3–5, qu. 1, art. 2, vol. 4, 345. The article is discussed on pp. 369–379.
230. Thomas Aquinas, *Sentences,* bk. 1, dist. 39, qu. 1, art. 2, vol. 1, 922.

could be moved from place to place without passing through the middle [point]."[231]

The eagerness to employ the powerful tools they had acquired may also explain why scholastic theologians so boldly applied logic and natural philosophy to the deepest mysteries of the Christian faith: the Trinity and the Eucharist.[232] In considering the Trinity in a question titled "whether the Father and the Son are one principle giving origin to the Holy Spirit," Gregory of Rimini applies logic in his analysis of the question, using supposition theory and concepts, such as "confused supposition" and "determinate supposition."[233] More than a century later, Gabriel Biel (d. 1495) used similar logical apparatus in the very same question in his *Sentence Commentary*.[234] In another question on the Trinity earlier in his commentary, Biel mentions four fourteenth-century theologians who provided a deeper understanding of the relationship between logic and the Trinity.[235]

To cope with these, and many similar, paradoxical questions, medieval theologians were compelled to resort to their logical and philosophical training. However, it seems plausible to assume that they introduced such questions because their backgrounds in logic and natural philosophy made it feasible to do so, and challenging as well. Inexorably, then, theology was transformed into a highly analytical pursuit.

But we have also seen that medieval theologians went well beyond the application of logico-mathematical techniques to theological questions. They introduced many questions into their *Sentence* commentaries that were solely devoted to natural philosophy and that had virtually nothing to do with theology and were not logico-mathematical. They were questions on a range of subjects, some of which natural philosophers routinely discussed in their questions on Aristotle's natural books, and some that they

231. Robert Holkot, *Sentences,* bk. 2, qu. 4, art. 5. The book is unpaginated, but the fifth article occurs on the page where AA appears in the right margin.

232. For the application of natural philosophy to the Eucharist, see Sylla, "Autonomous and Handmaiden Science," 361–372; for the application of logic to the Trinity, see A. Maierù, "Logique et théologie trinitaire dans le moyen-âge tardif: Deux solutions en présence," in Monika Asztalos, *The Editing of Theological and Philosophical Texts from the Middle Ages* (Stockholm: Almqvist & Wiksell International, 1986), 185–212.

233. "Utrum pater et filius sint unum principium spirans spiritum sanctum." Gregory of Rimini, *Sentences,* bk. 1, dist. 12, qu. 1, vol. 2, 191–192.

234. Gabriel Biel, *Collectorium circa quattuor libros Sententiarum,* bk. 1, dist. 12, qu. 1, dubium 4, vol. 1, 377–378.

235. Gabriel Biel, *Sentences,* bk. 1, dist. 5, qu. 1, art. 3, vol. 1, 271. The four are Henry Totting de Oyta, William Ockham, Robert Holkot, and Gregory of Rimini. See also, Maierù, "Logique et théologie trinitaire," 185.

did not routinely discuss, or even discuss at all. So deeply did natural philosophy permeate theology that on occasion, parts of *Sentence* commentaries were extracted and circulated as separate treatises on natural philosophy.[236] Medieval theology became an exercise in rigorous analysis for its professional practitioners. It was often more natural philosophy than theology. Medieval theologians seem to have been compulsively driven to explain all aspects of their faith in rational terms. In doing so, they seem to have emptied theology of spiritual content.

Many of the theological questions that have been cited in this chapter seem strange, even bizarre. Why did medieval theologians think it important to know "whether God could make a creature exist for only an instant", or "whether angels could have foreknowledge of their fall in the first instant of their existence?" And why would they ask whether God can speak falsely, or whether God could erase the past, or whether God could make someone sin? And why were they anxious to show that not even God can overcome certain problems, as, for example, when Robert Holkot showed that not even God can know whether a person's last act was sinful or meritorious if that person is alternately sinful and meritorious in every proportional part of the last hour of his life. One can easily cite a host of similar questions in which medieval theologians sought to determine whether God could do this or that specific act.

It is obvious that they thought they could resolve such questions by the application of reason in the guise of logic and natural philosophy. But why did they believe it important to raise such questions in the first instance? In part, the answer must be that they had the tools – logic and natural philosophy – to cope with such questions and were determined to use them. Indeed, possession of such powerful analytic tools, and the desire to use them, probably prompted theologians to raise the kinds of theological questions I have just described. But did the Church regard the resolution of paradoxical questions about God, angels, and faith as contributing to the well-being of the faith? Judging from numerous attempts to curb the zeal-

236. John Murdoch explains ("From Social into Intellectual Factors," 276) that "genuine parts of fourteenth-century theological tracts ... successfully masqueraded as straightforward tracts in natural philosophy. Thus, Gerard of Odo's examination of the problem of the composition of continua was detached from the *Sentence Commentary* to which it belongs and circulated separately. So totally without theological relevance (it was shorn of its introduction), it appears exactly as if it could be the initial *questio* of Book VI of a commentary on Aristotle's *Physics*. Even more interesting is the opening *questio* of the *Commentaria sententiarum* of Roger Rosetus: Its first article enjoyed an extensive separate career as a *Tractatus de maximo et minimo*."

ous use of logic and natural philosophy in *Sentence Commentaries* (see next section on the reaction to analytic philosophy), the answer seems to be in the negative.

Indeed, one wonders what theologians themselves thought about their efforts to do theology by the application of logic and natural philosophy to ostensible theological problems. Did they believe that they were contributing positively to knowledge and understanding about God and the faith? Did they regard the application of quantitative and analytic methods to theological problems as, in some sense, enhancing their spiritual understanding of the faith? And did they regard it as important to determine what God could or could not do, or what He could or could not know? The theologians themselves fail to shed light on such questions. But somehow, in addition to the personal pleasure they might have derived from the effort to resolve challenging, if bizarre, questions by analytical means, we must, I believe, assume that medieval theologians regarded their efforts as in some sense advancing and buttressing their faith. To think otherwise would signify that they knowingly engaged in meaningless and empty puzzle solving, analytic exercises that had no relevance to their faith. But in what sense they may have regarded their contributions as meaningful for the faith escapes my understanding.

By the fourteenth century, medieval theologians were as much logicians and natural philosophers as they were theologians. They made theology a mix of logic and natural philosophy. Consequently, the theology they produced was virtually unintelligible to those who lacked training in logic and natural philosophy. Nothing like the theology of the late Middle Ages had ever been seen before; and after its demise in the seventeenth century, nothing like it has been seen since.

The Reaction to Analytic Theology

The kind of theology that I have described and that was taught and written about in the late Middle Ages was nothing less than analytical theology. It was not without opponents. From the early thirteenth century, some Church officials were uneasy about it. As early as 1228, Pope Gregory IX tried to stop the infiltration of natural philosophy into theology:

> Gregory accuses theology of being dominated by philosophy, which should be its obedient servant, and of committing adultery with philosophical doctrines. The theologians, he says, have moved the limits set by the church fathers on the utilization of philosophy for the study of sacred Scripture. While trying to support faith

with natural reason, they weaken faith, for nature believes what it has understood, whereas faith understands what it has believed. The pope finally commands the masters to teach pure theology without any leavening of worldly science.[237]

In 1231, in a sermon by John of St. Giles, a Dominican master, we catch another glimpse of the uneasiness that some theologians felt about the infiltration of natural philosophy into theology. John criticizes career theologians who "when they come to theology ... can hardly part from their science [*scientia sua,* i.e., philosophy], as is clear in certain persons, who cannot part from Aristotle in theology, bringing with them brass instead of gold, that is to say philosophical questions and opinions."[238]

Concern about the intrusion of natural philosophy into theology continues into the fourteenth century, when Pope Clement VI (1342–1352) criticized theologians for absorbing themselves in philosophical questions.[239] But it was in 1366, with the promulgation of new statutes for the University of Paris, that the authorities made a serious effort to separate natural philosophy and theology as much as was feasible. Those who were teaching the *Sentences* were admonished to avoid the intrusion of philosophy and logic into their commentaries. But no ban on the use of philosophy in theology was issued, because the university authorities felt obliged to concede that it was permissible to introduce philosophy and logic if it was deemed essential to the resolution of an argument. This became a common approach in various universities.[240] It appears that theologians almost always found it necessary to include logic and natural philosophy. The impulse to explain most theological questions by the use of logic and natural philosophy was by then too deeply entrenched to be significantly weakened, or thwarted.[241]

From a statement in the sixteenth century by the eminent theologian John Major, we get a strong sense that official protestations were of little avail. For Major informs us that "for some two centuries now, theologians have not feared to work into their writings questions which are purely

237. Asztalos, "The Faculty of Theology," 421.

238. Ibid., 422.

239. Ibid., 432.

240. See ibid., 434.

241. W. G. L. Randles describes how Albertus Magnus and Thomas Aquinas used Aristotle's natural philosophy to explain how the Blessed in the empyrean heaven could continue to enjoy the use of their five senses. See Randles, *The Unmaking of the Medieval Christian Cosmos, 1500–1760: From Solid Heavens to Boundless Aether* (Aldershot Hants, Eng.: Ashgate, 1999), 12–24.

physical, metaphysical, and sometimes purely mathematical."[242] During Major's lifetime, however, the practices he describes drew a strong reaction from Martin Luther, who, even as he was ushering in the Protestant Reformation, found time to attack scholasticism, especially the way his fellow theologians commented upon the *Sentences* of Peter Lombard.

Luther's assault on medieval theology and scholasticism was but one – albeit one of great importance – of a growing chorus of opposition against the methods and content of scholastic learning. In time, that increasingly scornful opposition would alter the perception of the medieval contribution to reason and reasoned discourse. Instead of being judged as a period in which an enormous emphasis was placed on reason and analytical skills, as a period in which the firm foundations of a rational approach to nature were laid, the Middle Ages would be caricatured as a period of intellectual sterility, superstition, and even irrationality. We must now describe this extraordinary historical process.

242. Translated by Walter Ong, *Ramus, Method, and the Decay of Dialogue: From the Art of Discourse to the Art of Reason* (Cambridge, MA: Harvard University Press, 1958), 144.

7

THE ASSAULT ON THE MIDDLE AGES

THE EMPHASIS ON REASON – ITS DEVELOPMENT AND APPLICATION – that was so characteristic of medieval Western Europe, and which we have described, must now briefly be viewed in the broader context of subsequent history. It is essential to do this because I have claimed that the Age of Reason *began* in the Middle Ages. If it did, what connection does the latter have with the former? To make the connection, we must first arrive at some sense of what the phrase "Age of Reason" signifies.

THE MEDIEVAL AND EARLY MODERN "AGES OF REASON"

Many rightly regard the seventeenth century as a century of momentous change because it produced a "Scientific Revolution," an expression that is commonly used to characterize the science of that century.[1] The designation is appropriate because of the contributions of a series of extraordinary figures – Galileo Galilei, Johannes Kepler, René Descartes, Christiaan Huyghens, Robert Boyle, Isaac Newton, and many others – who produced scientific theories, experiments, and treatises that reflected a new approach to nature that radically transformed the way science had been done within the earlier context of medieval Aristotelian natural philosophy. Galileo captured a fundamental aspect of the dramatic change when he declared:

> Philosophy is written in this grand book, the universe, which stands continually open to our gaze. But the book cannot be understood unless one first learns to comprehend the language and read the letters in which it is composed. It is written in the language of mathematics, and its characters are triangles, circles, and other

1. For a monumental study of that revolution, see H. Floris Cohen, *The Scientific Revolution: A Historiographical Inquiry* (Chicago: University of Chicago Press, 1994).

geometric figures without which it is humanly impossible to understand a single word of it; without these, one wanders about in a dark labyrinth.[2]

The contributions of Sir Isaac Newton marked the culmination of the Scientific Revolution. His most famous treatise, *The Mathematical Principles of Natural Philosophy* (1687), which was based upon his universal theory of gravitation, furnished a new way of interpreting the structure and operation of the cosmos. After Newton, the application of mathematics to physics became routine. Dramatic changes in science were not confined to physics. Biology, physiology, and medicine also saw significant advances associated with such great names as William Harvey (1578–1657), Robert Hooke (1635–1702), Marcello Malphigi (1628–1694), Antoni van Leeuwenhoek (1632–1723), and others. Along with monumental scientific achievements, the seventeenth century witnessed major new departures in philosophy. The transformation of that traditional discipline was most dramatically affected by René Descartes (1596–1650) and John Locke (1632–1704), but other philosophers, such as Francis Bacon (1561–1626), Thomas Hobbes (1588–1679), Baruch Spinoza (1632–1677), and Gottfried Leibniz (1646–1716), also made momentous contributions.

Eighteenth-century philosophers and scientists built upon the achievements of their seventeenth-century predecessors. What they built has been called "The Age of Enlightenment," a descriptive phrase that has become virtually synonymous with reason and rationality. Enlightenment thought in science and philosophy, and eventually in the social sciences, was based solidly on seventeenth-century thought, especially the contributions of three Englishmen: Bacon, Locke, and Newton, who have been called "the patron saints of the Enlightenment."[3] The scientists and philosophers who together contributed the science and philosophy that constitute the Enlightenment are too numerous to mention here. Among the most significant scientists were Antoine-Laurent Lavoisier (1743–1794), Carl Linnaeus (1707–1778), Joseph Priestley (1733–1804), Leonhard Euler (1707–1783), Joseph Louis Lagrange (1736–1813), and Marquis de Laplace (1749–1827). Among the most important philosophers were George Berkeley (1685–1753), David Hume (1711–1776), and Immanuel Kant (1724–1804), with additional

2. From Galileo's *Assayer (Il Saggiatore)* as translated in Stillman Drake, *Discoveries and Opinions of Galileo* (Garden City, NY: Doubleday Anchor Books, 1957), 237–238.

3. See Lloyd Spencer and Andrzej Krauze, *Introducing the Enlightenment* (New York: Totem Books, 1997), 41.

major contributions from Voltaire (1694–1778), Denis Diderot (1713–1784), and Jean Le Rond d'Alembert (1717–1783).

When taken together, the Scientific Revolution of the seventeenth century and the Enlightenment of the eighteenth century may be said to constitute the Age of Reason, a momentous epoch in the history of Western civilization, and in the history of all civilizations. With regard to reason and rationality, what is the most striking similarity and the most dramatic difference between the Age of Reason and the late Middle Ages? The most striking similarity is that reason was applied on a large scale to religion and theology in both periods. The most striking difference is that those who applied reason to theology in the Middle Ages did not – indeed, could not – challenge the ultimate supremacy of the truths of revelation, whereas scholars in the Age of Reason began to do so. They could do so because of the emergence of natural theology, a subject that had barely begun in the Middle Ages. What is natural theology or natural religion? It is generally conceived as an attempt to establish fundamental religious truths, especially truths about God, by means of reason applied to nature, God's creation, without invoking, or appealing to, revealed truths. This was an approach that developed over the course of the seventeenth century, with not much of a prehistory in the Middle Ages, except for Thomas Aquinas, who was doing natural theology in his five famous proofs for the existence of God, and perhaps a few others.[4] Medieval theologians believed that the Christian faith was based upon revealed truths that were, in the final analysis, beyond the scope of reason. It would not have occurred to them to explain the Christian faith solely in terms of reason, or in terms of nature's harmony. And they certainly would not have pursued the path of reason if it somehow challenged revealed truth.

The emergence of natural theology in the seventeenth century was made possible by a series of monumental events, among which the most important were the Protestant Reformation and the Scientific Revolution. According to Richard Westfall:

> Christianity was forced to take notice of the new natural science.... That science did affect Christianity so profoundly was due in part to the fact that science reached maturity in an age when orthodoxy was shaken and Christian thought in

4. For the five proofs, see Thomas Aquinas, *Summa Theologiae,* prima pars, qu. 2, art. 3: "Whether there is a God?" in St. Thomas Aquinas, *Summa Theologiae* (Blackfriars, 1968), vol. 2, pp. 12–17; also see Norman Kretzmann, *The Metaphysics of Creation: Aquinas's Natural Theology in Summa contra gentiles II* (Oxford: Oxford University Press, 1999).

flux. Because old beliefs were unsettled and natural science offered a new criterion of certainty, Christianity felt the impact of science through its whole frame.[5]

Revealed truths came under scrutiny in the seventeenth century, when some concluded that revelation could be better arrived at by human reason than by traditional means. John Locke, for example, declared that

> [i]n all Things of this Kind, there is little need or use of *Revelation,* GOD having furnished us with natural and surer means to arrive at the Knowledge of them. For whatsoever Truth we come to the clear discovery of, from the Knowledge and Contemplation of our own *ideas,* will always be certainer to us than those which are conveyed to us by Traditional Revelation.[6]

And elsewhere Locke declares that God might offer the light of Revelation in such a way that our reason can only give "a probable determination." In such instances, Revelation

> *must carry … against the probable Conjectures of Reason.* But yet it still belongs to *Reason,* to judge of the Truth of its being a Revelation, and of the signification of the Words, wherein it is delivered. Indeed, if any thing shall be thought *Revelation,* which is contrary to the plain Principles of Reason, and the evident Knowledge the Mind has of its own clear and distinct *Ideas;* there *Reason* must be hearkened to, as to a Matter within its Province.[7]

Indeed, Locke, and many of his contemporaries, viewed revelation as a supplement, or reinforcement, to natural religion. John Tillotson offered a typical interpretation when he declared that "[n]atural religion is the foundation of all revealed religion, and revelation is designed simply to establish its duties."[8]

But revelation itself would come to be seen as unnecessary. Isaac Newton, and many of his fellow *virtuosi* – that is, contemporaries who were

5. Westfall, *Science and Religion in Seventeenth-Century England* (N.p.: Archon Books, 1970; first published by Yale University Press, 1958), 10
6. From *John Locke An Essay Concerning Human Understanding,* edited with an Introduction by Peter H. Nidditch (Oxford: Clarendon Press, 1975), bk. IV, ch. 18, sec. 4, 690–691.
7. Locke, ibid., bk. IV, ch. 18, sec. 8, 694. For more on Locke's rational approach to religion, see Andrew Fix, *Prophecy and Reason: The Dutch Collegiants in the Early Enlightenment* (Princeton: Princeton University Press, 1991), 14.
8. John Tillotson, *Works* (ed. 1857), II, 333. Cited from John Herman Randall, *The Making of the Modern Mind: A Survey of the Intellectual Background of the Present Age,* rev. ed. (Boston: Houghton Mifflin Co., 1940), 289.

interested in, and pursued, science – laid the firm groundwork for this move, one that would reach its full development in the eighteenth century, as Westfall explains:

> Little separated Newton's religion from the 18th century's religion of reason – only the name "Christianity" and an attitude which the name implied. The virtuosi had taken up natural religion originally in defense of Christianity, and this attitude still remained dominant in Newton. In removing the fragments of irrationality he was saving Christianity from itself and defending it from skepticism. The virtuosi's concentration on natural religion meant that they treated only those aspects of Christianity to which rational proofs might apply, while they ignored the spiritual needs to which Christianity had ministered through the centuries. In defending Christianity in this manner, they prepared the ground for the deists of the Enlightenment – the mechanical universe run by immutable natural laws, the transcendent God removed and separated from His creation, the moral law which took the place of spiritual worship, the rational man able to discover the true religion without the aid of special revelation.[9]

Already during the years 1650 to 1700, a small group of Dutch Protestant thinkers, known as the Collegiants, adopted a surprisingly secular view of the world that was largely based on reason. Influenced by Baruch Spinoza, most Collegiants rejected revelation and adopted "reason as the new standard for religious truth" and thus "moved toward the natural religion of the Enlightenment."[10] In the course of the eighteenth century, supporters of natural religion, often in the guise of Deism, rejected revelation and abandoned miracles as superstitious beliefs. In 1730, Matthew Tindal published *Christianity as Old as the Creation,* one of the most important of Deist treatises. In contrast to Locke, Tillotson, and many others, Tindal argued that revelation can add nothing to natural religion because the latter is perfect. Revelation is therefore superfluous.[11] The mysteries of the Christian religion were thus abandoned. Reason alone was necessary since God had created a rationally ordered world and would not disrupt its operations by the intrusion of miracles. Indeed, after creating the world, God left it to operate by itself in accordance with the rational laws He had conferred upon it. Many Deists did not regard themselves as Christians, preferring to view their Deism as a universal religion embodying a common core of rational beliefs. The rituals and dogmas of the great variety of religions were viewed as irrelevant to true belief.

9. Westfall, *Science and Religion in Seventeenth-Century England,* 218–219.
10. Fix, *Prophecy and Reason,* 256.
11. See Randall, *The Making of the Modern Mind,* 291, and Gerald R. Cragg, *The Church and the Age of Reason 1648–1789* (New York: Atheneum, 1961), 161–163.

Natural religion, in turn, also came under attack. In the mid–eighteenth century, David Hume not only attacked belief in miracles but also criticized natural religion in two treatises: in his *Essay on Providence and a Future State* (1748) and in his *Dialogues Concerning Natural Religion,* composed in 1751 but published posthumously. Hume undercut the basis for natural religion by rejecting the notion that there could be a rational basis of faith. Relying on empiricism, rather than rationalistic argument, Hume's "method was to ask how much of traditional religious beliefs could be actually derived from facts observable in nature; and his verdict was, very little."[12] Not long after, in 1770, Baron Paul Henri Thiry d'Holbach (1723–1789) went beyond Hume and took the ultimate step, arguing for atheism and materialism in his anonymously published *System of Nature (Système de la nature, ou des lois du monde physique et du monde moral).* He rejected both the argument from a first cause and the argument from design.[13]

Immanuel Kant (1724–1804) rejected the basic premise of natural religion, namely, that reason could legitimize religion. He demonstrated that "reason and science were valid only within a certain field, and that outside this field faith – Kant called it "practical reason" – could still establish the tenets of natural religion, God, freedom, and immortality."[14] Kant destroyed the idea of a natural religion based on reason. He did this in his *Critique of Pure Reason* (1781), where he declared:

> *There are only three possible ways of proving the existence of God by means of speculative reason.*
>
> All the paths leading to this goal begin either from determinate experience and the specific constitution of the world of sense as thereby known, and ascend from it, in accordance with laws of causality, to the supreme cause outside the world; or they start from experience which is purely indeterminate, that is, from experience of existence in general; or finally they abstract from all experience, and argue completely *a priori,* from mere concepts to the existence of a supreme cause. The first proof is the *physico-theological* [that is, the argument from design], the second the *cosmological* [that is, the argument from a first cause], the third the *ontological.* There are, and there can be, no others.[15]

12. See Randall, *The Making of the Modern Mind,* 300.
13. Although it was probably not widespread, atheism found a voice in the late Middle Ages, as is made evident in a recent collection of essays, *Atheismus im Mittelalter und in der Renaissance,* eds. Friedrich Niewöhner and Olaf Pluta (Wiesbaden: Harrassowitz, 1999).
14. Randall, *The Making of the Modern Mind,* 304.
15. *Immanuel Kant's Critique of Pure Reason,* translated by Norman Kemp Smith; unabridged ed. (New York: St Martin's Press, 1965; first published 1929), Chapter III, Section 3, 499–500. I have added the bracketed phrases.

The legacy of Kant and the eighteenth century "made the foundation of religion upon the principles of scientific reason henceforth impossible."[16] Thus did the Age of Reason gradually arrive at a religion based on reason, first with revelation, then without it. Atheism emerged to argue that there could be no rational religion, since reason could not demonstrate God's existence, because God does not exist. And, finally, like the atheists, Kant applied reason to the proofs of God's existence and found that not only was the argument from design, which was the foundation of natural religion, untenable, but so also were the other two possible proofs for the existence of God. But if the foundation of religion could not be based on pure reason, Kant argued that it could be based upon faith, or practical reason. Thus were the connections between traditional religion and reason severed.

After all this, I must return to the fundamental claim of my book: that the Age of Reason began in the Middle Ages. But in what sense can the Middle Ages be regarded as the beginning of the Age of Reason? The response to this question must be seen in terms of historical process and evolution. There can be no doubt that reason played a pervasive role in the intellectual life of medieval universities. The masters in the arts faculties, teaching and writing primarily about logic and natural philosophy, and the masters in the theological faculties, teaching and writing primarily about theology, placed a high value on reason, as we saw. That reason is central to logic is self-evident; that it is the driving force in the study of natural philosophy is apparent from the nature of Aristotle's natural philosophy, which formed the basis of the curriculum of the arts faculty and remained a highly rationalistic discipline as it was expanded during the Middle Ages beyond anything that Aristotle had envisioned (see Chapter 5). But it is by no means obvious that reason would permeate theology in the manner it did during the Middle Ages. Medieval theologians imported logic and natural philosophy into theology to such an extent that they effectively secularized it. Theology became an analytical subject because it relied heavily on logico-mathematical techniques. Although reason was ubiquitous in medieval theology, theologians were not free to use their reason to arrive at conclusions that were contrary to revealed truths and Church dogma. They could only analyze revealed truths by using logic and natural philosophy, but they could not subvert them (see also Chapter 6). Nevertheless, the major contribution of medieval theology to the periods that followed, especially the seventeenth century, was the bold,

16. Randall, *The Making of the Modern Mind*, 306.

intense, and widespread application of reason to theological subjects that seemed, by their very nature, unlikely candidates for such treatment.

What medieval scholastics started, their successors in the Age of Reason completed. The emergence of natural religion gradually reversed the roles between reason and revelation, as it became the next step in the interplay between the two. Whereas reason could be applied to revealed truths in the Middle Ages, but never challenge them, the development of natural religion made reason at first equal to revelation, then elevated it above revelation, retaining the latter in a supplementary role. And, finally, revelation was regarded as superfluous, at best, and eventually as untenable, until reason itself fell victim to the critiques of David Hume and Immanuel Kant, thus allowing faith to reemerge and operate beyond the reach of rational analysis.

The Age of Reason may justifiably be said to have begun in the Middle Ages because in the latter period, Europeans had become accustomed to the self-conscious application of reason to problems in all university subjects: natural philosophy, theology, law, and medicine. By the seventeenth and eighteenth centuries, much had changed from the fourteenth and early fifteenth centuries. The Protestant Reformation and Scientific Revolution altered Europe dramatically. In the Middle Ages, reason was employed in an abstract and often a priori manner, and was frequently applied to hypothetical arguments and examples with little relevance to the real world. By contrast, nonscholastic, or better antischolastic, scholars, in the seventeenth century, beginning with Francis Bacon, and continuing on through a stellar list of natural philosophers and scientists, laid great emphasis on empirical evidence as a control on pure reason. Although we saw (in Chapter 5) that empiricism formed the basis of the medieval theory of knowledge and that considerable lip service was paid to it, we also saw that it was largely an empiricism without observation. During the late Middle Ages, reason was clearly dominant over empiricism. But this approach to nature, which satisfied medieval natural philosophers for four centuries, was found to be inadequate for revealing nature's operations and laws. One can scarcely doubt that reason was applied more fruitfully in the Age of Reason than in the Middle Ages. But it would be rash to conclude that natural philosophers in the seventeenth century, and in the Age of Reason generally, were therefore "more rational" than their medieval predecessors. Medieval scholastic theologians and natural philosophers were as dedicated to the use of reason in the disciplines they discussed and analyzed as were the scientists and natural philosophers who developed the new science in the Age of Reason.

But the disciplines, and attitudes toward those disciplines, had certainly changed. Indeed, logic as it was done in the Middle Ages was largely gone, although a few of its logical texts were still in use. Logic had been replaced by mathematics and mathematical physics, both by-products of the Scientific Revolution. Great philosophers like John Locke paid virtually no attention to logic. Moreover, by the seventeenth century, theology was no longer done in the medieval manner. Commentaries on the *Sentences* of Peter Lombard played no role in Protestant countries, and were playing a diminishing role in Catholic lands. A separate class of professional theologians, who possessed the sole right to do theology during the Middle Ages, no longer existed, at least in Protestant countries, and probably not in a Catholic country such as France. Anyone could write on theology, and numerous natural philosophers included some theological discussions within their natural philosophy. Many held combined interests in natural philosophy and religion, using the former to provide a vague kind of exhortative support for the latter, usually by showing how natural philosophy reveals God's wisdom and foresight in the natural world. Robert Boyle and Isaac Newton were two who used their skills in natural philosophy to buttress religion.

Despite the numerous changes that occurred, one can make a strong case for regarding the Middle Ages as the beginning of the Age of Reason. One of the weightiest reasons for characterizing the Middle Ages as the true beginning of the Age of Reason lies in the fact that medieval theologians regularly applied reason, in the form of logic and natural philosophy, to theology and the mysteries of the faith. One expects reason to be used extensively in logic and natural philosophy, but not in theology and religion. And yet it became routine to apply analytic techniques and rigorous argumentation to all aspects of theology, including the Trinity and the sacraments. This was done deliberately and consistently for some four centuries, despite significant opposition to the practice that surfaced from time to time.

Although I have focused here on logic, natural philosophy, and theology, reason was also applied regularly in law and medicine, and, therefore, to all university subjects. The use of reason in a self-conscious manner began in the twelfth century and has continued, without interruption, to the present. The Middle Ages was itself an Age of Reason and marks the real beginnings of the intense, self-conscious use of reason in the West. But the catch phrase "Age of Reason" is not applied to the Middle Ages, but only to the seventeenth and eighteenth centuries. Why did this happen? Was it because the medieval understanding of the relationship between reason and revelation was repudiated in the seventeenth century and replaced by natural religion, which was based primarily on reason, with revelation playing a secondary role, or no role at all? In other words, was the Middle Ages denied credit for emphasizing reason because it laid too much emphasis on revelation, whereas, by contrast, the Age of Reason was so-called because it exalted reason over revelation? However plausible this explanation might seem, it is not the historical reason for calling the seventeenth and eighteenth centuries an Age of Reason. That title was conferred because the use of reason in those two centuries was believed to stand in stark contrast with the general absence of reason in the Middle Ages or, at best, a sterile use of reason. The grossly mistaken understanding of the Middle Ages was the result of general ignorance of medieval accomplishments, accompanied by a deep bias against the medieval period as a whole.

As a corrective to this gross historical distortion, a strong counterargument can be made to regard the medieval emphasis on reason as laying a foundation for the even more extensive application of reason in what would eventually be designated the Age of Reason. Without the long medieval period of reasoned argument in logic, natural philosophy, and theology, the Age of Reason could not have occurred. Although scholars and authors in the Age of Reason were largely, if not completely, unaware of any obligation to the Middle Ages – indeed, their scorn and contempt for it were deep rooted (this will be considered later in this chapter) – they could not have launched their critiques in favor of natural religion and against revelation without the medieval background of reasoned argument in theology. Logic had been employed in the analysis of theological claims for centuries; natural philosophy had also been used to explicate theology for centuries. A tradition of five centuries of reasoned argument cannot be ignored.

What if the great emphasis on logic in the medieval universities had not occurred? What if the tradition of reasoned argument in Aristotelian natural philosophy had not been a primary aspect of a medieval university education? And, finally, what if the long-standing tradition of applying logic and natural philosophy to theology in the theology faculties of medieval universities had not occurred? In sum, would an Age of Reason have occurred if the scholars of that age had not inherited from the Middle Ages a tradition of reasonably free and rational inquiry in the crucial subjects of logic, natural philosophy, and theology? The answer must surely be in the negative. It is unlikely that an Age of Reason would have developed in the seventeenth and eighteenth centuries, because it is doubtful that a Scientific Revolution would have occurred in the seventeenth century. The Middle Ages passed on to their intellectual heirs a tradition of reasoned discourse in subjects vital to the investigation of natural phenomena and for the investigation of arguments and claims in theology.[17]

The positive achievements of medieval intellectual life developed from a confident reliance on reason. Indeed, medieval scholars unequivocally recognized and assumed "the autonomy of reason in the discussion of philosophical and theological problems."[18] Few in the seventeenth and eighteenth centuries, the Age of Reason, appreciated or suspected, any of this. Why? Because, for the most part, scholars in that period repudiated medieval logic, Aristotelian natural philosophy, and scholastic theology. With the long struggle against these medieval disciplines over, there was little inclination to say anything good or positive about the Middle Ages. Any contributions to the history of Western civilization that may have been made by scholastic thinkers were either ignored or ridiculed and scorned. A cloud of ignorance about the Middle Ages and its contributions settled over Europe until the late nineteenth and early twentieth centuries. Modern scholars have altered their opinions somewhat, but still seem reluctant to accord any significant praise for the genuine contributions of medieval scholars.

We will better appreciate and understand the reasons for this when we see the manner in which the Middle Ages in general, and medieval scholasticism in particular, were viewed by early modern and modern scholars. Indeed, our story really begins in the Middle Ages itself, when harsh criticisms were leveled at logic and logicians.

THE ONSLAUGHT AGAINST SCHOLASTICISM AND THE MIDDLE AGES

In Chapter 2, I had occasion to note opposition to the application of logic to theology in the eleventh and twelfth centuries; and, in Chapter 6, I mentioned opposition to the importation of logic and natural philosophy into theology by Church and university officials. These criticisms, however, were by scholars and officials who were familiar with the traditions and practices they were criticizing. They were, in a sense, internal critics. The opinions and viewpoints that I shall focus on for the remainder of this

17. I have argued this at greater length in Grant, *The Foundations of Modern Science in the Middle Ages: Their Religious, Institutional, and Intellectual Contexts* (Cambridge: Cambridge University Press, 1996), 191–205.
18. R. W. Southern, *The Making of the Middle Ages* (New Haven: Yale University Press, 1953), 184.

chapter are almost exclusively from those who were outside the scholastic system (an exception is Juan Luis Vives) and who found it detrimental to intellectual activities they regarded as important. In writing on the idea of progress, Robert Nisbet speaks about the attitude of Renaissance humanists toward scholasticism. He explains that "in their reaction to medieval scholasticism, which was supremely rationalist and objectivist, the humanists were necessarily carried to place an emphasis on the emotions, passions, and other nonrational affective states which were scarcely compatible with any theory of progress."[19]

Continental Critics from the Middle Ages to the Seventeenth Century

Francesco Petrarch (Petrarca) (1304–1374), the great poet, humanist, and ardent champion of classical learning, had an attitude toward logic that was similar to that of John of Salisbury (see Chapter 2). He regarded logic as useful but did not feel that it was an end in itself. A friend, Tomasso da Messina, informed Petrarch that his attitude toward logic had antagonized an old logician, who thought that Petrarch wished to condemn logic. To this Petrarch replied:

> Far from it; I know well in what esteem it was held by that sturdy and virile sect of philosophers, the Stoics, whom our Cicero frequently mentions, especially in his work De Finibus. I know that it is one of the liberal studies, a ladder for those who are striving upwards, and by no means a useless protection to those who are forcing their way through the thorny thickets of philosophy. It stimulates the intellect, points out the way of truth, shows us how to avoid fallacies, and finally, if it accomplishes nothing else, makes us ready and quick-witted.
>
> All this I readily admit, but because a road is proper for us to traverse, it does not immediately follow that we should linger on it forever. No traveller, unless he be mad, will forget his destination on account of the pleasures of the way; his characteristic virtue lies, on the contrary, in reaching his goal as soon as possible, never halting on the road. And who of us is not a traveller? We all have our long and arduous journey to accomplish in a brief and untoward time, – on a short, tempestuous, wintry day as it were. Dialectics may form a portion of our road, but certainly not its end: it belongs to the morning of life, not to its evening. We may have done once with perfect propriety what it would be shameful to continue. If as mature men we cannot leave the schools of logic because we have found pleasure in them as boys, why should we blush to play odd and even, or prance upon a shaky reed, or be rocked again in the cradle of our childhood?...

19. See Robert Nisbet, *History of the Idea of Progress* (New York: Basic Books, Inc., 1980), 105.

Who would not scorn and deride an old man who sported with children, or marvel at a grizzled and gouty stripling? What is more necessary to our training than our first acquaintance with the alphabet itself, which serves as the foundation of all later studies; but, on the other hand, what could be more absurd than a grandfather still busy over his letters?

Use my arguments with the disciples of your ancient logician. Do not deter them from the study of logic; urge them rather to hasten through it to better things. Tell the old fellow himself that it is not the liberal arts which I condemn, but only hoary-headed children. Even as nothing is more disgraceful, as Seneca says, than an old man just beginning his alphabet, so there is no spectacle more unseemly than a person of mature years devoting himself to dialectics. But if your friend begins to vomit forth syllogisms, I advise you to take flight, bidding him argue with Enceladus. Farewell.[20]

Although Petrarch was harsh on those who pursued logic zealously, he exempted Aristotle, whom he lauded as "a man of the most exalted genius, who not only discussed but wrote upon themes of the very highest importance."[21] It was not Aristotle, but Aristotle's followers – Aristotelians – whom he disdained. Thus, he mentions friends who "are so captivated by their love of the mere name 'Aristotle' that they call it a sacrilege to pronounce any opinion that differs from his on any matter."[22] And elsewhere Petrarch says: "I snarl at the stupid Aristotelians, who day by day in every single word they speak do not cease to hammer into the heads of others Aristotle whom they know by name only."[23]

Italian humanists played a significant role in generating a hostile reaction to scholastic learning. Lorenzo Valla (1407–1457) was perhaps the most

20. From James Harvey Robinson, ed. and trans., *Francesco Petrarca: The First Modern Scholar and Man of Letters* (New York: G. P. Putnam, 1898), 221–223. Enceladus is "a ferocious giant of the ancient myth whom Jupiter buried under the flaming masses of Mount Etna." See the translation of the same letter ("A Disapproval of an Unreasonable Use of the Discipline of Dialectic") by Hans Nachod, in Ernst Cassirer, Paul Oskar Kristeller, and John Herman Randall, Jr., eds., *The Renaissance Philosophy of Man* (Chicago: University of Chicago Press, 1948), 137–139.

21. Robinson, *Francesco Petrarca,* 219.

22. From Petrarch's "On His Own Ignorance and That of Many Others" *(De sui ipsius et multorum ignorantia),* translated by Hans Nachod, in Cassirer, Kristeller, and Randall, *The Renaissance Philosophy of Man,* 102.

23. Petrarch, "On His Own Ignorance and That of Many Others," 107. For a good account of the reaction against dialectics, or logic, among the ancients, theologians, and humanists, see Lawn, *The Rise and Decline of the Scholastic 'Quaestio Disputata' With Special Emphasis on its Use in the Teaching of Medicine and Science* (Leiden: E. J. Brill, 1993), Chapter 9, 101–128. On Petrarch, see p. 107.

striking and important. He sought to reform dialectic and natural philosophy by refuting Aristotle and the Aristotelians.[24] In the process, he made some harsh remarks about Aristotle, but reserved his sharpest comments for Aristotle's scholastic followers. In what is surely a maliciously false description, Valla declared that "it is embarrassing to relate the initiation rites of his disciples. They swear an oath never to contradict Aristotle – a superstitious and foolish lot, who do a disservice to themselves. They deprive themselves of an opportunity to investigate the truth." Elsewhere in the same treatise, Valla proclaims that" modern Peripatetics are intolerable. They deny a person who does not adhere to any school the right to disagree with Aristotle."[25] And, finally, in the same vein, Valla attacks them for their narrow-minded attitude, because

> [t]hey regard all other philosophers as nonphilosophers and embrace Aristotle as the only wise man, indeed the wisest – not surprisingly, since he is the only writer they know. If one can call it 'knowing,' for they read him, not in his own language, but in a foreign, not to say corrupt, language. Most of his works are wrongly translated, and much that is well said in Greek is not well said in Latin.[26]

In the sixteenth century, attacks on the scholastics intensified. Medieval logic, natural philosophy, and theology all came under fire. One of the earliest attacks from northern Europe came from Desiderius Erasmus of Rotterdam (ca. 1469–1536), the greatest of Renaissance humanists. In 1511, Erasmus published his famous *Praise of Folly*, a work he had written, he says, in the course of a week in 1509 to amuse Thomas More, whose houseguest he was in England. Although Erasmus did not regard *Praise of Folly* as a major work, he revised it numerous times until the final edition of 1532 (published in Basle). Much to Erasmus's surprise, it was enormously successful, as evidenced by the fact that it was translated into numerous languages and saw 36 Latin editions before Erasmus died in 1536.[27] In his

24. On the details of Valla's reforms in dialectic and natural philosophy, see Charles Trinkaus, "Lorenzo Valla's Anti-Aristotelian Natural Philosophy," in *I Tatti Studies* (1993): 279–325.
25. These quotations from Valla appear in Erika Rummel, *The Humanist-Scholastic Debate in the Renaissance & Reformation* (Cambridge, MA: Harvard University Press, 1995), 160. Rummel's translations are from Valla's *Retractio totius dialectice cum fundamentis universe philosophie,* which is a third redaction of Valla's earlier *Repastinatio dialectice et philosophie.* The work has been edited under the latter title by G. Zippel (Padua, 1982).
26. Rummel, ibid., 160. For more on Valla, See Lawn, *The Rise and Decline of the Scholastic 'Quaestio Disputata,'* 108–109.
27. Erasmus, *Praise of Folly*, trans. Betty Radice; revised with a new introduction and notes by A. H. T. Levi (London: Penguin Books, 1993; first published 1971), xi–xii.

edition of 1514, Erasmus turned Folly's attention briefly to philosophers, and then, at much greater length, to scholastic theology and theologians. The philosophers, or better, natural philosophers,

know nothing at all, yet they claim to know everything. Though ignorant even of themselves and sometimes not able to see the ditch or stone lying in their path, either because most of them are half-blind or because their minds are far away, they still boast that they can see ideas, universals, separate forms, prime matters, <quiddities, ecceities> things which are all so insubstantial that I doubt if even Lynceus could perceive them.[28]

It is, however, against the theologians, whom Erasmus regarded as "a remarkably supercilious and touchy lot," that he directs his sharpest shafts. Theologians, Erasmus explains,

interpret hidden mysteries to suit themselves: how the world was created and designed; through what channels the stain of sin filtered down to posterity; by what means, in what measure, and how long Christ was formed in the Virgin's womb; how, in the Eucharist, accidents can subsist without a domicile. But this sort of question has been discussed threadbare. There are others more worthy of great and enlightened theologians (as they call themselves) which can really rouse them to action if they come their way. What was the exact moment of divine generation? Are there several filiations in Christ? Is it a possible proposition that God the Father could hate his Son? Could God have taken on the form of a woman, a devil, a donkey, a gourd, or a flintstone? If so, how could a gourd have preached sermons, performed miracles, and been nailed to the cross? And what would Peter have consecrated <if he had consecrated> when the body of Christ still hung on the cross? Furthermore at that same time could Christ have been called a man? Shall we be permitted to eat and drink after the resurrection? We're taking due precaution against hunger and thirst while there's time.[29]

To Erasmus and his fellow humanists, the kinds of questions theologians considered in their commentaries on the *Sentences* were absurd. In the chapter on theology (Chapter 6), we saw that scholastic theologians were fascinated with all sorts of questions that tested God's powers. Erasmus regards such questions as silly. Indeed, after citing similar paradoxical questions, Erasmus asks "Who *could* understand all this unless he has frittered away thirty-six whole years over the physics and metaphysics of Aristotle and Scotus?"[30]

28. Erasmus, ibid., 85.
29. Ibid., 86–87. See the valuable note (n. 107) to this paragraph by A. H. T. Levi on pp. 87–88.
30. Erasmus, ibid., 91.

After describing some "theological minutiae," Erasmus declares that those theologians are

> so busy night and day with these enjoyable tomfooleries, that they haven't even a spare moment in which to read the Gospel or the letters of Paul even once through. And while they're wasting their time in the schools with this nonsense, they believe that just as in the poets Atlas holds up the sky on his shoulders, they support the entire church on the props of their syllogisms and without them it would collapse. Then you can imagine their happiness when they fashion and refashion the Holy Scriptures at will, as if these were made of wax, and when they insist that their conclusions, to which a mere handful of scholastics have subscribed, should carry more weight than the laws of Solon and be preferred to papal decrees.[31]

And near the end of his attack on theologians, Erasmus explains:

> They also set up as the world's censors, and demand recantation of anything which doesn't exactly square with their conclusions, explicit and implicit, and make their oracular pronouncements: "This proposition is scandalous; this is irreverent; this smells of heresy; this doesn't ring true." As a result, neither baptism nor the gospel, neither Paul, Peter, St Jerome, Augustine, or even Thomas, 'the greatest of Aristotelians', can make a man Christian unless these learned bachelors have given their approval, such is the refinement of their judgment. For who could have imagined, if the savants hadn't told him, that anyone who said that the two phrases "chamber-pot you stink" and "the chamber-pot stinks", or "the pots boil" and "that the pot boils" are equally correct can't possibly be a Christian?[32] Who could have freed the church from the dark error of its ways when no one would ever have read about these if they hadn't been published under the great seals of the schools? And aren't they perfectly happy doing all this? They are happy too while they're depicting everything in hell down to the last detail, as if they'd spent several years there, or giving free rein to their fancy in fabricating new spheres and adding the most extensive and beautiful of all in case the blessed spirits lack space to take a walk in comfort or give a dinner-party or even play a game of ball.[33]

31. Erasmus, ibid., 93–94.
32. In a note to this sentence (n. 112), Levi believes that it "mocks at an Oxford dispute about correct Latin grammatical usage. The usage of accents in the original text suggests that abstruse fun is being made of a technical dispute. Folly replaces the traditional examples with those involving 'chamber-pot'." It is more likely, however, that Erasmus was poking fun at the way scholastics used word order to convey different meanings to two sentences that humanists would ordinarily have translated in the same way. See Chapter 4 where I discuss word order and cite Barbara and Norman Kretzmann, who regard such changes in word order as a rational procedure that foreshadows the use of formal, logical notation.
33. Erasmus, *Praise of Folly*, 94–95. In the last sentence, mention of a new sphere that would enable "the blessed spirits to take a walk in comfort," and engage in other activities, is probably a reference to the empyrean heaven.

Friends of Erasmus shared his hostility to scholasticism and the Middle Ages. Guillaume Budé, a great admirer of Greek culture, declares himself "thankful and a little surprised that anything of merit had been saved from the deluge of more than a thousand years; for a deluge indeed, calamitous of life, had so drained and absorbed literature itself and the kindred arts worthy of the name, and kept them so dismantled and buried in barbarian mud that it was a wonder they could still exist."[34]

Few critics of medieval scholastic logic were more knowledgeable than Juan Luis Vives (1493–1540), a Spaniard born in Valencia, but trained in Paris in scholastic logic in the early sixteenth century, although he soon after fervently embraced humanism. In 1512, Vives went to live in Bruges, where he spent the rest of his life, except for trips to Louvain and England. During a trip to the former in 1515, at the age of 23, he met Erasmus, who was then 50 and who made a strong impression on young Vives. Later, he traveled to England, where he met Thomas More.[35]

Vives was renowned for his writings on education, but was perhaps the major critic of the medieval logical tradition in the sixteenth century. In 1520, Vives published *Against the Pseudodialecticians,* an intense attack against the obscurities of logic as it was taught in the universities. Since Vives uses the language of logic, his criticisms are usually difficult to understand, but the savage tone of it is unmistakable. For example, with scorn and humor, he attacks the way medieval logicians analyzed the terms "begins," "ceases," and "instants" (see Chapter 4):

> By now one feels ashamed even to mention 'begins' and 'ceases'. Whoever taught us such subtle rigor, such subtle 'instants', such dull frivolities? In what language were these thought out? In Greek, or Latin? In Spanish, or French? Who ever said that a child cannot begin to learn an hour after he is brought to school? But the dialecticians [that is, the logicians] deny that he can, because many instants have flowed by after the first one in which he began to learn. Then they also say this statement is false, 'That spring now begins to appear two or three hours after the water first began to flow'. And they do not concede that 'This tree ceases to bloom a short while before it finally stops producing flowers', and 'A fountain ceases to flow half an hour before it dries up'. And in this hair-splitting the meaning of the words

34. See Russell Fraser, *The Dark Ages & the Age of Gold* (Princeton: Princeton University Press, 1973), 4. The passage is from Budé's *De studio literarum,* 1527.

35. Here, I rely on Rita Guerlac, *Juan Luis Vives Against the Pseudodialecticians: A Humanist Attack on Medieval Logic. The Attack on the Pseudodialecticians* and *On Dialectic,* Book III, v, vi, vii from *The Causes of the Corruption of the Arts with an Appendix of Related Passages by Thomas More,* the Texts, with translation, introduction, and notes by Rita Guerlac (Dordrecht, Holland: D. Reidel Publishing Co., 1979), 24–25.

'begins' and 'ceases' has shrunk until now they can be of no use at all. I think that according to their rules it can be said of absolutely nothing that it either begins or ceases either to be or to act.[36]

It was by citing frequent examples, or perhaps exaggerated, or even distorted, examples, that Vives drove home his point in the most effective manner – by satire, mockery and irony. He tells his readers that

they [the logicians, or "sophisters", as he frequently calls them] have invented for themselves certain meanings of words contrary to all civilized custom and usage, so that they may seem to have won their argument when they are not understood.

For when they are understood, it is apparent to everyone that nothing could be more pointless, nothing more irrational. So, when their opponent has been confused by strange and unusual meanings and word-order, by wondrous suppositions, wondrous ampliations, restrictions, appellations, they then decree for themselves, with no public decision or sentence, a triumph over an adversary not conquered but confused by new feats of verbal legerdemain. Truly, would not Cato, Cicero, Sallust, Livy, Quintilian, Pliny, and Marcus Varro (recognized as the first Latin writers on logic) be utterly at a loss to hear one of these sophisters make statements like these:

When he is full of drink, swear on the stone Jove that he has not drunk wine, because he has not drunk wine that is in India.

When he sees the King of France attended by a great retinue of servants, say This King does not have servants because he does not have those who wait on the King of Spain.

Though Varro is a man, yet he is not a man because Cicero is not himself Varro.

That a head no man has even though no man lacks a head.

There are more non-Romans than Romans in this hall, in which there are a thousand Romans and two Spaniards ...

Socrates and this donkey are brothers.

Two contradictory statements are also in a contradictory sense true ...

And what about these?

Any donkey as it were of a man *c* is *b* not an animal.

A man *a* and any man of any kind not Sortes [that is, Socrates] are necessarily both another man and *d p*.

So that *a, b, c, d* can make those suppositions confused, determinate, and a mixture of both.

Indeed you can add more commixtions than any quack pharmacist ever made – *e, f, g, h, i, j, k* – so that some of these men already have recourse to letters down as far as the tenth letter of the second alphabet, dreaming up and combining wonder-

36. Guerlac, trans., ibid., 71.

ful kinds of suppositions. It is clear that these men envied the mathematicians, because they alone seemed to use letters. Therefore, these, too, have taken over the whole alphabet for their own use, so that no one who observes them can deny that men of this kind are extremely 'literate'. But when they proceed to mathematics ... they are much dismayed, because they do not quite understand what those letters mean. I hear that one of them, having taken up the study of geometry, decided that a line designated as *b* was posited determinately while the one marked *a* was in the same state as himself, to wit, merely confused. For *a* and *b* are of such potency that one misplaced *b* can render separate and determined the whole confused and inseparable order of the lower regions or of the ancient Chaos. And on the contrary, a single *a* can invert and confuse the perfect order of the heavens.[37]

Thus did Vives lampoon the methods and ways of medieval logicians by using their own terminology – "supposition," "determinately," and "merely confused" (see Chapter 4) – and examples against them. On the face of it, what Vives presents seems absurd, incomprehensible, and laughable. The use of letters to designate quantities, groups, and individuals goes back to Aristotle and was widespread in the Middle Ages in logic and natural philosophy. Vives's illustrations, however, are intended to show the absurd ways logicians used them. Indeed, he would even have us believe that they would foolishly apply the jargon of logic to letters designating lines in geometry.

Against the Pseudodialecticians is filled with such examples and attacks. "And because of its brevity and force," Rita Guerlac explains, "it was widely read, and because of its eloquence and acerbity, frequently quoted."[38] Thus did Vives play a significant role in the undoing of scholastic logic in partic-

37. Guerlac, trans., ibid., 57–61. With reference to Vives's criticism, E. J. Ashworth explains that "[t]he one original development found in the works of the Parisian logicians and later satirised by Vives concerned the use of 'a' and 'b' as special signs of supposition, to be used especially in the analysis of such propositions as 'Of every man some donkey is running' and 'Every man has a head', which posed special problems for the theory of supposition. A simple example of how the signs were used is 'Every man is b animal' which, unlike 'Every man is animal', signifies, by virtue of the special sign 'b', that every man is identical to one and the same animal." See Ashworth, "The Eclipse of Medieval Logic," in Norman Kretzmann, Anthony Kenny, and Jan Pinborg, eds., *The Cambridge History of Later Medieval Philosophy* (Cambridge: Cambridge University Press, 1982), Chapter 42, 792. With propositions like "Of every man some donkey is running" and "Every man is *b* animal," it is hardly surprising that Vives, and many others, resorted to ridicule, satire, and straightforward denunciation against the subject of logic or dialectic. And yet, to those who were trained in the subject, these sentences made sense and modern scholars have been able to decipher them.
38. Guerlac, *Vives,* 3. For more on Vives, see Lawn, *The Rise and Decline of the Scholastic 'Quaestio Disputata,'* 117–120.

ular, and of scholasticism in general. Indeed, he also skewered scholastic physicians by observing that they also were ensnared in the desire to dispute about logical subtleties, carrying

> as it were, huge wagon loads of material for disputation, concerning the intension and remission of forms, rarity and density, proportional parts, instants, and things which neither are nor ever will be, airing their dreams and meanwhile forsaking their fight with the diseases in the locality, which afflict and kill people.[39]

The kinds of scholastic logical propositions that were attacked by Vives (and by Thomas More, as we shall see) were drawn from formal logic as it was taught in the arts courses of medieval universities up through the sixteenth century. Humanists rebelled against this approach, regarding it as sterile and mostly unintelligible. A new approach to dialectic was fashioned by Lorenzo Valla and Rudolph Agricola. They were no longer concerned with certainty based on formal syllogisms but, rather, emphasized techniques of persuasion in argument and probable and plausible arguments:

> What was needed in the way of dialectic for the humanist arts course was a simple introduction to the analysis and use of ordinary language (elegant Latin) for formal debating and clear thinking. Such an analysis was provided in the 'alternative' dialectics deriving from the work of Valla and Agricola. These were committed to 'plausibility' as the measure of successful argumentation. Their authors impressively argued the case for giving serious consideration to non-syllogistic forms of argument, and strategies which support or convince rather than prove, as an intrinsic part of the 'art of discourse.'[40]

To meet these new needs, medieval logic, with its incredibly complex terminology and concepts, was pushed into the background, and gradually ceased to be understood or readily available.

Even as scholastic logic was being severely criticized and ridiculed, so also did theology come under fire. We saw (Chapter 6) that official admonitions to theologians to desist from an overemphasis on logic and natural philosophy had always failed. But an attack on medieval theology from Martin Luther in the sixteenth century had a more profound impact. In

39. From Lawn, *The Rise and Decline of the Scholastic 'Quaestio Disputata,'* 119. The passage is from Vives's *De causis corruptarum artium,* bk. 5. In note 61, Lawn gives the Latin text on which he based his translation. Vives had an unerring eye for the topics that captivated medieval logicians. Most of these are mentioned in Chapter 4 of this study.

40. Lisa Jardine, "Humanism and the Teaching of Logic," in Kretzmann et al., eds., *The Cambridge History of Later Medieval Philosophy,* 806.

1517, Luther wrote a treatise titled *Against Scholastic Theology* in which, as the title suggests, he attacked scholastic theology, and also, we may add, attacked Gabriel Biel, who had been the first professor of theology at the new University of Tübingen. Luther repudiated and denounced all of the features and characteristics that made medieval theology a rationalistic enterprise. Among 97 criticisms against the scholastics, Luther included the following:[41]

43. It is an error to say that no man can become a theologian without Aristotle. This in opposition to common opinion.
44. Indeed, no one can become a theologian unless he becomes one without Aristotle.
45. To state that a theologian who is not a logician is a monstrous heretic – this is a monstrous and heretical statement. This in opposition to common opinion.
46. In vain does one fashion a logic of faith, a substitution brought about without regard for limit and measure. This in opposition to the new dialecticians.
47. No syllogistic form is valid when applied to divine terms. This in opposition to the Cardinal.[42]
48. Nevertheless it does not for that reason follow that the truth of the doctrine of the Trinity contradicts syllogistic forms. This in opposition to the same new dialecticians and to the Cardinal.
49. If a syllogistic form of reasoning holds in divine matters, then the doctrine of the Trinity is demonstrable and not the object of faith.
50. Briefly, the whole Aristotle is to theology as darkness is to light. This in opposition to the scholastics.

By speaking against "the whole Aristotle," Luther, who was himself a doctor of theology (1512), makes it clear that his attack was aimed not only against the application of logic to theology but also against the extensive use of natural philosophy. Theology as it was done in the Middle Ages would soon fall victim to the Protestant Reformation. In Protestant lands, theology would no longer be the exclusive preserve of professional theolo-

41. *Luther's Works,* Vol. 31: *Career of the Reformer: I,* ed. Harold J. Grimm (Philadelphia: Muhlenberg Press, 1957), 12.

42. In a note (*Luther's Works,* vol. 31, 12), the editor explains that Luther is here referring to "the Cardinal of Cambrai, Pierre d'Ailly (1350–1420), a French theologian, a commentator on the *Sentences* of Peter Lombard and guiding spirit of the conciliar movement which led to the calling of the Council of Constance (1414–1418)."

gians. Moreover, with the gradual abandonment of Peter Lombard's *Sentences* as the primary theological text – first in Protestant universities and then in Catholic universities, where it was replaced by the *Summa theologiae* of St. Thomas Aquinas – the tradition of lectures and commentaries on the *Sentences* was basically over by the end of the sixteenth century, although the *Sentences* continued to be used into the seventeenth century.

One of the most effective critics of Aristotelians and Aristotelianism was Peter Ramus (1515–1572), who was famous for his works on method. Ramus criticized scholastic preoccupation with the syllogism, arguing that the purpose of logic was not to do syllogisms but to sway audiences with rhetorical arguments. He was also severely critical of Aristotle's *Physics* and of his natural philosophy. He found traditional commentaries on Aristotle's natural philosophy of no value because they contained "nothing that points to nature: nothing, if you regard the truth of nature, that is not confused, muddied up, contaminated, and distorted."[43]

Medieval natural philosophy did not suffer the same fate as logic and theology in the sixteenth century. Few critical assaults were launched against it until the seventeenth century, but when unleashed, they were deadly. For

> it was ... the early seventeenth-century critique that finally assured Aristotle's doom as *Maestro di color che sanno* (the Master of those who know). Several of the major philosophers and scientists of the generation of Bacon and Galilei on to that of Hobbes and Descartes sealed the fate of Aristotelianism as a coherent philosophy, at least from an intellectual if not wholly from a historical point of view.[44]

One of the most potent attacks came from the pen of Galileo (1564–1642), who was undoubtedly the most devastating, single foe Aristotelian natural philosophy had ever confronted. With the dissemination of Galileo's writings, and the great fame he achieved, not only was medieval natural philosophy mortally wounded, but his ridicule and sarcasm also created a negative image of the Middle Ages that has remained remarkably persistent to the present.

In his famous work of 1613, *History and Demonstrations Concerning Sunspots and Their Phenomena,* which consists of three letters to Mark

43. Cited in Fraser, *The Dark Ages & the Age of Gold,* 4. For a good article on Ramus, see Michael S. Mahoney, "Ramus, Peter," in *Dictionary of Scientific Biography,* vol. 11 (1975), 286–290.
44. See Charles Schmitt, *Aristotle and the Renaissance* (Cambridge, MA: Published for Oberlin College by Harvard University Press, 1983), 5.

Welser, a wealthy merchant of Augsburg with an interest in science, Galileo had occasions to remark on Aristotelian natural philosophy and its practitioners. The famous telescopic discoveries he made of celestial phenomena were rejected by most Aristotelian commentators, who tried to explain them away in terms of Aristotelian natural philosophy. Some insisted that his telescopic discoveries lacked real existence in the heavens – they were mere illusions – produced by the telescope itself. In his third letter, Galileo replied to those who held such opinions by declaring:

I believe that there are not a few Peripatetics on this side of the Alps who go about philosophizing without any desire to learn the truth and the causes of things, for they deny these new discoveries or jest about them, saying that they are illusions. It is about time for us to jest right back at these men and say that they likewise have become invisible and inaudible. They go about defending the inalterability of the sky, a view which perhaps Aristotle himself would abandon in our age.[45]

Near the end of the third letter, and the treatise itself, Galileo, after declaring that he will examine Peripatetic arguments at another time, chastises Aristotelians for their defense of false conclusions. "It appears to me," he argues,

not entirely philosophical to cling to conclusions once they have been discovered to be manifestly false. These men are persuaded that if Aristotle were back on earth in our age, he would do the same – as if it were a sign of more perfect judgment and a more noble consequence of deep learning to defend what is false than to learn the truth! People like this, it seems to me, give us reason to suspect that they have not so much plumbed the profundity of the Peripatetic arguments as they have conserved the imperious authority of Aristotle. It would be enough for them, and would save them a great deal of trouble, if they were to avoid these really dangerous arguments; for it is easier to consult indexes and look up texts than to investigate conclusions and form new and conclusive proofs. Besides, it seems to me that we abase our own status too much and do this not without some offense to Nature (and I might add to divine Providence), when we attempt to learn from Aristotle that which he neither knew nor could find out, rather than consult our own senses and reason. For she, in order to aid our understanding of her great works, has given us two thousand more years of observations, and sight twenty times as acute as that which she gave Aristotle.[46]

Galileo's most devastating attacks on scholastic natural philosophers appear in his great cosmological work, *Dialogue on the Two Chief World*

45. Galileo, *Letters on Sunspots,* in Drake, *Discoveries and Opinions of Galileo,* 140–141.
46. Galileo, ibid., 142–143.

Systems, Ptolemaic and Copernican, published in 1632. It was this work that eventually brought the condemnation of the Church down on Galileo. In writing his treatise, Galileo was quite aware that Copernicus's *On the Revolutions of the Heavenly Orbs* had been condemned by the Congregation of the Index on March 5, 1616. It was condemned because Copernicus had presented his heliocentric theory as if it were true, rather than merely as a hypothesis, which would have been acceptable. If the heliocentric system were true, it meant that the traditional Aristotelian geocentric and geostatic system was false. Although Copernicus's treatise created relatively little religious tension when it was published in 1543 – it was unopposed by the Church and was even dedicated to Pope Paul III – much had happened to alter attitudes by 1616, when it was condemned until corrected.

When he wrote his *Dialogue on the Two Chief World Systems,* Galileo realized that he had to give the appearance of upholding the Church's official position, even though he rejected the traditional Aristotelian cosmos and believed in the truth of the heliocentric system as presented by Copernicus. To achieve this objective, Galileo used a dialogue form involving three speakers who discoursed over three days. One speaker, named Simplicio, defended the traditional Aristotelian-Ptolemaic geocentric and geostatic view of the cosmos; another, whom he called Salviati, defended the heliocentric view of the world espoused by Nicholas Copernicus in his great work of 1543; the third speaker was Sagredo, who was ostensibly neutral and who listened to the arguments of both sides and only then arrived at an opinion. Within this format, the Copernican spokesman, "in many places throughout the book, usually at the end of a particular topic," sought to convince readers that Galileo was not trying to demonstrate the truth of the Copernican system, but was merely trying to dispense "information and enlightenment" and "not to decide the issue, which is a task reserved for the proper authorities."[47] For complex reasons that cannot be presented here, Galileo's strategy did not work and his book was condemned in 1633.

In the document of condemnation, issued June 22, 1633, the inquisitors-general, who were cardinals commissioned by the pope to investigate Galileo and his book, came to the following judgment:

> We say, pronounce, sentence, and declare that you, the above-mentioned Galileo, because of the things deduced in the trial and confessed by you as above, have rendered yourself according to this Holy Office vehemently suspected of

47. This is the interpretation of Maurice A. Finocchiaro, ed. and trans., *The Galileo Affair: A Documentary History* (Berkeley/Los Angeles: University of California Press, 1989), 34.

heresy, namely of having held and believed a doctrine which is false and contrary to the divine and Holy Scripture: that the sun is the center of the world and does not move from east to west, and the earth moves and is not the center of the world, and that one may hold and defend as probable an opinion after it has been declared and defined contrary to Holy Scripture. Consequently you have incurred all the censures and penalties imposed and promulgated by the sacred canons and all particular and general laws against such delinquents. We are willing to absolve you from them provided that first, with a sincere heart and unfeigned faith, in front of us you abjure, curse, and detest the above-mentioned errors and heresies, and every other error and heresy contrary to the Catholic and Apostolic Church, in the manner and form we will prescribe to you.

Furthermore, so that this serious and pernicious error and transgression of yours does not remain completely unpunished, and so that you will be more cautious in the future and an example for others to abstain from similar crimes, we order that the book *Dialogue* by Galileo Galilei be prohibited by public edict.[48]

This unfortunate condemnation not only boomeranged against the Church but also reverberated into the past to produce a negative reaction against the Middle Ages. Those who condemned the Church's action in the seventeenth century as antiscientific and a blow against freedom of thought did not trouble to distinguish between the Church of the seventeenth century and the medieval Church. Whatever hostile and negative descriptions were applied to the actions of the Catholic Church in the seventeenth century were also applied to the Church of the Middle Ages, as well as to all aspects of medieval intellectual life. Although many, if not most, of these characterizations are untrue for the Middle Ages, few troubled to differentiate the periods. From the day of publication, Galileo's work was a success – by the time it was banned, it had been sold out.[49] And after it was banned from circulation, Galileo's *Dialogue,* not surprisingly became enormously successful in those parts of Europe where the Church's writ did not extend.

The second profound impact Galileo's work had on the Middle Ages derived from his effort to lend whatever support he could to the Copernican system. One effective way of achieving this objective was to subvert Aristotle and his scholastic followers. And this he did brilliantly. Galileo's favorite theme, one that he never tired of emphasizing, was the slavish, thoughtless devotion of scholastic natural philosophers to the words of Aristotle.

48. Finocchiaro, ibid., 291.

49. See Galileo Galilei, *Dialogue Concerning the Two Chief World Systems – Ptolemaic & Copernican,* translated by Stillman Drake, foreword by Albert Einstein (Berkeley/Los Angeles: University of California Press, 1962), xxiii.

In a striking passage, Galileo has Sagredo, the neutral observer, relate the following experience:

> One day I was at the home of a very famous doctor in Venice, where many persons came on account of their studies, and others occasionally came out of curiosity to see some anatomical dissection performed by a man who was truly no less learned than he was a careful and expert anatomist. It happened on this day that he was investigating the source and origin of the nerves, about which there exists a notorious controversy between the Galenist and Peripatetic doctors. The anatomist showed that the great trunk of nerves, leaving the brain and passing through the nape, extended on down the spine and then branched out through the whole body, and that only a single strand as fine as a thread arrived at the heart. Turning to a gentleman whom he knew to be a Peripatetic philosopher, and on whose account he had been exhibiting and demonstrating everything with unusual care, he asked this man whether he was at last satisfied and convinced that the nerves originated in the brain and not in the heart. The philosopher, after considering for awhile, answered: "You have made me see this matter so plainly and palpably that if Aristotle's text were not contrary to it, stating clearly that the nerves originate in the heart, I should be forced to admit it to be true."[50]

Sagredo remarks that the Peripatetic who assumed that Aristotle must be right about the nerves originating in the heart did not offer any evidence or cite any argument from Aristotle, but thought he could win the day by simply invoking Aristotle's name. Simplicio, still hoping to show that Aristotle might be right about the nerves originating from the heart, replies that Aristotle buried many of his ideas in different parts of his works and that the true student of Aristotle "must have a grasp of the whole grand scheme, and be able to combine this passage with that, collecting together one text here and another very distant from it. There is no doubt that whoever has this skill will be able to draw from his books demonstrations of all that can be known; for every single thing is in them."[51] To this Sagredo replies that he could do the same thing with the verses of Virgil and Ovid, cobbling them together to explain

> all the affairs of men and the secrets of nature. But why do I speak of Virgil, or any other poet? I have a little book, much briefer than Aristotle or Ovid, in which is contained the whole of science, and with very little study one may form from it the most complete ideas. It is the alphabet, and no doubt anyone who can properly join and order this or that vowel and these or those consonants with one another

50. Galileo, *Dialogue Concerning the Two Chief World Systems,* The Second Day, 107–108.
51. Ibid., 108.

can dig out of it the truest answers to every question, and draw from it instruction in all the arts and sciences.[52]

A few lines below, Galileo has another bizarre tale to tell about an Aristotelian who tried to attribute the invention of the telescope to Aristotle. Salviati recounts that

certain gentlemen still living and active were present when a doctor lecturing in a famous Academy, upon hearing the telescope described but not yet having seen it, said that the invention was taken from Aristotle. Having a text fetched, he found a certain place[53] where the reason is given why stars in the sky can be seen during daytime from the bottom of a very deep well. At this point the doctor said: "Here you have the well which represents the tube; here the gross vapors, from whence the invention of glass lenses is taken; and finally here is the strengthening of the sight by the rays passing through a diaphanous medium which is denser and darker."[54]

After a call for greater respect for Aristotle by Simplicio, Salviati replies to the ridiculous claim and says to Simplicio:

Tell me, are you so credulous as not to understand that if Aristotle had been present and heard this doctor who wanted to make him inventor of the telescope, he would have been much angrier with him than with those who laughed at this doctor and his interpretations? Is it possible for you to doubt that if Aristotle should see the new discoveries in the sky he would change his opinions and correct his books and embrace the most sensible doctrines, casting away from himself those people so weak-minded as to be induced to go on abjectly maintaining everything he had ever said? Why, if Aristotle had been such a man as they imagine, he would have been a man of intractable mind, of obstinate spirit, and barbarous soul; a man of tyrannical will who, regarding all others as silly sheep, wished to have his decrees preferred over the senses, experience, and nature itself. It is the followers of Aristotle who have crowned him with authority, not he who has usurped or appropriated it to himself. And since it is handier to conceal oneself under the cloak of another than to show one's face in open court, they dare not in their timidity get a single step away from him, and rather than put any alterations into the heavens of Aristotle, they want to deny out of hand those that they see in nature's heaven.[55]

Galileo was apparently convinced that virtually all scholastics were dubious about the validity of observations made by the telescope, which had

52. Ibid., 109.
53. Drake cites Aristotle's work *On Generation and Corruption,* 5.1.780b.21.
54. Galileo, *Dialogue,* The Second Day, 109.
55. Galileo, ibid., 110–111.

been invented in 1608. Simplicio is made to reinforce this attitude. After Sagredo makes reference to a few of Galileo's works involving telescopic observations, Simplicio declares that

[f]rankly, I had no interest in reading those books, nor up till now have I put any faith in the newly introduced optical device. Instead, following in the footsteps of other Perpatetic philosophers of my group, I have considered as fallacies and deceptions of the lenses those things which other people have admired as stupendous achievements.[56]

Galileo frequently proclaims his respect for Aristotle, but finds his "devoted" followers intolerable. Indeed, he emphasizes that their single-minded allegiance to Aristotle's texts actually discredits Aristotle. He puts this sentiment into the mouth of Salviati, who declares:

I often wonder how it can be that these strict supporters of Aristotle's every word fail to perceive how great a hindrance to his credit and reputation they are, and how the more they desire to increase his authority, the more they actually detract from it. For when I see them being obstinate about sustaining propositions which I personally know to be obviously false, and wanting to persuade me that what they are doing is truly philosophical and would be done by Aristotle himself, it much weakens my opinion that he philosophized correctly about other matters more recondite to me. If I saw them give in and change their opinions about obvious truths, I should believe that they might have sound proofs for those in which they persisted and which I did not understand or had not heard.[57]

Hearing these criticisms of Aristotle, Simplicio asks who could replace Aristotle if he were abandoned as a guide in philosophy? To whom could natural philosophers turn? Salviati replies:

We need guides in forests and in unknown lands, but on plains and in open places only the blind need guides. It is better for such people to stay at home, but anyone with eyes in his head and his wits about him could serve as a guide for them. In saying this, I do not mean that a person should not listen to Aristotle; indeed, I applaud the reading and careful study of his works, and I reproach only those who give themselves up as slaves to him in such a way as to subscribe blindly to everything he says and take it as an inviolable decree without looking for any other reasons.[58]

56. Ibid., The Third Day, 336.
57. Ibid., The Second Day, 111.
58. Ibid., 112–113.

Although Galileo did criticize Aristotle directly, he more often praised him, perhaps because that made it easier to denounce his followers, as the passage just quoted shows. The respect accorded to Galileo by all later scientists and natural philosophers, and the widespread knowledge of his relevant works, gave Galileo's frequent, incisive, and often amusing criticisms of Aristotelians and Aristotelian natural philosophy a far greater impact than they might otherwise have had. Of all the opponents of scholastic natural philosophy, Galileo was easily the most devastating. After Galileo, scholastic natural philosophy would rarely be evaluated with objectivity until the early twentieth century.

Before leaving Galileo, one important point must be made. Whether Galileo's reports about the absurd and bizarre behavior of Aristotelian natural philosophers were true or not is largely irrelevant. But they might well have been true, because Galileo's criticisms were made against Aristotelian natural philosophers in the seventeenth century, when scholasticism was under criticism from many quarters and was gradually succumbing to the new science that would almost completely displace it by the end of the century. Indeed, Aristotelian scholastic natural philosophy in the seventeenth century was challenged by new philosophies, such as Neoplatonism and Hermeticism, as well as by the developing Copernican cosmology that threatened the entire Aristotelian cosmos. When problems stemming from the Protestant Reformation are factored in, one can readily see that Aristotelian natural philosophers in the seventeenth century were under siege and had developed a siege mentality. Every criticism of Aristotle was seen as a potential threat to traditional interpretations of the world, which also embraced the Biblical account. Thus, the strange behavior that Galileo reports may have been true, or were exaggerated accounts of what he had witnessed and heard.

But a word of caution is in order. Galileo's reports should not be seen as also representative of behavior by scholastic natural philosophers in the late Middle Ages. Such an interpretation would be a gross distortion of the realities. It is well attested that medieval natural philosophers diverged from Aristotle on many points of natural philosophy.[59] Moreover, they frequently criticized his conclusions, basing their arguments on reason and the testimony of the senses. One significant reason for the numerous divergences from Aristotle during the medieval period may be attributable to the fact that no alternative to Aristotle's natural books existed. His works had no rivals. It was, therefore, quite accept-

59. In support of this claim, see Grant, "Medieval Departures from Aristotelian Natural Philosophy," in Stefano Caroti, ed., *Studies in Medieval Natural Philosophy* ([Florence:] Olschki, 1989), 237–256.

able to criticize Aristotle and to depart from his arguments and conclusions. Despite the high regard in which Aristotle was held, no one felt constrained to defend all of his opinions. But as new philosophies and outlooks emerged, beginning in the late fifteenth century with the works of Plato, and a body of literature attributed to Hermes Trismegistus and known as the Hermetic corpus, alternatives to Aristotle developed and Aristotelian natural philosophy came under severe attack.

A powerful critic of Aristotle and Aristotelian philosophy was Pierre Gassendi (1592–1655), who in 1624 published the first volume of his *Paradoxical Exercises Against the Aristotelians (Exercitationes Paradoxicae Adversus Aristoteleos).*[60] At the beginning of the second of these "exercises" *(Exercitationes*), Gassendi leaves no doubt about his attitude toward the followers of Aristotle. He titled the second exercise "That the Aristotelians have undeservedly taken away the freedom to philosophize" and titled the first article "Slothful distrust has truly seized the Aristotelians." The meaning of these titles becomes clear when Gassendi launches into the first article with these words:

> I have often asked myself what the source was, and from where did the current of corrupt philosophizing derive which has filled the Schools for so long a time. I have found no satisfactory answer other than a soft and slothful distrust by which Aristotelian philosophers, as we have seen above, are persuaded that the truth has been grasped by Aristotle in times past and therefore they cease to trouble to inquire further. Thus taking their predecessor, Aristotle, for a God who has descended from heaven [and] by whom the truth has been revealed, they have not dared to deviate from him by the thickness of a fingernail. And thus distrustful of their own powers, they have renounced the direct study of things and confined themselves to vain prattling about the writings and words of Aristotle.[61]

Gassendi devotes the whole of the *Exercitationes* to an analysis of Aristotle's works with the sole intent of demonstrating its gross inadequacy.

English Anti-Aristotelians

The harshest critiques of Aristotle and his followers came from England in the seventeenth century, when Aristotelianism had relatively few support-

60. See *Pierre Gassendi, Dissertations en Forme de Paradoxes contre les Aristotéliciens (Exercitationes Paradoxicae Adversus Aristoteleos) Livres I et II,* texte établi, traduit et annoté par Bernard Rochot (Paris: Librairie Philosophique J. Vrin, 1959).
61. Gassendi, *Dissertations,* ed. and trans. Rochot, 49 for the Latin text and 48 for Rochot's French translation.

ers. But already in the sixteenth century, Sir Thomas More, a friend of Erasmus, had occasion to ridicule logicians in a 1515 letter to Martin Dorp (1485–1525), a letter that Juan Luis Vives read and which inspired him to write against the medieval logicians, or pseudodialecticians, as he called them. Dorp had earned his doctorate in theology at the University of Louvain in 1515 and almost immediately became a member of the theology faculty at that institution. In two letters to Erasmus, one in 1514 and another in 1515, Dorp reported to Erasmus that the theologians at Louvain were displeased with his *Praise of Folly,* which, as we saw, was critical of the-ologians. More received a copy of the letters and sent a lengthy reply to Dorp on October 21, 1515. The letter was not published until 1563 in Basel,[62] although it circulated among humanists during the years prior to publication.

In his letter, More finds much that is absurd in what logicians, or dialec-ticians, do. As his point of departure, he cites a book called the *Little Logicals (Parva Logicalia),* and then declares:[63]

> All the same, that book of the *Little Logicals* (so named, I think, because it contains but little logic) is worth the trouble to look into for suppositions, as they call them, ampliation, restrictions, appellations. It also contains some pid-dling rules, not so much foolish as false, through which boys are taught to dis-tinguish among statements of this kind, 'The lion than an animal is stronger' and 'The lion is stronger than an animal' – as if they did not mean the same thing. Actually both statements are so clumsy that neither of them means much of anything, but if they do mean anything they doubtless mean the same. And these differ just as much, 'Wine twice I have drunk' and 'Twice wine I have drunk'; that is to say, a lot according to those logickers but in reality not at all. Now if a man has eaten meat roasted to burning, they want him to be speaking the truth if he puts it this way 'I raw meat have eaten', but not if he says 'I have eaten raw meat'. Again, if someone should take part of my money, leaving some for me, it seems I should be lying if I said, 'He has robbed me of my money'; but so that I will not be without words to complain before the judge, it will be all right to say 'Of my money he has robbed me'. In another case, assuming, as they say, the possibility, this will be true, 'The Pope I have beaten', while under the same circumstances this will be false, 'I have beaten the Pope', if, of course, he who is now Pope was beaten by me long ago as a boy. By Jove, it would serve

62. The letter appeared in a work by More titled *Lucubrationes.*

63. I have used Rita Guerlac's partial translation of More's letter to Martin Dorp in Guerlac, *Juan Luis Vives,* 171–173. For a complete translation of More's letter to Dorp, see Elizabeth Frances Rogers, *St. Thomas More: Selected Letters* (New Haven: Yale University Press, 1961), 6–64. The translation is by Rev. Dr. Marcus A. Haworth, S. J.

the old men right who now teach these things if they were beaten themselves every time they teach schoolboys![64]

In this passage, More, in the manner of Erasmus, ridicules the logician's use of word order to determine meaning (see Chapter 4). He then cites additional paradoxical propositions and indicates that dialecticians regard themselves as superior to poets because

[p]oets make up things, and tell lies, dialecticians never say anything but the truth, even when they say it is absolutely true that 'A dead man can celebrate Mass'. Although I dare not disbelieve them when they assert and practically swear to this (for it is not proper to contest so many incontestable doctors), yet so far as I remember I have never found anyone who said he had served Mass for a dead celebrant.... By heaven I wonder how rather intelligent men discerned that these statements should be understood in a way no one in the whole world except themselves understand them. These words do not belong to an art, its own private property, as it were, to be taken on loan by anyone who wishes to borrow them. Speech is truly common to all, except that these render back some words more corrupt than they received them from cobblers. They have taken their speech from the common people and misused the common meanings. But the rules they call logic say that such statements must be interpreted in those meanings. Can some damned rule made up in some corner by men who hardly know how to speak, impose new laws of speech upon the whole world?[65]

More explains that he does not oppose theologians who "treat human matters seriously or divine matters reverently, or both." He adds, however, that

I do not approve of theologians of the kind (and I speak with feeling) who not only grow old but wither away among petty questions of this kind. These are men who, whether hampered by some barrenness of talent or excited by the childish applause of the scholastics, have slighted the writings of all the ancients and even neglected the very Gospels of which they claim to be teachers. They have learned nothing at all except trivial questions that are empty in themselves and useless to men who are empty of all the rest of knowledge. Old men now, and for that reason to be pitied, they cannot discuss the Scriptures properly because they do not know the writings of the ancients, nor be equal to learning the things they should, because they are so deficient in the Latin language.[66]

64. Guerlac, *Vives*, 171–173.
65. Ibid., 175.
66. Ibid., 191.

Thus did Thomas More savagely criticize scholastic theologians, whom he regarded as grossly ignorant of Scripture because they lacked the humanistic skills that counted. They only seemed capable of analyzing strange propositions that were completely alien to ordinary people using ordinary language.

There was no shortage of critics in seventeenth-century England. One of the most severe was Francis Bacon (1561–1626), a strict inductionist, who advocated a purely empirical approach to nature in which one gathers facts by observation and arrives at probable generalizations.[67]

As compared to the inductionist method he advocated, Bacon judged Aristotle's syllogistic logic as useless and repudiated it in a series of aphorisms (XI–XIV) in the first book of his *New Organon (Novum Organum).* Although Aristotle is not mentioned by name, he is obviously intended:

> XI. As the present sciences are useless for the discovery of effects, so the present system of logic is useless for the discovery of the sciences.
> XII. The present system of logic rather assists in confirming and rendering inveterate the errors founded on vulgar notions than in searching after truth, and is therefore more hurtful than useful.
> XIII. The syllogism is not applied to the principles of the sciences, and is of no avail in intermediate axioms, as being very unequal to the subtilty of nature. It forces assent, therefore, and not things.
> XIV. The syllogism consists of propositions, propositions of words, words are the signs of notions. If, therefore, the notions (which form the basis of the whole) be confused and carelessly abstracted from things, there is no solidity in the superstructure. Our only hope, then, is in genuine induction.[68]

Bacon was rejecting not only Aristotle's syllogistic logic but also the way logic was used in natural philosophy by Aristotle and his medieval followers. Indeed, in Aphorism LIV, Bacon says that Aristotle "made his natural philosophy completely subservient to his logic, and thus rendered it little more than useless and disputatious."[69] And in Aphorism LXIII, he declares that Aristotle "corrupted natural philosophy by logic – thus he formed the world of categories" and that "Aristotle's physics are mere logical terms."[70] Bacon considers the causes of the errors that

67. See Stuart Hampshire, *The Age of Reason: The 17th Century Philosophers,* selected, with Introduction and Commentary (New York: Mentor Books, 1956), 20.
68. *The Physical and Metaphysical Works of Lord Bacon including "the Advancement of Learning" and "Novum Organum,"* ed. Joseph Devey (London: George Bell and Sons, 1904), 384–386.
69. Ibid., 396.
70. Ibid., 400–401.

have bedeviled the sciences and natural philosophy and concludes that (Aphorism LXXVIII)

> out of twenty-five centuries with which the memory and learning of man are conversant, scarcely six can be set apart and selected as fertile in science and favourable to its progress. For there are deserts and wastes in times as in countries, and we can only reckon up three revolutions and epochs of philosophy. 1. The Greek. 2. The Roman. 3. Our own, that is the philosophy of the western nations of Europe: and scarcely two centuries can with justice be assigned to each. The intermediate ages of the world were unfortunate both in the quantity and richness of the sciences produced. Nor need we mention the Arabs, or the scholastic philosophy, which, in those ages, ground down the sciences by their numerous treatises, more than they increased their weight. The first cause, then, of such insignificant progress in the sciences, is rightly referred to the small proportion of time which has been favourable thereto.[71]

Thus did Bacon denigrate the efforts in science and natural philosophy of both medieval Islam and the medieval West.

Bacon's antagonism to scholastic learning was already well honed in *The Advancement of Learning* of 1605, where he declared that the learning of the schoolmen "grew ... to be utterly despised as barbarous."[72] In a section in which he argues that "vain matter is worse than vain words," Bacon asserts that

> [i]t is the property of good and sound knowledge to putrify and dissolve into a number of subtle, idle, unwholesome, and, as I may term them, vermiculate questions, which have indeed a kind of quickness and life of spirit, but no soundness of matter or goodness of quality. This kind of degenerate learning did chiefly reign amongst the Schoolmen: who having sharp and strong wits, and abundance of leisure, and small variety of reading, but their wits being shut up in the cells of a few authors (chiefly Aristotle their dictator) as their persons were shut up in the cells of monasteries and colleges, and knowing little history, either of nature or time, did out of no great quantity of matter and infinite agitation of wit spin out unto those laborious webs of learning which are extant in their books.[73]

Bacon would later observe that the schoolmen also made their commentaries vaster than the original writings on which they commented. After

71. Ibid., 413–414.
72. Francis Bacon, *The Advancement of Learning,* edited with an Introduction by G. W. Kitchin (London: Everyman's Library, Dent, 1965), I.IV.2, 24.
73. Bacon, ibid., I.IV.5, 26.

Bacon, scholasticism was truly in ill repute. Many other Englishmen also found scholastic authors wanting and, like Francis Bacon, vented their displeasure.

One of the harshest critics of Aristotle and his scholastic followers was Thomas Hobbes (1588–1679), a critical philosopher and great political thinker. In *Leviathan, or the Matter, Form, and Power of a Commonwealth, Ecclesiastical and Civil,* published in 1651 and regarded as his most famous work, Hobbes, in the fifth chapter on reason and science, explains that only men are subject to the "privilege of absurdity":[74] "And of men, those are of all most subject to it, that profess philosophy."[75] He then identifies seven causes of absurdity, the seventh of which concerns "names that signify nothing; but are taken and learned by rote from the schools, as *hypostatical, transubstantiate, consubstantiate, eternal-now,* and the like canting of schoolmen."[76] The "schools" to which Hobbes refers are the universities in their medieval and early modern form. Indeed, Hobbes was a severe critic of Oxford, where he was educated, regarding it as a bastion of scholastic thought, a place where schoolmen held sway.

Hobbes continues his onslaught against the philosophers in Part 4, Chapter 46 ("Of Darkness from Vain Philosophy and Fabulous Traditions"), where he says, "And I believe that scarce any thing can be more absurdly said in natural philosophy, than that which now is called *Aristotle's Metaphysics;* nor more repugnant to government, than much of that he hath said in his *Politics;* nor more ignorantly, than a great part of his *Ethics.*"[77] Hobbes continues his assault against Aristotle by launching an attack on the universities with their faculties of theology, law, and medicine. Within this environment, philosophy serves as the handmaid "to the Roman religion," that is theology. "That which is now called an University," says Hobbes,

> is a joining together, and an incorporation under one government of many public schools, in one and the same town or city. In which, the principal schools were ordained for the three professions, that is to say, of the Roman religion, of the Roman law, and of the art of medicine. And for the study of philosophy it hath no

74. *The English Works of Thomas Hobbes of Malmesbury,* Now First collected by Sir William Molesworth, 11 vols. (London: John Bohn, 1839–1845; reprint, Darmstadt: Scientia Verlag Aalen, 1966), Vol. 3: *Leviathan, or the Matter, Form, and Power of a Commonwealth Ecclesiastical and Civil,* pt. 1, ch. 5, 33.
75. Hobbes, *Works,* Vol. 3: *Leviathan,* pt. 1, ch. 5, 33
76. Hobbes, ibid., 34–35.
77. Ibid., pt. 4, ch. 46, 669.

otherwise place, than as a handmaid to the Roman religion: and since the authority of Aristotle is only current there, that study is not properly philosophy, (the nature whereof dependeth not on authors,) but *Aristotelity.* And for geometry, till of very late times it had no place at all; as being subservient to nothing but rigid truth. And if any man by the ingenuity of his own nature, had attained to any degree of perfection therein, he was commonly thought a magician, and his art diabolical.

Now to descend to the particular tenets of vain philosophy, derived to the Universities, and thence into the Church, partly from Aristotle, partly from blindness of understanding.[78]

The philosophy (presumably natural philosophy, moral philosophy, and metaphysics) that is subservient to theology is the philosophy of Aristotle, which Hobbes does not regard as "properly philosophy" but, rather, something he calls "Aristotelity," some perversion of philosophy that Aristotle passed on to the scholastics.

In the course of his criticism of Aristotelian scholastic thought, Hobbes denounces Aristotle's metaphysics, which he says is concerned with definitions of such terms as "body, time, place, matter, form, essence, subject, substance, accident, power, act, finite, infinite, quantity, quality, motion, action, passion, and divers others, necessary to the explaining of a man's conceptions concerning the nature and generation of bodies." Determining the meaning of such terms is

in the Schools called *metaphysics;* as being a part of the philosophy of Aristotle, which hath that for title. But it is in another sense; for there it signifieth as much, *as books written, or placed after his natural philosophy:* but the schools take them for *books of supernatural philosophy:* for the word *metaphysics* will bear both these senses. And indeed that which is there written, is for the most part so far from the possibility of being understood, and so repugnant to natural reason, that whosoever thinketh there is any thing to be understood by it, must needs think it supernatural.[79]

Aristotle's metaphysics, Hobbes explains, is concerned with "certain essences separated from bodies, which they call *abstract essences, and substantial forms.*"[80] After a brief attempt to explain "abstract essences," Hobbes acknowledges that someone might ask why he has included so much about Aristotle's metaphysics in a book about the "doctrine of government and obedience":

78. Ibid., 670–671.
79. Ibid., 671–672.
80. Ibid., 672.

It is to this purpose, that men may no longer suffer themselves to be abused, by them, that by this doctrine of *separated essences,* built on the vain philosophy of Aristotle, would fright them from obeying the laws of their country, with empty names; as men fright birds from the corn with an empty doublet, a hat, and a crooked stick. For it is upon this ground, that when a man is dead and buried, they say his soul (that is his life) can walk separated from his body, and is seen by night amongst the graves. Upon the same ground they say, that the figure, and colour, and taste of a piece of bread, has a being, there, where they say there is no bread. And upon the same ground they say, that faith, and wisdom, and other virtues are sometimes *poured* into a man, sometimes *blown* into him from Heaven, as if the virtuous and their virtues could be asunder; and a great many other things that serve to lessen the dependance of subjects on the sovereign power of their country. For who will endeavour to obey the laws, if he expect obedience to be poured or blown into him? Or who will not obey a priest, that can make God, rather than his sovereign, nay than God himself? Or who, that is in fear of ghosts, will not bear great respect to those that can make the holy water, that drives them from him? And this shall suffice for an example of the errors, which are brought into the Church, from the entities and essences of Aristotle.[81]

It is clear that Hobbes believed that Aristotle's metaphysics was a danger for both the state and religion.

Aristotle's natural philosophy, which in the following passage is called "physics," receives similar treatment:

Then for *physics,* that is, the knowledge of the subordinate and secondary causes of natural events; they render none at all, but empty words. If you desire to know why some kind of bodies sink naturally downwards toward the earth, and others go naturally from it; the Schools will tell you out of Aristotle, that the bodies that sink downwards, are *heavy;* and that this heaviness is it that causes them to descend. But if you ask what they mean by *heaviness,* they will define it to be an endeavour to go to the centre of the earth: so that the cause why things sink downward, is an endeavour to be below: which is as much as to say, that bodies descend, or ascend, because they do. Or they will tell you the centre of the earth is the place of rest, and conservation for heavy things; and therefore they endeavour to be there: as if stones and metals had a desire, or could discern the place they would be at, as man does; or loved rest, as man does not; or that a piece of glass were less safe in the window, than falling into the street.[82]...

And in many occasions they put for cause of natural events, their own ignorance; but disguised in other words: as when they say, fortune is the cause of things

81. Ibid., 674–675.
82. Ibid., 678.

contingent; that is, of things whereof they know no cause: and as when they attribute many effects to *occult qualities;* that is, qualities not known to them; and therefore also, as they think, to no man else. And to *sympathy, antipathy, antiperistasis, specifical qualities,* and other like terms, which signify neither the agent that produceth them, nor the operation by which they are produced.[83]

Hobbes illustrates the absurdity and obscurity of Aristotle's physics by the way Aristotle explained the fall of a heavy body toward the earth's center. Thus, in a book on sovereignty and government, Hobbes found numerous occasions to ridicule and denounce Aristotelian metaphysics and physics, or natural philosophy, and those who used it. He concludes, "If such *metaphysics,* and *physics* as this, be not vain philosophy, there was never any; nor needed St. Paul to give us warning to avoid it."[84]

Theology, which Hobbes describes as "metaphysics, which are mingled with the Scripture to make school divinity,"[85] is simply another example of "vain philosophy brought into religion by the doctors of School-divinity."[86] Hobbes declares that he will offer no examples to show how "vain philosophy" appears in theology, but says:

I shall only add this, that the writings of School-divines, are nothing else for the most part, but insignificant trains of strange and barbarous words, or words otherwise used, than in the common use of the Latin tongue; such as would pose Cicero, and Varro, and all the grammarians of ancient Rome. Which if any man would see proved, let him, as I have said once before, see whether he can translate any School-divine into any of the modern tongues, as French, English, or any other copious language: for that which cannot in most of these be made intelligible, is not intelligible in the Latin. Which insignificancy of language, though I cannot note it for false philosophy; yet it hath a quality, not only to hide the truth, but also to make men think they have it, and desist from further search.[87]

Because scholastic theology is full of "strange and barbarous words," it cannot be translated into vernacular languages; and if Latin discourse

83. Ibid., 679–680.

84. Hobbes, ibid., 680. Despite his hostile attitude, Hobbes was definitely influenced by scholastic natural philosophy. The extent to which Hobbes was influenced by scholastic Aristotelians has been thoroughly investigated in an excellent study by Cornelis Hendrik Leijenhorst, *Hobbes and the Aristotelians: The Aristotelian Setting of Thomas Hobbes's Natural Philosophy* (Ph.D. diss., University of Utrecht, 1998); see especially the conclusion, 255–263; see also Lawn, *The Rise and Decline of the Scholastic 'Quaestio Disputata,'* 81–82, and Grant, *The Foundations of Modern Science in the Middle Ages,* 197.

85. Hobbes, *Leviathan,* 672.

86. Ibid., 685.

87. Ibid., 686.

cannot be rendered intelligibly into the vernacular, it cannot be "intelligible in the Latin."

In the next chapter, Hobbes makes 12 points to show who benefits from obscurity ("darkness") of language and thought. In the twelfth, and final, point, he declares:

> Lastly, the metaphysics, ethics, and politics of Aristotle, the frivolous distinctions, barbarous terms, and obscure language of the Schoolmen, taught in the universities, which have been all erected and regulated by the Pope's authority, serve them to keep these errors from being detected, and to make men mistake the ignis fatuus [i.e., feeble glow] of vain philosophy, for the light of the Gospel.[88]

Thomas Hobbes was surely one of the most savage critics of the medieval tradition of Aristotelian metaphysics and natural philosophy, as well as the theology that relied so heavily on "vain philosophy."

The most significant English philosopher of the seventeenth century, John Locke, was less caustic than Hobbes, but every bit as antischolastic. Where Hobbes thought Aristotle's achievements largely worthless, Locke, although he also found much to criticize in Aristotle's philosophy, nevertheless professed great admiration for Aristotle, "whom I look on," he declared, "as one of the greatest Men amongst the Antients, whose large Views, acuteness and penetration of Thought, and strength of Judgment, few have equalled."[89] Locke was, however, convinced that the formal syllogism, which Aristotle had practically invented in his *Prior Analytics,* was unnecessary for reasoning. He explains that

> [i]f Syllogisms must be taken for the only proper instrument of reason and means of Knowledge, it will follow that before *Aristotle* there was not one Man that did or could know anything by Reason; and that since the invention of Syllogisms, there is not one of Ten Thousand that doth.
>
> But God has not been so sparing to Men to make them barely two-legged Creatures, and left it to *Aristotle* to make them Rational.[90]

It was not Aristotle but his scholastic followers that Locke found objectionable. Aristotle may have invented the syllogism, but it was scholastics who used and abused it. "Of what use then are *Syllogisms?*" asks Locke, to

88. Hobbes, *Leviathan,* pt. 4, ch. 47, 693. I have added the bracketed words.
89. John Locke, *An Essay Concerning Human Understanding,* edited with an Introduction by Peter H. Nidditch (Oxford: Clarendon Press, 1975), bk. IV, ch. XVII, sec. 4, 671.
90. Locke, ibid.

which he replies: "I answer, Their chief and main use is in the Schools, where Men are allowed without Shame to deny the Agreement of *Ideas,* that do manifestly agree; or out of the Schools to those, who from thence have learned without shame to deny the connexion of *Ideas,* which even to themselves is visible."[91]

In a chapter on the truth and certainty of universal propositions, Locke declares: "*All Gold is fixed,* is a Proposition whose Truth we cannot be certain of, how universally soever it be believed." And then he strikes at the scholastics: "For if, according to the useless Imagination of the Schools, any one supposes the term *Gold* to stand for a Species of Things set out by Nature, by a real Essence belonging to it, 'tis evident he knows not what particular Substances are of that Species; and so cannot, with certainty, affirm any thing universally of *Gold.*"[92] His low opinion of the scholastics is made apparent again, when, in a chapter on maxims, or axioms, he asks what use is made of maxims. The second usage for maxims is in disputes, "for the silencing of obstinate Wranglers," that is, quarrelsome disputants:

> The Schools having made Disputation the Touchstone of Mens Abilities, and the *Criterion* of Knowledge, adjudg'd Victory to him that kept the Field: and he that had the last Word was concluded to have the better of the Argument, if not of the Cause. But because by this means there was like to be no Decision between skilful Combatants ...[93] [the Schools introduced] certain general Propositions, most of them indeed self-evident.

These maxims were given the name of principles beyond which one could not go. Hence, these principles were mistakenly assumed to be the source of all knowledge "and the Foundations whereon the Sciences were built. Because when in their Disputes they came to any of these, they stopped there, and went no farther, the Matter was determined. But how much this is a mistake hath been already shewn."[94]

After a brief discourse on the role of maxims, Locke again scornfully turns his attention to the disputatious scholastics:

> But the Method of the Schools, having allowed and encouraged Men to oppose and resist evident Truth, till they are baffled, *i.e.* till they are reduced to contradict themselves, or some established Principle; 'tis no wonder that they should not in

91. Ibid., bk. IV, ch. XVII, sec. 4, 675.
92. Ibid., bk. IV, ch. VI, sec. 8, 582–583.
93. Ibid., bk. IV, ch. VII, sec. 11, 600.
94. Ibid.

civil Conversation be ashamed of that, which in the Schools is counted a Vertue and a Glory; *viz.* obstinately to maintain that side of the Question they have chosen, whether true or false, to the last extremity; even after Conviction. A strange way to attain Truth and Knowledge.[95]

A few lines later, Locke concludes his attack on the scholastics by way of an aside:

This, I think, that bating[96] those places, which brought the *Peripatetick* Philosophy into their Schools, where it continued many Ages, without teaching the World any thing but the Art of Wrangling; these Maxims were no where thought the Foundations on which the Sciences were built, nor the great helps to the Advancement of Knowledge.[97]

Locke undoubtedly denounced scholastics and scholastic thought in other of his works, but his criticisms in *An Essay Concerning Human Understanding* clearly reveal his intense dislike for their thought and methods.

In the course of the seventeenth century, scholasticism was subjected to devastating criticisms that had roots in humanist assaults in the sixteenth century by the likes of Gianfrancesco Pico della Mirandola (1469–1533) and are even traceable to Petrarch in the fourteenth century. Christia Mercer has observed:

Through the works of humanists like Petrarca and Pico, subsequent anti-Aristotelians acquired a set of stock complaints. The standard criticisms were that the Peripatetics are more committed to Aristotle than to the pursuit of the truth and, hence, are removed from the proper source of knowledge; talk about many things but understand little; do not even agree among themselves; and use obscure terminology which they neither properly define nor fully understand....

The writings of seventeenth-century philosophers like Gassendi, Descartes, and Leibniz contain exactly the same list of grievances. Descartes, for instance, remonstrates about the language of the scholastics, their lack of concern with the truth, their disagreements among themselves, and their obscurity.[98]

95. Ibid., bk. IV, ch. VII, sec. 11, 601.

96. That is, "leaving out of account."

97. Locke, *An Essay Concerning Human Understanding,* bk. IV, ch. VII, sec. 11, 601. In the expression "Art of Wrangling," wrangling signifies "contentious disputation."

98. Mercer, "The Vitality and Importance of Early Modern Aristotelianism," in Tom Sorrell, ed., *The Rise of Modern Philosophy: The Tension between the New and Traditional Philosophies from Machiavelli to Leibniz* (Oxford: Clarendon Press, 1993), 35. In support of Descartes's criticisms, Mercer cites *Oeuvres de Descartes,* ed. C. Adam and P. Tannery (rev. ed., 1964–1976), vol. 9, 25f., 33, 35.

From the fourteenth to the end of the seventeenth centuries, criticisms of Aristotelian philosophy grew ever louder and more widespread. Most would have agreed with Francis Bacon that "due to the persistent philosophical wrong-headedness of the schoolmen, they had become the objects of 'popular contempt.'"[99] As the seventeenth century moved along, the severe criticisms of Aristotelianism became repetitious, and it is obvious that many of the critics derived their criticisms from a storehouse of complaints that had accumulated over the centuries. In the course of the eighteenth century, fewer and fewer scholars had any direct knowledge of scholastic texts. Although the works of Aristotle may have been known directly to a few, the works of his medieval and early modern commentators had faded from history. Lack of direct knowledge of scholastic texts, however, proved no obstacle to further criticism. One had only to draw upon the readily available common legacy of hostile comments and insert them at will. By the end of the seventeenth century, medieval scholastic theology and natural philosophy had been severely denigrated and denounced. The *philosophes* of the eighteenth century added to the sum total of antischolastic rhetoric.

The Philosophes of the Age of Enlightenment

Voltaire could find nothing positive to say about scholastic philosophy and theology. He was convinced that in the thirteenth century, we passed "from savage ignorance to scholastic ignorance."[100] He was convinced that "the studies of the scholastics were then, and have remained to our day, systems of absurdities, such that, if we attributed them to the peoples of Ceylon (Taprobane), we would believe that we had insulted them,"[101] and that "scholastic theology, bastard daughter of Aristotle's philosophy, badly translated and misunderstood," had "caused more error for reason and good education than the Huns and Vandals had done."[102]

99. Mercer, ibid., 37.
100. "On passa, dans ce xiiie siècle, de l'ignorance sauvage à l'ignorance scolastique." *Essai sur les Moeurs, I,* in *Oeuvres complètes de Voltaire, nouvelle édition avec notices, préfaces, variantes, table analytique, les notes de tous les commentateurs et des notes nouvelles . . .,* 52 vols. (Paris: Garnier Frères, 1877–1885), vol. 11 (1878), 506.
101. "Les études des scolastiques étaient alors et sont demeurées, presque jusqu'à nos jours, des systèmes d'absurdités, tels que, si on les imputait aux peuples de la Taprobane, nous croirions qu'on les calomnie." Ibid., 507.
102. "La théologie scolastique, fille bâtarde de la philosophie d'Aristote, mal traduite et méconnue, fit plus de tort à la raison et aux bonnes études que n'en avaient fait les Huns et les Vandales." Ibid., vol. 12 (1878): *Essai sur les moeurs, II,* 61.

The French philosophes generally shared Voltaire's contempt for the Middle Ages and scholastic thought. One of the most important members of this group, Jean Le Rond d'Alembert (1717–1783), was the author of the *Preliminary Discourse to the Encyclopedia* edited by Denis Diderot (1713–1784).[103] Published in 1751, d'Alembert's *Preliminary Discourse* has been regarded as "the manifesto of the French Enlightenment," and it won the highest praise from Frederick the Great, Condorcet (1743–1794), and many others.[104] Because d'Alembert's introduction was widely read, we may properly assume that his criticisms of the Middle Ages and scholasticism had a greater impact than would similar sentiments uttered by lesser-known figures. But whether it is d'Alembert or Voltaire, or any other philosophe, the key words used to describe the Middle Ages and scholasticism are "ignorance" and some form of "barabarous" or "barbarism," as in d'Alembert's declaration that "Scholasticism, which constitutes the whole of so-called science of the centuries of ignorance, still was prejudicial to the progress of true philosophy in that first century of enlightenment,"[105] that is, the sixteenth century. And elsewhere, d'Alembert informs us that "Descartes dared at least to show intelligent minds how to throw off the yoke of scholasticism, of opinion, of authority – in a word, of prejudices and barbarism."[106] In denouncing eloquence through oratory, d'Alembert found occasion to scorn "those pedantic puerilities which have been honored by the name of Rhetoric," and to observe that they are "to the oratorical art what scholasticism is to true philosophy."[107] Thus did d'Alembert strike at scholasticism directly and by analogy.

Although Condorcet[108] was as hostile as any eighteenth-century French philosophe toward the Church and priests, he was surprisingly moderate and sensible toward scholasticism. Indeed, his views lend some support to the thesis of my book. Although he judged that "scholasticism neither encouraged the discovery of truth nor promoted better methods of evaluating and discussing evidence," he allowed that "it whetted men's intellects." Consequently,

103. See Jean Le Rond D'Alembert, *Preliminary Discourse to the Encyclopedia of Diderot,* translated by Richard N. Schwab (Chicago: The University of Chicago Press, 1963; revised Introduction, 1995).

104. D'Alembert, *Preliminary Discourse to the Encyclopedia of Diderot,* x, xi.

105. Ibid., 71.

106. Ibid., 80.

107. Ibid., 34.

108. His full name was Marie-Jean-Antoine-Nicolas Caritat, Marquis de Condorcet.

the taste for subtle distinctions and the need to sharpen ideas to the last refinement, to grasp the most fugitive shades of meaning and to clothe them in new expressions, the whole paraphernalia intended to confound one's opponent or escape his traps – all this was the first beginnings of that philosophical analysis which has since been the fruitful source of our progress.[109]

What most critics denounced as the vices of scholasticism, Condorcet characterized as "the first beginnings of that philosophical analysis which has since been the fruitful source of our progress." Condorcet also went on to mention that a debt was owed to the Schoolmen for presenting "more precise notions" about various terms and concepts and "the manner of distinguishing the various operations of the human mind, and the correct way of classifying such ideas as it can form of real objects and their properties."[110] But, although these contributions were important, and represented "the first beginnings of that philosophical analysis which has since been the fruitful source of our progress," Condorcet emphasized that "this same method could only retard the progress of the natural sciences in the schools." Thus, Condorcet recognized, and was willing to grant, that scholasticism had contributed something to the advance of progress, a concession that few, if any, philosophes and philosophers would make.

The Nineteenth and Twentieth Centuries

William Whewell's *History of the Inductive Sciences* was one of the best known nineteenth-century works on the history of science. He made his opinion of the Middle Ages apparent by the title he assigned to the fourth book: "History of Physical Science in the Middle Ages or, View of the Stationary Period of Inductive Science."[111] Straightaway, he declares:

We have now to consider more especially a long and barren period, which intervened between the scientific activity of ancient Greece, and that of modern Europe; and which we may, therefore, call the Stationary Period of Science.... We must

109. Antoine-Nicolas de Condorcet, *Sketch for a Historical Picture of the Progress of the Human Mind*, translated by June Barraclough, with an Introduction by Stuart Hampshire (London: Weidenfeld and Nicolson, 1955; first published 1795; first translated 1795), 94–95.
110. Condorcet, ibid., 95.
111. *The Historical and Philosophical Works of William Whewell*, 10 vols. Collected and edited by G. Buchdahl and L. L. Laudan. *History of the Inductive Sciences*, vols. 2–4 (London: Frank Cass & Co., 1967; first published in 1837; 3d edition, 1857). For Book IV, see vol. 2, 181–267.

endeavour to delineate the character of the Stationary Period, and as far as possible, to analyze its defects and errours; and thus obtain some knowledge of the causes of its barrenness and darkness.[112]

The medieval period in Europe was an age of commentaries on authoritative texts. At the outset of a chapter titled "Of the Dogmatism of the Stationary Period" (Chapter 4 of Book IV), Whewell explains:

In speaking of the character of the age of commentators, we noticed principally the ingenious servility which it displays; – the acuteness with which it finds ground for speculation in the expression of other men's thoughts; – the want of all vigour and fertility in acquiring any real and new truths. Such was the character of the reasoners of the stationary period from the first; but, at a later day, this character, from various causes, was modified by new features. The servility which had yielded itself to the yoke, insisted upon forcing it on the necks of others; the subtlety which found all the truth it needed in certain accredited writings, resolved that no one should find there, or in any other region, any other truths; speculative men became tyrants without ceasing to be slaves; to their character of Commentators, they added that of Dogmatists.[113]

After a discussion of what purports to be scholastic philosophy, theology, physics, and the role of Aristotle among the Schoolmen, Whewell begins Book V ("History of Formal Astronomy After the Stationary Period") with a backward glance, declaring:

We have thus rapidly traced the causes of the almost complete blank which the history of physical science offers, from the decline of the Roman empire, for a thousand years. Along with the breaking up of the ancient forms of society, were broken up the ancient energy of thinking, the clearness of idea, and steadiness of intellectual action. This mental declension produced a servile admiration for the genius of the better periods, and thus, the spirit of Commentation: Christianity established the claim of truth to govern the world; and this principle, misinterpreted and combined with the ignorance and servility of the times, gave rise to the Dogmatic System: and the love of speculation, finding no secure and permitted path on solid ground, went off into the regions of Mysticism.[114]

The cumulative impact of criticisms from the seventeenth through nineteenth centuries produced a powerfully negative image of medieval contributions to natural philosophy, science, and theology. In the late nineteenth

112. Whewell, *History of the Inductive Sciences,* vol. 2, 181.

113. Ibid., vol. 2, 237.

114. Ibid., vol. 2, 271.

and early twentieth centuries, medievalists made significant attempts to rectify these gross misperceptions and present a balanced picture. The most heroic effort was made by Pierre Duhem, a French physicist, who wrote multivolume works on medieval science and natural philosophy in the first fifteen years of the twentieth century. But many ignored these corrective works and were guided by the traditional judgments.

In a book widely used by historians of science and philosophy, E. A. Burtt presented a misleading sense of the medieval attitude toward religion and nature when he declared that

> the religious experience to the medieval philosopher was the crowning scientific fact. Reason had become married to mystic inwardness and entrancement.... In this graciously vouchsafed kinship of man with an eternal Reason and Love, lay, for medieval philosophy, a guarantee that the whole natural world in its present form was but a moment in a great divine drama which reached over countless aeons past and present and in which man's place was quite indestructible.[115]

Two histories of modern philosophy repeat biases that had become routine. Francis Bowen, professor of natural religion and moral philosophy at Harvard College, claimed that "Aristotelic premises were evoked to support theological conclusions. Novelty was shunned, because it immediately incurred suspicion of heresy."[116] Bowen offers another, more demeaning opinion in the following comparison: "The philosophy of the sixteenth century is rightly called by Cousin a necessary and useful transition from the absolute slavery of thought in mediaeval times to its absolute independence a century afterwards."[117]

Harald Höffding was in some ways sympathetic to the Middle Ages, remarking that "it would be erroneous to regard the Middle Ages as an age of utter darkness," and admitting that "the Middle Ages has rendered important contributions to intellectual development and was by no means the wilderness or the world of darkness which it is so often depicted as being."[118] But these moderate sentiments are counterbalanced by intemperate and insupportable remarks:

115. E. A. Burtt, *The Metaphysical Foundations of Modern Physical Science*, rev. ed. (London: Routledge & Kegan Paul, 1932; reprinted 1950), 6–7.
116. Francis Bowen, *Modern Philosophy from Descartes to Schopenhauer and Hartmann* (New York: Charles Scribner's Sons, 1885), 2.
117. Bowen, ibid., 3.
118. Harald Höffding, *A History of Modern Philosophy: A Sketch of the History of Philosophy from the Close of the Renaissance to Our Own Day*, vol. 1 (London: Macmillan and Co., Limited, 1935; first edition 1900), 4.

And when once the conviction had become established that the Aristotelian philosophy was especially suited to represent natural science in the construction of the scholastic system, and deviation from Aristotle – from Aristotle as he was known and interpreted in the Middle Ages – attracted attention and was regarded as heresy. In other words: thought and inquiry were arbitrarily checked in order that the structure which had been raised might not be shaken. The Aristotelian philosophy, which in its own time denoted such an enormous advance, was now set up as valid for all eternity.[119]

And in the next paragraph, Höffding goes on to say that in the Middle Ages,

[t]he dogmatist ever held watch over the mystic, so often carried by the tides of his inner life beyond the limits of the feeling sanctioned by the Church as right and true. The Church could no more venture to give inner experience its own way than she could allow free play to outer experience. She saw that it was dangerous for men to withdraw into themselves and thus come into contact with the highest, for so they might become independent of the Church. She suspected that self-knowledge, no less than knowledge of Nature, offered possibilities of spiritual freedom and opened the way to a very different conception of the world from that presented by theology.[120]

As late as 1941, Charles Singer published *A Short History of Science to the Nineteenth Century* in which he mentioned that "many attempts have been made to rehabilitate the intellectual achievement of the Middle Ages. So far as science is concerned they have been unsuccessful. There is no reason to reverse the decision that in this domain the period is one of intellectual degradation."[121]

The Larger Context of the Assault on Scholastic Thought

The harsh criticisms of medieval logic, philosophy, natural philosophy, and theology that have been described thus far were made within a larger context of hostility toward the Middle Ages, a hostility that was already evident from the very way the Middle Ages was named. Francesco Petrarch (1304–1374) divided history into two eras: the "ancient" *(antiqua),* which included Roman history to the adoption of Christianity by the Roman

119. Ibid., 8.
120. Ibid., 9.
121. Charles Singer, *A Short History of Science to the Nineteenth Century* (Oxford: Clarendon Press, 1941), 161.

Emperors; and the "modern" *(nova)*, which embraced the period from the conversion of the Roman emperors to Christianity to Petrarch's own day. Petrarch loved ancient Roman history and regarded it as the only period worthy of study. The second, or modern period, covering the history of Christian Europe, he viewed with complete disdain and contempt, characterizing it as barbarous, a period of "darkness" *(tenebrae)*.[122] Some generations later, humanists came to believe that they were not living in a period of darkness, but in a period of renewal following the darkness. In 1469, the humanist Giovanni Andrea (1417–1475) invented a term *media tempestas*, or "middle time," to identify the period of European history after the period of ancient Rome.[123]

But it was the reaction by humanists against the period that intervened between the conversion of Roman emperors to Christianity and their own time, in combination with the Protestant Reformation, that foreordained a dark future for the period we now call the Middle Ages. For "although they were often worlds apart in their values, humanist scholars and Protestant reformers both agreed, but for radically different reasons, that antiquity had ended by the fifth century and was followed by a thousand years of ignorance and, worse, degeneracy."[124] The third period, which followed the second degenerate period, was of course that in which the humanists lived. Not surprisingly, they regarded themselves as denizens of a radically altered world, one in which they were revitalizing Europe and bringing it forth from the darkness of the preceding age. By the seventeenth century "that middle period was increasingly known as medium aevum, 'the Middle Age,'"[125] a term that was apparently first used in a book title in 1688, when Christoph Keller published a work he called *A History of the Middle Ages from the times of Constantine the Great to the Capture of Constantinople by the Turks*.[126] From then on, the term "Middle Age," which eventually became the Middle Ages in English (it remains Middle Age in French, "le Moyen-Âge"), became a permanent feature of European history.

C. Warren Hollister, a distinguished medievalist, has brilliantly captured the attitude of many scholars for whom the Middle Ages represented the

122. Wallace K. Ferguson, *The Renaissance in Historical Thought: Five Centuries of Interpretation* (Cambridge, MA: The Riverside Press, 1948), 8. See also, T. E. Mommsen, "Petrarch's Conception of the 'Dark Ages,'" *Speculum* 17 (1942), 226–242.

123. For this, see Alfred J. Andrea, "The Myth of the Middle Ages," *The Historian* 55, no. 1 (Autumn 1992): 184.

124. Andrea, ibid., 184–185.

125. Ibid., 185.

126. Ibid.

nadir of European history. "The Middle Ages, stretching across a thousand years from the fifth century to the fifteenth," he declared,

> are still viewed by some as a long stupid detour in the march of human progress – a thousand years of poverty, superstition, and gloom that divided the golden age of the Roman Empire from the new golden age of the Italian Renaissance. During these thousand years, as a famous historian once said (in 1860), human consciousness "lay dreaming or half awake."[127] The Middle Ages were condemned as "a thousand years without a bath" by one well-scrubbed nineteenth-century writer. To others they were simply "the Dark Ages" – recently described (facetiously) as the "one enormous hiccup in human progress." At length, sometime in the fifteenth century the darkness is supposed to have lifted. Europe awakened, bathed, and began thinking and creating again. After a long medieval intermission, the Grand March resumed.[128]

The general hostility toward the Middle Ages that Professor Hollister describes has existed in Western culture since the Renaissance. To document his claim, I shall sample opinions from some of those he had in mind. The attitudes that are reflected in the hostile quotations and paraphrases that follow reveal more about their authors, and the times in which they wrote, than about the Middle Ages, as Norman Cantor, a distinguished medievalist, has observed:

> It is well known that the image of the Middle Ages which obtained at any given period in early modern Europe tells us more about the difficulties and dilemmas, the intellectual commitments of the men of the period than it does about the medieval world itself. The Renaissance thinkers who first branded the Middle Ages as "barbaric" and "dark" were men in rebellion against certain aspects of scholasticism and clerical authority. To justify their own departure from prevailing intellectual orthodoxies, they found it psychologically necessary and didactically useful to project their disdain backwards in time and pronounce a thousand years in the history of European man as boorish and intellectually retrograde. Similarly, the sarcastic opprobrium of the Enlightenment toward the medieval world was in large measure dictated by the guerilla warfare which the *philosophes* waged against the power and pretensions of the Church in their own day. Monsieur le curé was personally obnoxious to Voltaire, who found him bigoted, narrow-minded, parochial, and ignorant. Voltaire proceeded to apply this unflattering stereotype to an entire historical epoch.[129]

127. The historian is Jacob Burckhardt. For the passage in which the phrase occurs, see the quotation from Burckhardt later in this section.
128. C. Warren Hollister, *Medieval Europe: A Short History*, 7th ed. (New York: McGraw-Hill, 1994), 1.
129. Norman Cantor, *The Meaning of the Middle Ages: A Sociological and Cultural History* (Boston: Allyn and Bacon, 1973), 5.

In judging the comments that follow, it will be well to keep this passage in mind. But having said that, I hasten to add that simply because one may have certain biases, either of the age in which one lives, or because of some bias that may be sui generis to the individual making the comment, it does not follow that the comment is false, or unwarranted, or is a distortion of what seems to be the truth. Even a biased person may speak truthfully some of the time, or even much of the time. In the end, one must judge comments about the Middle Ages against the evidence that is relevant to the Middle Ages as a whole, or relevant to some particular aspect of medieval life or thought.

The criticisms and denunciations of the Middle Ages that I present here, however, were quite obviously prompted by deep animosity toward the Middle Ages, often motivated by hostility toward the Catholic Church. This is certainly true for Voltaire and most eighteenth-century philosophes, who were locked in a struggle with the ancien regime and the Church. Voltaire frequently gave expression to his loathing of the Middle Ages. "In the dismemberment of the Roman Empire in the West," he declared, "a new order of things began, which we call the *history of the Middle Ages:* a barbarous history of barbarous peoples, who became Christians but did not become better because of it."[130] Of the Middle Ages, Voltaire would declare that "it is necessary to know the history of that age only in order to scorn it."[131]

As the Middle Ages receded and knowledge of its accomplishments faded, opinions of its achievements as a civilization often became more scathing and extreme. This occurred even as scholars began to study the Middle Ages more dispassionately for its own sake during the nineteenth century. A devoted scholar of the Renaissance, John Addington Symonds, drew the starkest possible contrasts between the two periods. In the medieval world,

> [b]eauty is a snare, pleasure a sin, the world a fleeting show, man fallen and lost, death the only certainty, judgment inevitable, hell everlasting, heaven hard to win;

130. "Au démembrement de I'empire romain en Occident, commence un nouvel ordre de choses, et c'est ce qu'on appelle *l'histoire du moyen âge; histoire* barbare de peuples barbares, qui devenus chrétiens, n'en deviennent pas meilleurs." *The Complete Works of Voltaire,* W. H. Barber, general ed. (The Voltaire Foundation, Taylor Institution Oxford), Vol. 33 (1987), Jeroom Vercruysse, ed.: *Oeuvres alphabétiques* I, Articles pour I'Encyclopédie: "Histoire," 173.

131. "Il ne faut connaitre I'histoire de ces temps-là que pour la mépriser." *Essai sur les moeurs II, Oeuvres complètes de Voltaire,* nouvelle édition, vol. 12 (1878), 123. Translated by Ferguson, *The Renaissance in Historical Thought,* 89.

ignorance is acceptable to God as a proof of faith and submission; abstinence and mortification are the only safe rules of life: these were the fixed ideas of the ascetic medieval Church.

How different would life become in the Renaissance, when the unpleasant attitudes of the medieval Church were destroyed by

rending the thick veil which they had drawn between the mind of man and the outer world, and flashing the light of reality upon the darkened places of his own nature. For the mystic teaching of the Church was substituted culture in the classical humanities; a new ideal was established, whereby man strove to make himself the monarch of the globe on which it is his privilege as well as destiny to live. The Renaissance was the liberation of the reason from a dungeon, the double discovery of the outer and inner world.[132]

Symonds acknowledged that the Renaissance did not burst forth upon the world suddenly. There were a few stirrings in the Middle Ages – he mentions Peter Abelard, Roger Bacon, and Joachim of Flora. But

[t]he nations were not ready. Franciscans imprisoning Roger Bacon for venturing to examine what God had meant to keep secret; Dominicans preaching crusades against the cultivated nobles of Toulouse; Popes stamping out the seed of enlightened Frederick; Benedictines erasing the masterpieces of classical literature to make way for their own litanies or lurries, or selling pieces of the parchment for charms; a laity devoted by superstition to saints and by sorcery to the devil; a clergy sunk in sensual sloth or fevered with demoniac zeal: these still ruled the intellectual destinies of Europe. Therefore the first anticipations of the Renaissance were fragmentary and sterile.[133]

The Renaissance, however,

is not the history of arts or of sciences, or of literature, or even of nations.... It is no mere political mutation, no new fashion of art, no restoration of classical standards of taste. The arts and the inventions, the knowledge and the books, which suddenly became vital at the time of the Renaissance, had long lain neglected on the shores of the Dead Sea which we call the Middle Ages.[134]

In 1860, before Symonds had issued his works on the Renaissance, Jacob Burckhardt, a Swiss professor of history at Basel University, published *The*

132. The two passages appear in John Addington Symonds, *Renaissance in Italy: The Age of the Despots,* new ed. (London: Smith, Elder, & Co., 1901), 11.

133. Symonds, ibid., 8.

134. Ibid., 3.

Civilization of the Renaissance in Italy, a book that was destined to exert a powerful influence. Far more judicious and restrained than Symonds, Burckhardt was more appreciative of the Middle Ages than was Symonds. Nevertheless, in order to emphasize the originality of the Renaissance, he, like so many others, found it convenient, if not necessary, to denigrate the Middle Ages. In this vein, he declared:

> In the Middle Ages both sides of human consciousness – that which was turned within as that which was turned without – lay dreaming or half awake beneath a common veil. The veil was woven of faith, illusion, and childish prepossession, through which the world and history were seen clad in strange hues. Man was conscious of himself only as a member of a race, people, party, family, or corporation – only through some general category.[135]

By the fourteenth century, a social world had arisen in Italy "which felt that want of culture, and had the leisure and the means to obtain it. But culture, as soon as it freed itself from the fantastic bonds of the Middle Ages, could not at once and without help find its way to the understanding of the physical and intellectual world."[136] And, in one of his numerous comparisons between the newness of things in the Italian Renaissance as contrasted with the medieval North of Europe, Burckhardt found occasion to mention that in the North, hostility to innovation, and persecution of innovators, came from "the upholders of the received official, scholastic system of nature." Such an attitude, however, "was of little or no weight in Italy,"[137] where things were otherwise.

In Burckhardt's view, "the worldliness, through which the Renaissance seems to offer so striking a contrast to the Middle Ages, owed its first origin to the flood of new thoughts, purposes, and views, which transformed the medieval conception of nature and man."[138] The worldliness acquired by the Renaissance, "once gained, can never again be lost." But the investigations inspired by that worldliness have apparently led men away from God, and affected religious attitudes. Burckhardt asks, how soon will men return to God and how will their religious attitudes be affected by that return? Whatever the response to those questions, Burckhardt is certain of one thing: "The Middle Ages, which spared themselves the trouble of induction

135. Jacob Burckhardt, *The Civilization of the Renaissance in Italy* (London: Phaidon Press, 1950; reprint of the translation by S. G. Middleton, 1878), 81.
136. Burckhardt, ibid., 107.
137. Ibid., 175.
138. Ibid., 304.

and free inquiry, can have no right to impose upon us their dogmatical verdict in a matter of such vast importance."[139]

One of the most extreme assaults against the Middle Ages came from the pen of John William Draper, who characterized the period as the "Age of Faith." Since he felt no obligation to provide footnotes to support his numerous claims, Draper was free to make any charges he pleased, as is readily apparent from the following discussion:

> I have now completed the history of the European Age of Faith as far as is necessary for the purposes of this book. It embraces a period of more than a thousand years, counting from the reign of Constantine. It remains to consider the intellectual peculiarity that marks the whole period – to review briefly the agents that exerted an influence upon it and conducted it to its close.
>
> Philosophically, the most remarkable peculiarity is the employment of a false logic, a total misconception of the nature of evidence. It is illustrated by miracle-proofs, trial by battle, ordeal test, and a universal belief in supernatural agency even for objectless purposes. On the principles of this logic, if the authenticity of a thing or the proof of a statement be required, it is supposed to be furnished by an astounding illustration of something else. If the character of a princess is assailed, she offers a champion; he proves victorious, and therefore she was not frail. If a national assembly, after a long discussion, cannot decide "whether children should inherit the property of their father during the lifetime of their grandfather," an equal number of equal combatants is chosen for each side; they fight; the champions of the children prevail, and therefore the law is fixed in their favour.... In all such cases the intrinsic peculiarity of the logic is obvious enough; it shows a complete misconception of the nature of evidence. Yet this ratiocination governed Europe for a thousand years, giving birth to those marvellous and supernatural explanations of physical phenomena and events upon which we now look back with unfeigned surprise, half disbelieving that it was possible for our ancestors to have credited such things.[140]

In apparent ignorance of the long history of Church support for the university system during the Middle Ages, Draper declares: "The Roman ecclesiastical system, like the Byzantine, had been irrevocably committed in an opposition to intellectual development. It professed to cultivate the morals, but it crushed the mind."[141]

In France, Jules Michelet (1798–1874) was absorbed in the history of his country. He not only wrote a *History of France* in 17 volumes, the first pub-

139. Ibid., 304, 305.

140. Draper, *History of the Intellectual Development of Europe,* rev. ed., 2 vols. (New York: Harper & Brothers, Publishers, 1876), vol. 2, 111–112.

141. Draper, ibid., 129.

lished in 1833, the last in 1867, but he interrupted it in 1846 to publish a 7-volume *History of the French Revolution,* which he completed in 1853. In the latter, Michelet warns "the revolutionary reign of Terror" to "beware of comparing herself with the Inquisition," because the latter would only "have good cause to laugh." Why? In only two or three years, the revolution could hardly pay back what the Inquisition had taken 600 years to accomplish:

What are the twelve thousand men guillotined by the one, to the millions of men butchered, hanged, broken on the wheel, – to that pyramid of burning stakes, – to those masses of burnt flesh, which the other piled up to heaven.[142]

But one can apparently compare them in their attitudes toward suffering. For then:

History will inform us that in her most ferocious and implacable moments, the Revolution trembled at the thought of aggravating death, that she shortened the sufferings of victims, removed the hand of man, and invented a machine to abridge the pangs of death.

By the invention of the guillotine, the Revolution mercifully dispatched its victims quickly and efficiently.

In contrast to the humanitarian actions of the Revolution, history

will also inform us that the church of the middle ages exhausted herself in inventions to augment suffering to render it poignant, intense; that she found out exquisite arts of torture, ingenious means to contrive that, without dying, one might long taste of death – and that, being stopped in the path by inflexible nature, who, at a certain degree of pain, mercifully grants death, she wept at not being able to make man suffer longer.

Unfortunately, "the greater part of those grand butcheries can no longer be related," because the inquisitorial evildoers "have burnt the books, burnt the men, burnt the calcined bones over again, and flung away the ashes. When, for instance, shall I recover the history of the Vaudois, or of the Albigenses? The day when I shall have the star that I saw falling to-night. A world, a whole world has sunk, perished, both men and things." Michelet bemoans the fact that "our enemies triumph that they have rendered us

142. Jules Michelet, *History of the French Revolution,* translated by Charles Cocks; edited with an Introduction by Gordon Wright (Chicago: The University of Chicago Press, 1967), 32.

powerless, and at having been so barbarous that one cannot, with certainty recount their barbarities."[143]

The lack of evidence of which Michelet complains did not prevent him from inventing a fantastic claim about "the millions of men butchered, hanged, broken on the wheel," over a period of 600 years. If the numbers he provides about executions in the French Revolution are accurate, then the millions of executions he attributes to the Middle Ages might, more appropriately, be assigned to the Revolution. For if the executioners of the Revolution did away with men at the rate of 12,000 every three years, as Michelet suggests in the passage just cited, they would have executed 2,400,000 victims in 600 years; and if they did away with 12,000 every two years, the total number of victims over 600 years would be 3,600,000!

In a volume titled *Satanism and Witchcraft: A Study in Medieval Superstition,* Michelet argues that the Church, by its unfortunate behavior, has made Satan what he is. For Satan

> is not hard to please. Nothing rebuffs him; what Heaven throws in his way, he picks up with alacrity. For instance the Church has rejected Nature as something impure and suspect. Satan seizes on it, and makes it his pride and ornament.[144]...
>
> Another trifle the Church has cast away and condemned – Logic, the free exercise of Reason. Here again is an appetising dainty *the Enemy* snaps up greedily.
>
> The Church has built of solid stone and tempered mortar ... the *schools.* A few shavelings were let loose in it, and told "to be free"; they one and all grew halting cripples. Three hundred, four hundred years, only made them more helplessly paralysed. Between Abelard and Occam the progress made is – nil.[145]

So wretched were the conditions of life in the Middle Ages that "a man slips from free man to *vassal* – from vassal to *servant* – from servant to *serf*." This "is the great terror of the Middle Ages, the basis of its despair. There is no way of escape; one step, and the man is lost." Such conditions "drove men to give themselves to Satan."[146] In such a society, originality and creativity were utterly stifled. "For a thousand long, dreary, terrible years! For ten whole centuries," Michelet proclaims, "a languor no previous age has known oppressed the Middle Ages."[147]

143. For all of these quotations, see Michelet, *History of the French Revolution,* 32–33.

144. Jules Michelet, *Satanism and Witchcraft: A Study in Medieval Superstition,* translated by A.R. Allinson. (London: Tandem Books, 1965), 15.

145. Michelet, ibid.

146. Ibid., 29

147. Ibid., 27.

Not only did Michelet convey a grossly distorted and utterly bizarre picture of the Middle Ages, but his prose and tone also border on the hysterical. It was probably for these reasons that his book was selected for translation and publication in 1965. On a single introductory page, the translator or officers of Tandem Books Limited, the publishers, present a brief description of the book to catch the eye of potential readers. The description mirrors Michelet's frantic, hyperbolic style and achieves a similar degree of accuracy about the Middle Ages:

This was the age of fear and superstition when witchcraft became the great force in people's lives.

The age of the Black Mass, the reign of Satan, the weird rites of the damned ...

The age of luxury beyond imagination, unbridled sensuality, and unendurable squalor ...

The age of torture, summary decapitation and the brand of the witch on any young girl who could not survive the tests of immersion in water and boiling oil ...

The age of potions, poisons, incantations, wizards and spells, of feudal barons and the serf whose bride dare not fail to show favour to the overlord.

The age of Intolerance, the Inquisition and the Ordeal by Fire ...

This superb re-creation is the most brilliant book of its kind ever written about the Age of Darkness.[148]

Given routine public acceptance of such graphic and wildly erroneous descriptions of medieval life, this blurb was probably accepted as a good assessment of conditions in the Middle Ages. Although I mentioned it earlier in the Introduction to this study, it bears repeating: The witchcraft persecutions, and the accompanying torture, reached their greatest peak in the seventeenth century, not in the Middle Ages, when the witchcraft phenomenon was relatively mild by comparison to what was to come.

The Flat Earth

In his book *On the Heavens,* book 2, Chapters 13 and 14, Aristotle discusses the shape and position of the earth, and whether it is at rest or in motion. Near the end of the discussion, he concludes that the earth is spherical and insists that

148. None of the ellipses in this passage are mine. They were all inserted by the author, or authors, of this paragraph.

> [t]he evidence of the senses further corroborates this. How else would eclipses of the moon show segments shaped as we see them? As it is, the shapes which the moon itself each month shows are of every kind – straight, gibbous, and concave – but in eclipses the outline is always curved; and since it is the interposition of the earth that makes the eclipse, the form of this line will be caused by the form of the earth's surface, which is therefore spherical. Again, our observations of the stars make it evident, not only that the earth is circular, but also that it is a circle of no great size. For quite a small change of position on our part to south or north causes a manifest alteration of the horizon.... All of which goes to show not only that the earth is circular in shape, but also that it is a sphere of no great size; for otherwise the effect of so slight a change of place would not be so quickly apparent. Hence one should not be too sure of the incredibility of the view of those who conceive that there is continuity between the parts about the pillars of Hercules and the parts about India, and that in this way the ocean is one. As further evidence in favour of this they quote the case of elephants, a species occurring in each of these extreme regions, suggesting that the common characteristic of these extremes is explained by their continuity. Also, those mathematicians who try to calculate the size of the earth's circumference arrive at the figure 400,000 stades. This indicates not only that the earth's mass is spherical in shape, but also that as compared with the stars it is not of great size.[149]

In this momentous passage, Aristotle not only assumes the sphericity of the earth but also offers convincing arguments in its favor. Moreover, he suggests that perhaps the pillars of Hercules – the Straits of Gibraltar – and India are linked by a single ocean that makes them in some sense continuous, a possibility that gains some credence by the fact that elephants are found in both places.

All medieval students who attended a university knew this. In fact, any educated person in the Middle Ages knew the earth was spherical, or of a round shape. Medieval commentators on Aristotle's *On the Heavens,* or in their commentaries on a popular thirteenth-century work titled *Treatise on the Sphere* by John of Sacrobosco, usually included a question in which they inquired "whether the whole earth is spherical."[150] Scholastics answered this question unanimously: The earth is spherical, or round. No university-trained author ever thought it was flat. When Columbus sought backing for his voyage from Spain to India, he gathered evidential support from a scholastic treatise, *The Image [or Representation] of the World* (*Ymago*

149. Aristotle, *On the Heavens,* 2.14, 297b.24–298a.20. Translated by J. L. Stocks.
150. For a list of some of those who treated this question, see Grant, *Planets, Stars, & Orbs,* 737–738. In the list, the first eight authors lived before Columbus set sail for India.

mundi), written by the eminent theologian and natural philosopher Pierre d'Ailly. At the outset, d'Ailly asserts that the earth is spherical and, in Chapter 10, reports that Alfraganus (or al-Farghani) (d. after 861), an Arab astronomer, had recorded the measurement of a degree of the earth's circumference as 56 2/3 miles, which, when multiplied by 360 degrees, yields a value of 20,400 miles for the earth's circumference. In Chapter 11, d'Ailly explains that "[a]ccording to Aristotle and Averroës at the end of the second book of *De caelo et mundo,* the end of the habitable earth toward the east and the end of the habitable earth toward the west are very close and there is a small sea between."[151]

Pierre d'Ailly's *Image of the World* was one of the most popular early printed books and exercised a great influence in the late fifteenth, sixteenth and seventeenth centuries. Christopher Columbus owned a copy, which he heavily annotated.[152] This annotated copy is preserved in Seville, Spain. It reveals that Columbus was well aware of d'Ailly's report of the earth's circumference of 20,400 miles and d'Ailly's subsequent report from Aristotle and Averroës that Spain and India were only separated by a small sea. In a number of places, Columbus wrote 56 2/3 in the margin and even drew boxes around it for emphasis, also declaring that 56 2/3 was the exact measure of a degree of the earth's circumference. He also copied, in the margin of his book, d'Ailly's report about the separation of Spain and India by a small sea.

D'Ailly's work provided Columbus with two vital pieces of information: (1) the earth is a sphere and (2) its circumference is only 20,400 miles, an estimate that was much too small, but was advantageous for Columbus because it made the earth appear much smaller than it actually was, so that sailing west from Spain to India would require only a few days to cross the small sea separating them. Columbus thus had the data he needed to convince the court of Spain that the voyage was feasible. It is apparent that Columbus did not rescue Europe from a universal medieval belief and conviction that the earth is flat. Columbus himself made no such claim. In fact, he learned about the sphericity of the earth and its size from medieval sources.[153]

And yet the popular conception of Columbus's voyage is that he discovered that the world is round, thereby refuting the medieval view that the

151. Grant, *A Source Book in Medieval Science,* 639.

152. I have translated Columbus's marginal comments along with sections of d'Ailly's *Ymago mundi* in Grant, ibid., 630–639.

153. For a brief discussion of the size and shape of the earth in the Middle Ages, see Grant, *Planets, Stars, & Orbs,* 620–622.

earth is flat. The widespread conviction that all in the Middle Ages believed in a flat earth until Columbus showed otherwise was an invention of the nineteenth century, as Jeffrey Burton Russell has shown in his aptly titled book: *Inventing the Flat Earth: Columbus and Modern Historians.*[154] Russell observes that prior to the nineteenth century, no one really attributed the erroneous idea of a flat earth to the Middle Ages. As he correctly observes, in the eighteenth century, "the vehement anticlericals of the Enlightenment seldom made the Flat Error. They were concerned with attacking scholastics and their successors for being hidebound Aristotelians, and they were fully aware that sphericity was central to Aristotle's cosmology."[155]

The flat-earth error is probably traceable to the fact that in late antiquity, two Christians assumed that the earth was flat, Lactantius (ca. 245–325) and Cosmas Indicopleustes, who wrote a *Christian Topography* in which he argued that the earth was the flat floor of a cosmos that was shaped as a rectangular vaulted arch.[156] Lactantius thought it silly to assume sphericity because everything on the other side of the earth would be upside down. The influence of these two was negligible, and virtually all other educated early Christians and Church Fathers assumed the sphericity of the earth. Lactantius's opinion, however, was given prominence in the early modern period when Nicholas Copernicus, in the preface of his great treatise *On the Revolutions of the Heavenly Orbs,* declared that "Lactantius, otherwise an illustrious writer but hardly an astronomer, speaks quite childishly about the earth's shape, when he mocks those who declared that the earth has the form of a globe."[157] Whether Copernicus's mention of Lactantius was subsequently influential is unclear. It was surely not during the sixteenth to eighteenth centuries, when Copernicus's treatise was most frequently used. For it was in this period that scholastic authors were accused of slavish devotion to Aristotle, and, since Aristotle argued for a spherical earth, it was assumed that his medieval scholastic followers also did. Prior to the nineteenth century, therefore, medieval thinkers were only occasionally accused of believing in a flat earth, usually because some scholar extrapolated the flat-earth concept espoused by Lactantius or Cosmas, or both, and falsely applied it to the Middle Ages.[158]

154. New York: Praeger, 1991.

155. Russell, *Inventing the Flat Earth,* 71.

156. For a summary of their views, see Russell, ibid., 32–35.

157. Cited by Russell, ibid., 64. The quotation is from Edward Rosen's translation of Nicholas Copernicus's *On the Revolutions of the Heavenly Orbs* (London: Macmillan, 1978).

158. Among the exceptions were Tom Paine, Thomas Hobbes, and John Wilkins. See Russell, *Inventing the Flat Earth,* 61–62.

By the nineteenth century, however, Aristotle and medieval scholasticism had largely faded from history. One could, therefore, level the flat-earth charge against the Middle Ages with impunity. In 1837, William Whewell published his famous *History of the Inductive Sciences* and saw fit to cite Lactantius and Cosmas Indicopleustes as evidence that a flat earth was universally accepted in the Middle Ages.[159] Few had qualms about citing one author from the third century and another from the sixth century, and then extrapolating their opinions approximately one thousand years to extend over the entire Middle Ages. Since Whewell's book was influential, others would repeat this false charge.

According to Jeffrey Russell, however, the real culprits in establishing the flat-earth myth preceded Whewell's publication. Washington Irving laid the basis for it in his *History of the Life and Voyages of Christopher Columbus* in 1828. In the passage Russell cites as the source of the error, Irving does not explicitly state that in the Middle Ages there was a common belief in a flat earth. Rather, Irving conjured up an imaginary scene in which Columbus pleads his case before Church dignitaries and professors at the University of Salamanca. He informs us that Columbus was "assailed with citations from the Bible and the Testament." The citations were presumably of the kind that showed that the earth was flat, or that it could not be spherical or globular. Irving then says:

> Columbus, who was a devoutly religious man, found that he was in danger of being convicted not merely of error, but of heterodoxy. Others more versed [than the scripture quoters] in science admitted the globular form of the earth ... but ... maintained that it would be impossible to arrive there.... Such are specimens of the errors and prejudices, the mingled ignorance and erudition, and the pedantic bigotry, with which Columbus had to contend."[160]

Although Irving does imply that Columbus believed in a spherical earth and that most of his opponents did not, it is obvious that, in addition to Columbus, some were in attendance who also believed that the earth was round. If this vague and ambiguous passage was one of the primary causes of the flat-earth myth, it was so only because subsequent authors chose to make it so. And some did choose to do so.

Washington Irving's book on Columbus was instrumental in disseminating the flat-earth error on a popular level. Antoine-Jean Letronne

159. See ibid., 32.

160. Cited by Russell, ibid., 53. The source is Washington Irving, *The Life and Voyages of Christopher Columbus,* edited by John Harmon McElroy (Boston: Twayne Publishers, 1981), 47–51.

(1787–1848) was largely responsible for ensnaring scholars into a belief in the flat-earth myth.[161] Letronne set forth his views in 1834, in an article titled "On the Cosmological Views of the Church Fathers." Although he allowed that a few theologians thought the earth was round, Letronne argued that the majority were committed to a flat earth, following the opinion of Cosmas Indicopleustes. Of crucial importance was Letronne's sweeping claim that flat-earth theories dominated to the time of Columbus and Magellan, thus extending belief in a flat earth throughout the whole of the Middle Ages.

From these beginnings, the flat-earth theory that was attributed to the Middle Ages gradually became embedded in both popular and intellectual thought. Since approximately 1870, the flat-earth theory has become very nearly a quintessential example of the backwardness of the Middle Ages. Columbus was depicted as the rationalist who was confronted by irrational, obscurantist, medieval defenders of a flat earth. In 1887, one author placed Columbus before a commission at Salamanca that is made to terrorize Columbus with the following chilling lines:

> "You think the earth is round, and inhabited on the other side? Are you not aware that the holy fathers of the church have condemned this belief? ... Will you contradict the fathers? The Holy Scriptures, too, tell us expressly that the heavens are spread out like a tent, and how can that be true if the earth is not flat like the ground the tent stands on? This theory of yours looks heretical."
>
> Columbus might well quake in his boots at the mention of heresy; for there was that new Inquisition just in fine running order, with its elaborate bone-breaking, flesh-pinching, thumb-screwing, hanging, burning, mangling system for heretics. What would become of the Idea if he should get passed over to that energetic institution?[162]

Such false and fantastic accounts of medieval support for a flat earth and against a round one became commonplace in the late nineteenth and twentieth centuries. John William Draper spoke of Magellan's circumnavigation of the globe and, after praising the voyage, remarks that "it is to be remembered that Catholicism had irrevocably committed itself to the dogma of a flat earth, with the sky as the floor of heaven, and hell in the under-world. Some of the Fathers, whose authority was held to be paramount, had, as we have previously said, furnished philosophical and religious arguments

161. I follow Russell, *Inventing the Flat Earth*, 58–61.

162. Russell, ibid., 6. Cited from James Johonnot, compiler and arranger, *Ten Great Events in History* (New York, 1887), 123–130.

against the globular form. The controversy had not suddenly come to an end – the Church was found to be in error."[163]

Andrew Dickson White (1832–1918), the first president of Cornell University, published a famous two-volume work in 1896 titled *History of the Warfare of Science with Theology and Christendom.* In this work, White conceded that most in the Middle Ages accepted the sphericity of the earth – he specifically mentions Roger Bacon, Albertus Magnus, Thomas Aquinas, Dante, and Vincent of Beauvais. Despite this concession, White wrote as if the Church opposed the sphericity of the earth when he declared:

The warfare of Columbus the world knows well; ... how, even after he was triumphant, and after his voyage had greatly strengthened the theory of the earth's sphericity ... the Church by its highest authority solemnly stumbled and persisted in going astray.... In 1519 science gains a crushing victory. Magellan makes his famous voyage....Yet even this does not end the war. Many conscientious men oppose the doctrine for two hundred years longer.[164]

In 1929, the eminent historian of the Middle Ages, Lynn Thorndike wrote:

Only thirty years ago the following statement was made, not by some crude American unacquainted with the more recent findings of European scholarship, but by the Gallic author of a history of French literature in the sixteenth century: "The world has been discovered for only a trifle over three hundred years. The time has not been more than that during which men have known that the world is round, that it is small and the sky infinite. This [discovery] has changed all ideas." This utterance represents one of the worst slanders current against the fair name of the middle ages, when every astronomical textbook or lecturer taught that the earth was a sphere, and this was well known to any educated layman such as Dante.[165]

In the twentieth century, the flat-earth myth continues on. It is often taught in the public schools and appears in numerous publications,

163. Draper, *History of the Conflict Between Religion and Science,* 8th ed. (New York, 1884; 1st ed., New York, 1874), 294.

164. Cited in Russell, *Inventing the Flat Earth,* 45–46, from White, *History of the Warfare of Science with Theology in Christendom,* vol. 1, 108–109.

165. Thorndike, *Science and Thought in the Fifteenth Century: Studies in the History of Medicine and Surgery. Natural and Mathematical Science, Philosophy and Politics* (New York: Columbia University Press, 1929), 20. The French work is *Seizième siècle* (1898) by E. Faguet. The translation of Fauget's text is by Thorndike.

although some encyclopedias and publishers of school texts have corrected the error.[166] The authors of a prestigious college textbook assert that "the fact that the earth is round was known to the ancient Greeks but lost in the Middle Ages."[167] In a play, titled *Christopher Columbus,* the author, Joseph Chiari, includes the following dialogue between Columbus and a prior:

Columbus: The Earth is not flat, Father, it's round!
The Prior: Don't say that!
Columbus: It's the truth; it's not a mill pond strewn with islands, it's a sphere.
The Prior: Don't, don't say that; it's blasphemy.[168]

Although some progress has been made in rectifying the egregious historical error that a flat earth was commonly assumed in the Middle Ages, the error lives on. Perhaps it is because, as Russell plausibly suggests, "the idea of the dark Middle Ages is still fixed in the popular consciousness," and, consequently, "no caricature is too preposterous to be accepted."[169]

CONTEMPORARY ATTITUDES TOWARD "MEDIEVAL" AND "MIDDLE AGES"

Recent articles in the press offer an excellent opportunity to gauge the extent to which modern attitudes about the Middle Ages have been shaped by three centuries of harsh, prejudicial, and adverse criticism. In reacting to the bombing of Iraq by the United States and Britain, one journalist found occasion to mention that "the Nazis combined the worst of medieval cruelty with twentieth-century technology and created a terrifying synthesis."[170] The author of this piece was searching for the most recognizable, extreme symbol for cruelty and could think of nothing better than "medieval." He did not say: "the Nazis combined the worst of ancient Greek and Roman cruelty"; or "the worst of Renaissance cruelty." Indeed, for medieval, he might have substituted "Age of Reason," or "seventeenth

166. For a list, see Russell, *Inventing the Flat Earth,* 4 and n. 11, 80. On p. 3, Russell cites a passage from a 1983 textbook for fifth graders in which the flat earth is attributed to Columbus's contemporaries.
167. Russell, ibid., 3.
168. Ibid. Russell (p. 80, no. 10) says that the play was published in New York in 1979.
169. Ibid., 67.
170. The article, dated December 23, 1998, is by Jimmy Reid, titled: "A bid to oil the wheels of a runaway power," and appears in the *Herald* (Glasgow), which is part of Scottish Media Newspapers Limited. This citation and those that follow from other newspapers were found by searching Lexis-Nexis.

century," the period in which witchcraft persecutions reached their horrible climax. On any objective cruelty meter, all of these periods are approximately equal to the Middle Ages. But the author knew that the association of "medieval" with "cruelty" would strike a responsive chord with his readers. Jimmy Reid, the author, was not yet finished with things medieval. Later in the article he asks whether the United States thinks it has the exclusive prerogative to proclaim sanctions against other countries? What if some country calls for sanctions against the United States? "Will those who dare use sanctions against it be bombed back to the Middle Ages?" he asks. Obviously, Mr. Reid could think of nothing more primitive to be "bombed back to" than the Middle Ages. Once again, the author used a name that he knew would instantly resonate with his readers. Obviously, Middle Ages equals: as primitive and crude as it gets.

Thousands of Asians – more than 50,000 – from China, the Philippines, Bangladesh, and Thailand have been recruited for sweatshops on the island of Saipan and promised good wages. Instead of good wages, they "wind up in sweatshops that 'would make medieval conditions look good,'" said the plaintiffs' attorney as cited in a 1999 Associated Press article.[171] To make his or her point, the anonymous author could think of no worse working conditions than those in the Middle Ages, until the clothing manufacturers of Saipan came upon the scene. Although the Renaissance and Middle Ages overlap, and the same working conditions obtained in both, only the latter – never the former – would be invoked in this dreadful context.

From a 1998 Lifestyle article in the *San Diego Union-Tribune*, we learn that approximately 4 million Americans are infected with hepatitis C, a chronic disease that causes approximately 10,000 deaths every year. "Fortunately," the author of the article informs us, "it's not the Middle Ages," so that "while hepatitis C is incurable, a recently approved drug combination reduces blood levels of the virus in nearly half of patients, and other treatments are under study."[172] In this context, "Middle Ages" clearly signifies "age of ignorance" wherein almost any disease or ailment, not only hepatitis C, was incurable. Although the author could have chosen almost any period prior to the nineteenth century and the medical treatment would not have differed much, if at all, from the Middle Ages, the author chose "Middle Ages" because she knew that most of her readers were condi-

171. Associated Press article, dated January 14, 1999, in *Newsday* (New York) titled: "Citing America's Worst Sweatshop/Suit Targets U.S. Retailers, Saipan."

172. Article by Michele Lesie, *San Diego Union-Tribune*, October 19, 1998, Lifestyle section, D–1.

tioned to react negatively to that term, and that it would instantly convey the desired sense of backwardness and hopelessness better than any other period description.

Derogatory remarks about the Middle Ages turn up in many contexts. David Goodman, an Associated Press writer, described the antics of Dr. Jack Kevorkian, noted for his role in assisted suicides. Goodman first explains that "Dr. Jack Kevorkian arrived in homemade stocks and a ball and chain today for his arraignment on assisted suicide charges." This prompted Kevorkian's attorney, Geoffrey Fieger, to explain to reporters that "Kevorkian is accepting his medieval punishment." A sign that Kevorkian wore on his front read: "Common law of Middle Ages. What's next, the Inquisition?"[173] "Medieval" and "Middle Ages" served Dr. Kevorkian very well. He, and his attorney, could think of nothing more telling than to depict the authorities, who were seeking to jail him, as throwbacks to the Middle Ages, implying thereby that they were ignorant, cruel, and barbaric. To my knowledge, the stocks in which Dr. Kevorkian confined himself was not a medieval instrument of punishment but a device used in the seventeenth century by Puritans in the American colonies. Its purpose was the public humiliation of its hapless victims.

Humorists also reveal to us the deep-seated, almost instinctive sense of contempt for the Middle Ages. Dave Barry, the syndicated humor columnist who writes for the *Miami Herald*, wrote a column about the Renaissance, in which he declares:

> The Renaissance – as you recall from not spelling it correctly even one single time in your academic career – was the historical period that started in the 15th century at approximately 3:30 p.m. when humanity, after centuries of being cooped up in the Dark Ages, finally stumbled out into the light and got a whiff of its own armpits and said, "Whoa! Time to invent cologne!" This was followed by tremendous advances in science, philosophy, literature and paintings of naked women.[174]

Playful and humorous as they are, Dave Barry's remarks are illuminating, for he was in fact giving instinctive utterance to ideas about the Middle Ages that he had probably absorbed, perhaps unwittingly, beginning in elementary school and in his subsequent education.

These instances reveal the way journalists reflect cultural attitudes toward the Middle Ages and things medieval. Their biases may be taken as a reliable measure of popular opinion about the Middle Ages. Although we

173. Article by David Goodman, Associated Press writer, September 14, 1995.

174. Article, "If you can't spell 'Renaissance,' at least wear a squeaking codpiece," by Dave Barry as printed in the *Indianapolis Star*, March 8, 1999.

may choose to ignore contemporary popular opinion, it is more difficult to disregard the opinions of a professional historian of the caliber of William Manchester, who had occasion to present his views on the Middle Ages in a book he published in 1992, titled: *A World Lit Only by Fire: The Medieval Mind and the Renaissance: Portrait of an Age.*[175] In speaking of the ten centuries of the Middle Ages, Manchester declares that

> [i]n all that time nothing of real consequence had either improved or declined. Except for the introduction of waterwheels in the 800s and windmills in the late 1100s, there had been no inventions of significance. No startling new ideas had appeared, no new territories outside Europe had been explored. Everything was as it had been for as long as the oldest European could remember.[176]

The Renaissance and great changes were imminent, but medieval Europeans were unaware of them and did not even believe such things could occur:

> Shackled in ignorance, disciplined by fear, and sheathed in superstition, they trudged into the sixteenth century in the clumsy, hunched, pigeon-toed gait of rickets victims, their vacant faces, pocked by smallpox, turned blindly toward the future they thought they knew – gullible, pitiful innocents who were about to be swept up in the most powerful, incomprehensible, irresistible vortex since Alaric had led his Visigoths and Huns across the Alps, fallen on Rome, and extinguished the lamps of learning a thousand years before."[177]

In these passages, Professor Manchester presents the Rip Van Winkle view of the Middle Ages: a thousand years of troubled sleep before the great awakening brought on by the Renaissance.[178] It is difficult to imagine how Manchester managed to miss all the weighty and abundant evidence that contradict his claim that, apart from the introduction of waterwheels and windmills, nothing of real consequence occurred in the Middle Ages.

REDRESSING THE BALANCE

Medievalists, however, have revealed a radically different picture, showing unmistakably that the Middle Ages in Western Europe was a very creative period. As one of them put it:

175. Boston: Little, Brown and Company. Manchester is professor emeritus of history at Wesleyan University, Middletown, Connecticut. He is the author of numerous books, including biographies and histories.
176. Manchester, *A World Lit Only by Fire,* 27.
177. Ibid., 27.
178. I derived the Rip Van Winkle characterization from C. Warren Hollister, *Medieval Europe,* 1.

By the close of the Middle Ages – by about 1500 – Europe's technology and political and economic organization had given it a decisive edge over all other civilizations on earth. Columbus had discovered America; the Portuguese had sailed around Africa to India; Europe had developed the cannon, the printing press, the mechanical clock, eyeglasses, distilled liquor, and numerous other ingredients of modern civilization.[179]

If medieval Europe had been asleep for a thousand years – from approximately 500 to 1500 – these momentous events could not have occurred. But medieval Europeans contributed much more than what is mentioned in the brief passage just cited. They did not merely invent waterwheels and watermills, as Manchester implies; innovative medieval technicians showed their ingenuity and inventiveness by adapting waterwheels and windmills to serve numerous functions. Thus, the watermill was used for brewing beer, fulling cloth, sawing wood, and making paper.[180]

Medieval civilization was able to adapt machinery to a variety of useful purposes because early in the Middle Ages, anonymous laborers invented the crank, a device unknown to the Greeks and Romans. In the later Middle Ages, this incredibly useful device made possible the construction of power machines that could produce reciprocal and rotary motion, for which reason the crank has been regarded as "an invention second in importance only to the wheel itself."[181] Medieval civilization emphasized technology and machine power to such an extent that Lynn White was moved to declare that

[t]he chief glory of the later Middle Ages was not its cathedrals or its epics or its scholasticism: it was the building for the first time in history of a complex civilization which rested not on the backs of sweating slaves or coolies but primarily on non-human power.[182]

The invention of labor-saving and time-keeping machinery was only one facet of the medieval European contribution to Western civilization. Even the raids and ravages of the Vikings from around 750 to 1050 had their positive side. The Vikings were Europe's earliest explorers, adventurers, and

179. Hollister, ibid.

180. See Edward Peters, *Europe and the Middle Ages*, 2nd ed. (Englewood Cliffs, NJ: Prentice Hall, 1989), 310.

181. See Lynn White, Jr., *Medieval Religion and Technology: Collected Essays* (Berkeley: University of California Press, 1978), 17; also see p. 18.

182. Ibid., 22.

settlers of previously unknown lands. Not only were they "successful ship builders who engaged in ever-widening trade, east to Russia and south to Rome and Baghdad," but "in their Iceland colony at the end of the 10th century, these people created the first democratic parliament."[183] A collection of some 800 artifacts from a Viking camp, however, shows beyond a doubt that the Vikings reached America around A.D. 1000 And so it was that in both the early and late Middle Ages, medieval Europeans exhibited an exploratory spirit that seems to have been unmatched anywhere else.

If the Vikings fashioned the earliest parliamentary democracy, their European descendants would lay the permanent foundations for democratic government a few centuries later. In 1215, Magna Carta was issued by King John of England. Like other great documents in the centuries to follow, most of the articles applied initially to the aristocratic classes. By the mid–fourteenth century, however, their protections were extended to all subjects in the kingdom. The thirty-ninth article would have been a monumental achievement by itself, since it declared that "[n]o free man shall be taken or imprisoned or disseised or outlawed or exiled or in any way ruined, nor will we go or send against him, except by the lawful judgment of his peers, or by the law of the land."[184] Thus did article 39 compel the king to follow "due process" and thereby prevent him from taking arbitrary actions against his subjects.

The parliaments of Europe were founded in the late Middle Ages of which the most notable emerged in England. The English parliament evolved over the thirteenth and fourteenth centuries and was solidly established by the late fifteenth century.[185] With England leading the way, European parliaments served as the instrumentality for the gradual establishment of democratic government in Western Europe.

From what has been said in this study, medieval Europe invented other institutions that proved vital to its ultimate advancement, as is obvious in the university and the corporation (for both, see Chapter 3). But the late Middle Ages also paved the way for amazing progress in the study of medi-

183. My source is an article by John Noble Wilford, "Ancient Site Offers Clue to Vikings in America," which appeared as the lead article on the front page of the Science Times section of the *New York Times* on May 9, 2000. The article was occasioned by an exhibition about the Vikings at the National Museum of Natural History in Washington, DC.

184. Quoted from Ralph V. Turner, "Magna Carta," in *Dictionary of the Middle Ages,* vol. 8 (1987), 41–42.

185. For a brief summary of the development of the English parliament in the Middle Ages, see William Huse Dunham, Jr., "Parliament, English," in *Dictionary of the Middle Ages,* vol. 9 (1987), 422–434.

cine, when in the thirteenth century, post mortems were made in Italy, followed shortly thereafter by the introduction of anatomical dissection in the medieval medical schools. One can scarcely imagine what modern medicine would be like without the knowledge that was subsequently derived from the study and analysis of the dissected human body.

I could cite additional achievements, but I have mentioned enough impressive contributions to convince unbiased readers that William Manchester's assessment of the Middle Ages is untenable if he seeks to convince us that "in all that time nothing of real consequence had either improved or declined" and that "no startling new ideas had appeared, no new territories outside Europe had been explored. Everything was as it had been for as long as the oldest European could remember." We should conclude – indeed, we must conclude – that the Middle Ages was a period in which the broad foundations of the Western world were firmly laid. Indeed, without those foundations, the modern world could not have come into being.

Despite the enormity of these achievements, and others, the Middle Ages is still all too frequently held in low regard. The dissemination of a more faithful and accurate depiction of medieval achievements was slow and laborious. The effort to counter the many false interpretations about the Middle Ages began in earnest in the nineteenth century and was pursued with considerable vigor in the twentieth century. Until the end of the eighteenth century, the Middle Ages was regarded by most scholars and educated people as a priest-ridden age of superstition and ignorance. By the end of the eighteenth century, the Romantic Age began to develop. Romantics rejected the harsh indictments of the Middle Ages made "by Renaissance humanists, Protestant Reformers, and Enlightenment philosophes," and "went to the opposite extreme and imbued the medieval past with special, almost magical, qualities,"[186] perhaps best exemplified by Sir Walter Scott's *Ivanhoe.* Much of what the Romantics wrote was unhistorical and largely invented. But historians who were imbued with the Romantic outlook became fascinated with the Middle Ages and reinterpreted it. "That much abused era now became the Age of Faith, a kind of golden age of innocence, of childlike trust and emotional security, now lost forever,"[187] wrote Wallace Ferguson. It was largely because of the Romantic interpretation of the Middle Ages that it came to be viewed as an Age of Faith, rather than an Age of Reason.

186. Andrea, "The Myth of the Middle Ages," 185.
187. Ferguson, *The Renaissance in Historical Thought,* 121.

Another powerful impetus toward the study of the Middle Ages was nineteenth-century nationalism. As European historians sought the deeper roots of their respective countries, they found it necessary to probe more deeply into the Middle Ages to locate the beginnings of the institutions, laws, languages, customs, religion, and politics of their respective countries. Romantic historians stressed "the unconscious and irrational, the typical and corporate qualities in medieval society at the expense of the conscious, rational, and individual."[188] This newly awakened positive interest in the Middle Ages motivated many historians to investigate the medieval period in a careful, scholarly way. In doing so, they began to produce critical studies of the Middle Ages, first in German universities, then in other parts of Europe, and finally in the United States, where medieval studies have flourished.[189] What they found is that during the Middle Ages, a new Europe emerged that laid the foundations for the modern Western world. Medievalists showed that the Middle Ages was nothing like the false picture painted of it during the preceding centuries. But they could not overcome the pejorative image of the Middle Ages, an image that seems almost genetically embedded in our culture.

To appreciate how deeply ingrained are the prejudicial images and perceptions of the Middle Ages, one need only juxtapose "Middle Ages" (or "the medieval period") alongside "Renaissance." They seem antithetical: The former evokes a sense of what is dark, gloomy, unenlightened; the latter conjures up positive feelings about progress and brilliant achievements in the arts and sciences. Indeed, the Renaissance was originally so-called because it signifies a "revival" or "renewal of learning." The description clearly implies that immediately prior to the "renewal of learning," there wasn't any learning worthy of the name. What was really meant was that learning was expanded by the addition of ancient Greek and Latin works on history and literature, and to a lesser extent by the addition of scientific and philosophical works. By any standard of measure, however, the Middle Ages had available a large body of ancient learning, which they absorbed and made their own, and to which they added much. The negative perceptions of the Middle Ages seem to be an unconscious, but integral, part of our culture, buried so deep within it, that they may never be extracted and eliminated. An enduring example of this negativism is the falsely ascribed belief in a flat earth, described earlier.

188. Ibid., 124.

189. For a brief description of twentieth-century scholarship in medieval studies, see Ferguson, ibid., 330–385, which is a section called "Revolt of the Medievalists."

Most misperceptions of the Middle Ages were formed in the sixteenth to nineteenth centuries, when the discipline of history was evolving, before it became a professional activity in the nineteenth century. A significant part of the problem derives from the period labels that European historians routinely use: "Antiquity," "Middle Ages," "Renaissance," "Age of Reason," "Age of Enlightenment," and so on. One sees at a glance that the nondescript title "Middle Ages" is followed by three very positive period labels: "Renaissance," "Age of Reason," "Age of Enlightenment." These labels have become almost a reflexive part of our thought patterns. Despite the convenience of using shorthand catch-phrases to represent historical periods, these traditional labels are a source of significant bias and distortion. If I say that you have a "medieval" mind, you, and almost everyone else, would regard that as an insult. It implies that you are superstitious, crude, backward, and reactionary, the kind of person who fails to use reason where it is clearly required, and so on. If your hygiene is "medieval," all would be well advised to hold their noses and distance themselves from your smelly, unwashed body. But if I identified you as a "Renaissance" person, you, and almost everyone else, would regard that as a major compliment, since it clearly implies that you are a well-rounded individual with broad-ranging interests and talents, a person with an open, inquisitive mind.

Ironically, the Middle Ages and Renaissance in Europe overlap in the fourteenth and fifteenth centuries. Numerous scholars in Italy are culled out and assigned to the Renaissance, with all other Italian and non-Italian scholars allotted to the Middle Ages. In the fourteenth century, Petrarch and Boccaccio go to the Renaissance, Dante to the Middle Ages. In any period that embraces a large geographic area, as Western Europe, there are bound to be different currents and trends evolving simultaneously. It is futile to treat them in isolation, assigning to each a name, and then pretending that each group acted independently of the others. The study of history is only obscured by such a strategy. In comparing the Middle Ages and Renaissance, Charles Haskins rejected any sharp distinction between the two periods, declaring that "the continuity of history rejects such sharp and violent contrasts between successive periods, and that modern research shows us the Middle Ages, less dark and less static, the Renaissance less bright and less sudden, then once was supposed."[190] But, as we saw, they were not really successive periods. The Renaissance and Middle Ages overlapped for some three centuries, with the former beginning in the latter

190. Haskins, *The Renaissance of the Twelfth Century* (Cleveland/New York: Meridian Books, World Publishing Co., 1964; first published 1927), v.

period, gaining dominance as time moved on. But important features of the Middle Ages remained even as the high Renaissance emerged in the fifteenth century and beyond. Moreover, even as the Italian Renaissance reached its peak in the fifteenth century, medieval culture remained a powerful force in northern Europe. The relationship between the Middle Ages and Renaissance will always remain, at best, muddled. But the stereotypes of the two periods will live on: The Middle Ages will be regarded as static, dark, and barbarous; the Renaissance enlightened and progressive.

To avoid Pavlovian reactions to periodizations, it would be preferable if historians inquired about events within a numerically bounded time span, say 1100 and 1700, or any subdivision thereof, or any extension of years beyond 1100 or 1700. Such neutral temporal designations would be much more preferable to an inquiry about what happened on this or that theme during the Middle Ages, or what happened in the Renaissance, and so on.

A few scholars have had sufficient insight and understanding to see that the Middle Ages was a period in which reason – both in the large sense and in the narrower analytic sense of logic – played a powerful and positive role. In 1795, Condorcet, as we saw, boldly allowed that scholasticism had made a contribution to the advance of civilization because it contained within itself "the first beginnings of that philosophical analysis which has since been the fruitful source of our progress."[191] Until the twentieth century, such a concession was rare in the literature on medieval civilization. Even in the twentieth century, scholastic rationality is rarely mentioned or recognized. It is, therefore, surprising to find it in a history textbook on the modern world. But, in the spirit of Condorcet, R. R. Palmer and Joel Colton acknowledged the debt of the modern world when they declared that

> scholastic philosophy laid foundations on which later European thought was to be reared. It habituated Europeans to great exactness, to careful distinctions, even to the splitting of hairs. It called for disciplined thinking. And it made the world safe for reason. If any historical generalization may be made safely, it may be safely said that any society that believes reason to threaten its foundations will suppress reason.[192]

Historians of medieval philosophy are, of course, unavoidably aware of the contributions of medieval scholasticism to the cause of reason and rationality. In this book, I have sought to make the role of reason in the intellectual life of the Middle Ages known to a much wider public. It

191. Condorcet, *Sketch for a Historical Picture of the Progress of the Human Mind*, 95.

192. R. R. Palmer, *A History of the Modern World*, 2nd ed., revised with the collaboration of Joel Colton (New York: Alfred A. Knopf, 1957; first published 1950), 39.

therefore seems fitting to close this final chapter with a quotation from an eminent historian of medieval philosophy, Etienne Gilson, who rightly declared:

> It is necessary ... to relegate to the domain of legend the history of a Renaissance of thought succeeding to centuries of sleep, of obscurity, and error. Modern philosophy did not have to undertake the struggle to establish the rights of reason against the Middle Ages; it was, on the contrary, the Middle Ages that established them for it, and the very manner in which the seventeenth century imagined that it was abolishing the work of preceding centuries did nothing more than continue it.[193]

193. Translated by Ferguson, *The Renaissance in Historical Thought,* 340, from Etienne Gilson, *La philosophie au Moyen Age* (Paris, 1922), p. 311f.

CONCLUSION

The Culture and Spirit of "Poking Around"

WHAT MADE IT POSSIBLE FOR WESTERN CIVILIZATION TO develop science and the social sciences in a way that no other civilization had ever done before? The answer, I am convinced, lies in a pervasive and deep-seated spirit of inquiry that was a natural consequence of the emphasis on reason that began in the Middle Ages. With the exception of revealed truths, reason was enthroned in medieval universities as the ultimate arbiter for most intellectual arguments and controversies. It was quite natural for scholars immersed in a university environment to employ reason to probe into subject areas that had not been explored before, as well as to discuss possibilities that had not previously been seriously entertained.

Reason and the spirit of inquiry appear to be natural companions. This spirit of inquiry can be aptly described as the spirit of "poking around," a spirit that manifests itself through an urge to apply reason to almost every kind of question and problem that confronts scholars of any particular period. Indeed, a vital aspect of "poking around" involves an irresistible urge to raise new questions, which eventually give rise to even more questions. I regard the spirit of poking around as nothing less than the spirit of scientific inquiry.

When scholars in the eleventh and twelfth centuries began the process of opposing reason to authority, they also began what may be appropriately called "the culture of poking around," or the irrepressible urge to probe into many things. Although there may be no necessary connection between the application of reason to various problems and the spirit of poking around, it is difficult to imagine a society in which reason would be highly valued, but in which there is no particular drive or urge to pose questions about basic problems that concern religion, nature, and even hypothetical or imaginary conditions relevant to religion and nature.

As we saw in Chapter 2, scholars in the eleventh and twelfth centuries began to challenge authority by invoking the use of reason. Once reason was used to challenge authority, it was quite natural to raise questions about received doctrines and explanations. Scholars like Adelard of Bath and Peter Abelard raised many questions opposing authoritative solutions to problems and sought to resolve the questions they raised by the use of reason. It was not until the thirteenth century, however, that the question, with its formal structure, became the basic instrument for applying reason to natural philosophy and theology. The question format remained in extensive use for some four centuries. By the time it was abandoned in the seventeenth century, to be replaced by treatises that were given topical treatment, the question approach to natural philosophy was ingrained in Western thought. Western Europe had been transformed into a society in which scholars routinely sought to answer all sorts of questions, which they could raise somewhere in their texts, or indicate by the substance of their arguments. In the course of the seventeenth and eighteenth centuries, questions were regularly raised and resolved in the newly developing sciences and social sciences. As an approach to these new disciplines, the questioning method became ever more widespread and effective and has continued to the present day. Indeed, this culture, or spirit, of poking around has been the driving force of Western civilization ever since the thirteenth century. Without it, the dazzling developments in science, social science, philosophy, and technology that have become characteristic of Western thought and achievement would not have occurred.

That the probing culture of Western civilization emerged from the Middle Ages should occasion no great surprise when one realizes that all the ingredients were there to produce it. The most important ingredient is reason, which, as we have seen, became characteristic of medieval intellectual life. Once reason became a major feature of medieval thought, scholars seem to have concluded that the best way to employ and exhibit reason was to express themselves by way of responding to questions. The most basic form of medieval teaching and scholarship came to be based on the posing of questions to which careful responses were given.

Thus was reason joined to an analytic questioning technique that was ubiquitous in medieval university education and, therefore, widespread among the literate class. Questions were posed in natural philosophy that asked about the structure and operation of the physical world that Aristotle had described. Questions were also posed in theology about every aspect of faith and revelation. But the probing character of medieval questions went far beyond the straightforward and routine. Scholastic natural philosophers and theologians asked questions not only about what is but also about what could be, but probably wasn't. Theologians exercised their logical talents by inquiring about what God could and could not do, or what He could and could not know. Every question in the scholastic arsenal produced pro and contra arguments that were intended to include all plausible and feasible positions.

By virtue of this questioning approach to the world, many interesting and even strange questions were formulated. Most of the questions were about problems in Aristotle's works. Scholars asked about natural phenomena of all kinds, prefacing their questions with the interrogative "whether," as, for example,

whether the whole earth is habitable;
whether spots appearing in the moon arise from differences in parts of the moon or from something external;
whether the earth is spherical;
whether by their light the celestial bodies are generative of heat;[1]
whether a compound is possible;
whether there are four elements, no more nor less;
whether any element is pure;
whether one element could be generated directly from another, so that water could be generated directly from air without something else being generated from it previously; and [that same question can be asked] of the other elements.[2]
whether it is possible for an actual infinite magnitude to exist;
whether the existence of a vacuum is possible;
whether a continuum is composed of indivisibles; for example, as a line might be composed of points, time of instants, and motion of mutations [that is, instantaneous changes];[3]
whether the mass of the whole earth – that is, its quantity or magnitude – is much less than certain stars;
whether a comet is of a celestial nature or [whether it is] of an elementary nature, say, of a fiery exhalation;
whether lightning is fire descending from a cloud;

1. The first four questions are drawn from John Buridan's *Questions on Aristotle's On the Heavens* in *Ioannis Buridani Quaestiones super libris quattuor De caelo et mundo*, ed. Ernest A. Moody (Cambridge, MA: The Mediaeval Academy of America, 1942). See Grant, *A Source Book in Medieval Science* (Cambridge, MA: Harvard University Press, 1974), 204–205.
2. The next four questions are from Albert of Saxony's *Questions on Aristotle's On Generation and Corruption.* See Grant, *Source Book,* 206.
3. The next three questions are from Albert of Saxony's *Questions on Aristotle's Physics.* See Grant, ibid., 201–202.

whether the colors appearing in the rainbow are where they seem to be and are true colors;

On the supposition that a rainbow can occur by reflection of rays, we inquire whether such reflection occurs in a cloud or whether it occurs in tiny dewdrops or raindrops.[4]

In Chapter 6, we saw the kinds of questions theologians asked. Many were drawn directly from natural philosophy. The theological questions were a mixture of theology, logic, and natural philosophy. Many of the questions in theology probed the domain of God's powers: what God could do and not do, and what God knows and does not know. Typical questions, in addition to those cited in Chapter 6, are

whether God could have made the world before He made it;[5]
whether God knew that He would create a world from eternity;[6]
whether God could do evil things;[7]
whether the Creator could have created things better than He did;[8]
whether God could make a better world than this world;[9]
whether God could make the humanity of Christ better than it is;[10]
whether, without any change in Himself, God could not want something that at some [earlier] time He had wanted.[11]
whether if God wants something new that He did not want from eternity, would this constitute a real change [in God].[12]

In both natural philosophy and theology, hypothetical questions were frequently proposed, as, for example,

4. The last five questions are from Themon Judaeus's *Questions on Aristotle's Meteorology.* See Grant, ibid., 207–209. For a list of the rest of the questions from the four treatises mentioned in the last four notes, see Grant, *Source Book,* 199–210. For 400 questions on cosmology, see Grant, *Planets, Stars, & Orbs,* 681–741.
5. In notes 5 to 20 of this Conclusion, I give only short titles; for the complete titles, see the bibliography or the reference cited. Richard of Middleton, *Sentences,* bk. 1, dist. 44, art. 1, qu. 3, vol. 1, 391, col. 1.
6. Robert Holkot, *Sentences,* bk. 2, qu. 2, Q ii.
7. Richard of Middleton, *Sentences,* bk. 1, dist. 42, qu. 2, vol. 1, 372, col. 2.
8. From *De universo* of William of Auvergne; see Grant, *Planets, Stars, & Orbs,* 693, n. 71.
9. See Gabriel Biel, *Sentences,* bk. 1, dist. 44, qu. unica, vol. 1, 749.
10. Thomas Aquinas, *Sentences,* bk. 1, dist. 44, qu. 1, art. 3, vol. 1, 1021.
11. "Utrum deus posset non velle aliquid quod aliquando voluit sine sui mutatione." Johannes Bassolis, *Sentences,* bk. 1, dist. 45, qu. 2, art. 3, fols. 217r, col. 2, for enunciation of article; fols. 218v, col. 2–219r, col. 2, for discussion of article.
12. See Thomas of Strasbourg, *Sentences,* bk. 1, dists. 47 and 48, qu. 1, fol. 121r, col. 1.

whether the elements could operate if the heavens exerted no influence on them;[13]

if there were several worlds, whether the earth of one would be moved naturally to the middle [or center] of another;[14]

whether beyond this world, God could make another earth of the same species as this world;[15]

whether it is possible that a body moved rectilinearly could be infinite;[16]

whether, if a vacuum existed, up or down could be in it;[17]

whether, if a vacuum did exist, a heavy body could move in it.[18]

Scholastic natural philosophers frequently asked questions involving the ultimate limit: infinity. For example, they asked

whether an infinite magnitude could be traversed in a finite time; and whether a finite magnitude could be traversed in an infinite time;

whether it is possible that a mover of finite power move [or act] through an infinite time;

whether an infinite power [or force] could exist in a finite magnitude;

whether motion could be accelerated to infinity;[19]

whether charity could be increased, or be made greater intensively [all the way] to infinity.[20]

Hundreds of questions like these in both natural philosophy and theology were regularly discussed for more than four centuries. Since the question format required that two alternative sides to each question be presented, the intent was to vent each question as thoroughly as possible. The tools used in scholastic responses to natural and theological questions were usually the same: logic, with its technical terminology, and reasoned argument, with occasional appeals to empirical evidence. Theologians and

13. From Richard of Middleton's *Sentence Commentary.* See Grant, *Planets, Stars, & Orbs,* 721, n. 207.

14. From Buridan's *Questions on On the Heavens;* see Grant, *Planets, Stars, & Orbs,* 692, qu. 66.

15. From Godfrey of Fontaines as cited in Grant, *Planets, Stars, & Orbs,* 692, qu. 65.

16. From Buridan's *Questions on On the Heavens;* see Grant, *Planets, Stars, & Orbs,* 693, qu. 72.

17. From Benedictus Hesse's *Questions on the Physics* as cited in Grant, *Planets, Stars, & Orbs,* 730, qu. 324.

18. From Albert of Saxony's *Questions on the Physics;* see Grant, *Source Book,* 201.

19. These four questions appear in Albert of Saxony's *Questions on the Physics;* see Grant, *Source Book in Medieval Science,* 202–203.

20. Gregory of Rimini, *Sentences,* bk. 1, dist. 17, qu. 6, vol. 2, 460.

natural philosophers used much the same approach, with theologians having less occasion for empirical appeals.

What do all of these categories of questions have in common? They were meant to probe into all aspects of the world: nature, the supernatural, and an imaginary world of the hypothetical and possible. Collectively, they represent something of fundamental importance. They reveal a questioning approach toward a great range of problems. The questions are an attempt to resolve a host of particular problems that were deemed important in medieval intellectual life. Not only did medieval scholastics seek to determine answers to problems about nature and the supernatural, but they also sought to answer questions that were about hypothetical and imaginary conditions, about the way things might have been. They wanted to know how the world would operate if it had been created with different structural and operational properties, which they conjured up from their fertile imaginations. To ascertain such things to their satisfaction, they applied the best available contemporary knowledge – largely, Aristotelian natural philosophy – to these hypothetical conditions.

All this questioning about the real world and about hypothetical conditions within that world or beyond it – as exemplified in the sample questions just cited, and in Chapters 4 to 6 – indicate an approach that may be appropriately characterized as "probing and poking around." What makes this of great significance is the fact that the questioning method – the probing and poking around – was institutionalized in the medieval universities, where it was practiced for more than four centuries. There was undoubtedly some probing and poking around in other civilizations before the Latin West institutionalized it. But whatever was done before and elsewhere was minuscule by comparison to what would be done in the medieval Latin West, where probing and poking around became a routine way of approaching problems in all aspects of life. The myriad questions that were raised reflected the desires of an intellectual class that sought to know as much as they could by reason alone. The structural form of the question as it was used in the medieval universities was meant to provide a definitive answer to each question raised, although scholars might arrive at different, and conflicting, answers. Even if modern critics judge the questions and their responses to be trivial, or of little utility, those who posed the questions and answered them regarded their efforts as of great importance. They were, after all, solving questions that ostensibly informed their contemporaries about the inner and outer workings of the world, as these were understood at the time. Not only did they provide their audience with answers to such questions, but they also included refutations of the arguments they found wanting.

Why did medieval scholastics provide elaborate responses to innumerable questions? What did they hope to achieve? They clearly did not seek to acquire power over nature – very few during the Middle Ages had such thoughts. Their purpose was fairly straightforward: the desire to know, and to profit from their knowledge. Those who came to know the answers to the questions raised in university courses and texts, and went on to acquire baccalaureate and masters degrees, earned the respect of their society and were rewarded, in some way or other, for their efforts.

Was the question method of expression in natural philosophy and theology of any historical value? Did the veritable blizzard of questions that poured forth for more than four centuries produce anything of lasting value? Indeed it did. The probing and poking around that medieval scholastic natural philosophers and theologians did routinely with their questioning approach to the world blossomed into a powerful, dynamic mode of inquiry that is perhaps the most characteristic aspect of Western civilization. Having begun in approximately the year 1200, it has been a feature of Western intellectual life for 800 years. Although the probing questions have varied greatly over time, place, and subject matter, probing by questions has been a constant feature of Western thought since the thirteenth century. Science, philosophy, and technology depend on it, even though, after the Middle Ages, the mode of presentation ceased to be by way of formal questions, except occasionally, as when Isaac Newton attached his famous *Queries* to the end of his *Opticks* in 1704. "Though he couched them as questions," however, "no one is apt to mistake the positive answers Newton intended,"[21] no more so than we are likely to mistake the positive answers to the overwhelming majority of scholastic questions.

The hypothetical questions and the questions about what God could or could not do, or what He knows or does not know, which were so characteristic of the Middle Ages, were largely abandoned by the natural philosophers who produced the Scientific Revolution. But just as surely as their medieval predecessors, they proceeded by way of questions. But the questions were now often only in their minds to guide them in their research and inquiries. The results they published might not explicitly include the questions that guided the researcher and led to those *results.* Moreover, the questions they posed to themselves and to others were rarely about hypothetical, or imaginary, conditions, or about God's power to do or not to do some particular act, but were about the real world. Also noteworthy is the

21. Richard S. Westfall, *Never at Rest: A Biography of Isaac Newton* (Cambridge: Cambridge University Press, 1980), 642.

fact that natural philosophers in the seventeenth century answered the questions they posed to nature by appeals to observation, or by means of experiments, or by the application of mathematics. This became the way scientists would proceed to the present day.

Although scientists in the various sciences have evolved different techniques and procedures for answering the never-ending parade of questions they generate, and without which modern science could not exist, the spirit of inquiry remains essentially what it was in the Middle Ages: an effort to advance a subject by probing and poking around with one or more questions to which answers are sought, after which more questions are posed, in a process that never ends. Questions for which satisfactory answers are unobtainable are left open. If they are important, they remain active until an acceptable response appears. Some questions cannot be answered definitively and receive various responses over time. We are a questioning society that constantly seeks answers to queries about virtually everything, especially about nature, religion, government, and society.

The questioning method is the driving force in science, social science, and technology. Ironically, it is absent from modern theology, which no longer raises the kinds of questions that theologians in the Middle Ages characteristically posed. It would be difficult to imagine modern theologians asking about the limits of God's power and determining those limits by application of the law of noncontradiction. Not only did the scholars in the Middle Ages lay the basis for our probing society by means of an unending stream of questions; they used reason as the fundamental criterion for arriving at their answers. By the seventeenth century, natural philosophers saw that "pure" reason alone was often inadequate, and they devised the experimental method to furnish evidence that reason alone could not provide. It was in this spirit that Isaac Newton began his work on the *Opticks* by proclaiming to his readers that "My Design in this Book is not to explain the Properties of Light by Hypotheses, but to propose and prove them by Reason and Experiments."[22]

All of Newton's work, however, was in response to questions that both puzzled and inspired him. And so it is with modern science. If one were to peruse articles in the weekly "Science Times" of the *New York Times,* it is immediately obvious that virtually all of the articles are about results that are answers to questions that are easy, or difficult, or still unresolved. In the latter category, an article of April 20, 1999, describes the efforts of a physicist, Dr. Abhay Ashtekar (Pennsylvania State University), to merge

22. Quoted by Westfall, ibid., from p. 1 of Newton's *Opticks.*

Albert Einstein's theory of general relativity with the laws of quantum mechanics.... Between them, these theories seem to explain the observed universe, but they express profoundly different conceptions of matter, time and space. That philosophical schism also leaves physicists in deep doubt over how to deal with phenomena in which both theories should be valid – in the realms of the very small and the very energetic, like the Big Bang in which the universe was allegedly born.

Over the years, physicists have believed, from time to time, that they had reconciled the two theories only to be proven wrong and compelled to try again. The large question faced by Dr. Ashtekar and his colleagues is obvious. If a sustainable solution is ever forthcoming, it will be because a host of smaller questions will have been proposed and answered to provide the basis for resolving the big question.

In the Middle Ages, scholars also kept questions alive by offering new answers to old questions. If modern science has progressed almost unrecognizably beyond anything known or contemplated in the natural philosophy and science of the Middle Ages, modern scientists are, nonetheless, heirs to the remarkable achievements of their medieval predecessors. The idea, and the habit, of applying reason to resolve the innumerable questions about our world, and of always raising new questions, did not come to modern science from out of the void. Nor did it originate with the great scientific minds of the sixteenth and seventeenth centuries, from the likes of Copernicus, Galileo, Kepler, Descartes, and Newton. It came out of the Middle Ages from many faceless scholastic logicians, natural philosophers, and theologians, in the manner I have described in this study. It is a gift from the Latin Middle Ages to the modern world, a gift that makes our modern society possible, though it is a gift that may never be acknowledged. Perhaps it will always retain the status it has had for the past four centuries as the best-kept secret of Western civilization.

BIBLIOGRAPHY

Abelard, Peter. *The Story of Abelard's Adversities, A Translation with Notes of the "Historia Calamitatum."* Translated by J. T. Muckle, with a preface by Etienne Gilson. Toronto: The Pontifical Institute of Mediaeval Studies, 1964.

Peter Abailard Sic Et Non, A Critical Edition. Edited by Blanche Boyer and Richard McKeon. Chicago: The University of Chicago Press, 1976, 1977.

A Dialogue of a Philosopher with a Jew, and a Christian. Translated by Pierre J. Payer. Toronto: Pontifical Institute of Mediaeval Studies, 1979.

Adelard of Bath. See Gollancz, Hermann.

Albert of Saxony. *Questiones et decisiones physicales insignium virorum. Alberti de Saxonia in octo libros Physicorum; tres libros De celo et mundo; duos libros De generatione et corruptione; Thimonis in quatuor libros Meteororum; Buridani in tres libros De anima; librum De sensu et sensato; librum De memoria et reminiscentia; librum De somno et vigilia; librum De longitudine et brevitate vite; librum De iuventute et senectute Aristotelis. Recognite rursus et emendatae summa accuratione et iudicio Magistri Georgii Lokert Scotia quo sunt tractatus proportionum additis.* Paris, 1518.

Expositio et Quaestiones in Aristotelis "Physicam" ad Albertum de Saxonia Attributae. Édition critique. Benoît Patar. 3 vols. Louvain-Paris: Éditions de l'Institut Supérieur de Philosophie. Éditions Peeters, 1999.

Albertus Magnus. *Alberti Magni Opera Omnia.* Ed. Bernhard Geyer.

Vol. 4, Part 1: *Physica,* bks. 1–4. Ed. Paul Hossfeld. Aschendorff: Monasterii Westfalorum, 1987.

Vol. 4, Part 2: *Physica,* bks. 5–8. Ed. Paul Hossfeld. Aschendorff: Monasterii Westfalorum, 1993.

Vol. 5: *De caelo et mundo.* Ed. Paul Hossfeld. Aschendorff: Monasterii Westfalorum, 1971.

Alembert, Jean Le Rond d'. *Preliminary Discourse to the Encyclopedia of Diderot.* Translated by Richard N. Schwab, with the collaboration of Walter E. Rex, with an Introduction and Notes by Richard N. Schwab. Chicago: The University of Chicago Press, 1963; revised Introduction, 1995.

Alexander of Hales. *Summa Theologica.* Tomus II: Prima Pars secundi libri. Florence: Collegium S. Bonaventurae, 1928.

Andrea, Alfred J. "The Myth of the Middle Ages." *The Historian* 55, no. 1 (Autumn 1992): 183–188.

Anselm, St. *Proslogium: Monologium: An Appendix in Behalf of the Fool Guanilon; and Cur Deus Homo.* Translated from the Latin by Sidney Norton Deane, with an Introduction, Bibliography, and Reprints of the Opinions of Leading Philosophers and Writers on the Ontological Argument. La Salle, IL: The Open Court Publishing Co., 1944.

Ariew, Roger. See Duhem, Pierre.

Aristotle. *On the Heavens.* Trans. W. K. C. Guthrie. London: William Heinemann Ltd; Cambridge, MA: Harvard University Press (Loeb Classical Library), 1960.

The Complete Works of Aristotle. The Revised Oxford Translation. 2 vols. Ed. Jonathan Barnes. Princeton: Princeton University Press, 1984.

Categories, trans. J. L. Ackrill.

Metaphysics, trans. W. D. Ross.

Posterior Analysis, trans. Jonathan Barnes.

On the Soul, trans. J. A. Smith.

Nicomachean Ethics, trans. W. D. Ross, revised by J. O. Urmson.

Parts of Animals, trans. W. Ogle.

On Generation and Corruption, trans. H. H. Joachim.

Sophistical Refutations, trans. W. A. Pickard-Cambridge.

Politics, trans. B. Jowett.

On the Heavens, trans. J. L. Stocks.

Physics, trans. R. P. Hardie and R. K. Gaye.

Armstrong, A. H., ed. *The Cambridge History of Later Greek and Early Medieval Philosophy.* Cambridge: Cambridge University Press, 1970.

Ashworth, E. J. "The Eclipse of Medieval Logic." In Norman Kretzmann, Anthony Kenny, and Jan Pinborg, eds., *The Cambridge History of Later Medieval Philosophy from the Rediscovery of Aristotle to the Disintegration of Scholasticism 1100–1600.* Cambridge: Cambridge University Press, 1982, Chapter 42, 787–796.

Asztalos, Monika. "The Faculty of Theology." In H. de Ridder-Symoens, ed. *A History of the University in Europe.* Vol. 1: *Universities in the Middle Ages.* Cambridge: Cambridge University Press, 1992, 409–441.

Audi, Robert, ed. *The Cambridge Dictionary of Philosophy.* Cambridge: Cambridge University Press, 1995.

A. R. M. (Alfred R. Mele). "theoretical reason," 796.

J. A. B. (José A. Bernadete). "infinity," 372–373.

Augustine, St. *St. Augustine The Literal Meaning of Genesis: De Genesi ad litteram.* Ed. and trans. John Hammond Taylor. 2 vols. These are vols. 41–42 in Johannes Quasten, Walter J. Burghardt, and Thomas Comerford Lawler, eds., *Ancient Christian Writers: The Works of the Fathers in Translation.* New York: Newman, 1982.

The Works of Saint Augustine. A Translation for the 21st Century. Ed. John E. Rotelle, O. S. A. Vol. 11: *Teaching Christianity (De Doctrina Christiana),* introduction, translation, and notes by Edmund Hill, O. P. Hyde Park, NY: New City Press, 1996.

Aureoli, Peter. *Petri Aureoli Verberii Ordinis Minorum Archiepiscopi Aquensis S. R. E. Cardinalis Commentariorum in primum [-quartum] librum Sententiarum pars prima [-quarta].* 2 vols. Rome: Aloysius Zannetti, 1596–1605.

Bacon, Francis. *The Physical and Metaphysical Works of Lord Bacon including "The Advancement of Learning" and "Novum Organum."* Edited by Joseph Devey. London: George Bell and Sons, 1904.

The Advancement of Learning. Edited with an Introduction by G. W. Kitchin. London: Everyman's Library, Dent, 1965. First included in Everyman's Library, 1915. First edited by G. W. Kitchin, 1861.

Bacon, Roger. *Opera hactenus inedita Rogeri Baconi, fasc. IV: Liber secundus Communium Naturalium Fratris Rogeri: De celestibus.* Partes quinque. Ed. Robert Steele. Oxford: Clarendon Press, 1913.

Opera hactenus inedita Rogeri Baconi, fasc. XIII: Questiones supra libros octo Physicorum Aristotelis. Ed. Ferdinand M. Delorme, with the assistance of Robert Steele. Oxford: Clarendon Press, 1935.

The Opus Majus of Roger Bacon, a translation by Robert Belle Burke. 2 vols. New York: Russell and Russell, 1962.

Roger Bacon's Philosophy of Nature. A Critical Edition, with English Translation, Introduction, and Notes, of 'De multiplicatione specierum' and 'De speculis comburentibus.' David C. Lindberg, ed. and trans. Oxford: Clarendon Press, 1983

Roger Bacon and the Origins of 'Perspectiva' in the Middle Ages, A Critical Edition and English Translation of Bacon's 'Perspectiva.' Ed. and trans. David C. Lindberg, with Introduction and Notes. Oxford: Clarendon Press, 1996.

Barnes, Jonathan. "Boethius and the Study of Logic." In Margaret Gibson, ed., *Boethius: His Life, Thought and Influence.* Oxford: Basil Blackwell, 1981, 73–89.

Aristotle. In *Past Masters.* Oxford: Oxford University Press, 1982.

"Life and Work." In Jonathan Barnes, ed. *The Cambridge Companion to Aristotle.* Cambridge: Cambridge University Press, 1995, 1–26.

Bassolis, John. *Opera Joannis Bassolis Doctoris Subtilis Scoti (sua tempestate) fidelis Discipuli, Philosophi, ac Theologi profundissimi: In Quatuor Sententiarum Libros (credite) Aurea. Quae nuperrime Impensis non minimis. Curaque & emendatione non mediocri. Ad debite integritatis sanitatem revocata. Decoramentisque marginalibus ac Indicibus adnotata. Opera denique et Arte Impressionis mirifica Dextris Syderibus elaborata fuere.* Venundantur a Francisco Regnault et loanne Fellon. Paris, [n.d.].

Becker, Carl L. *The Heavenly City of the Eighteenth-Century Philosophers.* New Haven: Yale University Press, 1952; first published, 1932.

Berman, Harold J. *Law and Revolution: The Formation of the Western Legal Tradition.* Cambridge, MA: Harvard University Press, 1983.

Bernard of Clairvaux. *The Life and Letters of St. Bernard of Clairvaux.* Newly translated by Bruno Scott James. London: Burns Oates, 1953.

Biel, Gabriel. *Collectorium circa quattuor libros Sententiarum.* 4 vols. (in five parts); plus index. Ed. Wilfredus Werbeck and Udo Hofmann. Tübingen: J. C. B. Mohr (Paul Siebeck), 1973–1992.

Black, Max. "Ambiguities of Rationality." In Newton Garver and Peter H. Hare, eds., *Naturalism and Rationality.* Buffalo, NY: Prometheus Books, 1986, 25–40.

Blasius of Parma. *Questio Blasij de Parma De tactu corporum durorum* in *Questio de modalibus Bassani Politi: Tractatus proportionum introductorius ad calculationes Suisset: Tractatus proportionum Thoma Bradwardini: Tractatus proportionum Nicholai Oren: Tractatus de latitudinibus formarum Blasij de Parma: Auctor sex inconvenientum.* Venice, 1501. Blasius's treatise *De tactu corporum durorum* is not mentioned on the title page of the edition cited here, but appears as the last treatise in the volume.

Boehner, Philotheus. *Medieval Logic: An Outline of its Development from 1250–c. 1400.* Manchester: Manchester University Press, 1952.

Boethius. *Boethius: The Theological Tractates,* with an English translation by H. F. Stewart, E. K. Rand, and S. J. Tester; *The Consolation of Philosophy,* with an English translation by S. J. Tester. London: William Heinemann Ltd.; Cambridge MA: Harvard University Press, 1973.

Boh, Ivan. "Consequences." In Norman Kretzmann, Anthony Kenny, and Jan Pinborg, eds., *The Cambridge History of Later Medieval Philosophy from the Rediscovery of Aristotle to the Disintegration of Scholasticism 1100–1600.* Cambridge: Cambridge University Press, 1982, Chapter 15, 300–314.

Bolgar, R. R. *The Classical Heritage and Its Beneficiaries.* Cambridge: Cambridge University Press, 1954.

Bonaventure, Saint. *Opera omnia.* Ad Claras Aquas [Quaracchi]: Collegium S. Bonaventurae, 1882–1901. Vol. 2 (1885): *Commentaria in quatuor libros Sententiarum Magistri Petri Lombardi: In secundum librum Sententiarum.*

See also, Vollert, Cyril.

Bowen, Francis. *Modern Philosophy from Descartes to Schopenhauer and Hartmann.* New York: Charles Scribner's Sons, 1885.

Bowersock, Glen W. "The Vanishing Paradigm of the Fall of Rome." *Bulletin of the American Academy of Arts and Science* 49 (May 1996): 29–43.

Bradwardine, Thomas. *Thomas of Bradwardine His "Tractatus de proportionibus." Its Significance for the Development of Mathematical Physics.* Ed. and trans. H. Lamar Crosby. Madison: The University of Wisconsin Press, 1955.

Broadie, Alexander. *Introduction to Medieval Logic.* Oxford: Clarendon Press, 1993.

Brown, Peter. *Augustine of Hippo, A Biography.* London: Faber, 1957.

Burckhardt, Jacob. *The Civilization of the Renaissance in Italy.* London: Phaidon Press, 1950; reprint of the translation by S. G. Middleton, 1878.

Buridan, John. *Ioannis Buridani Quaestiones super libris quattuor De caelo et mundo.* Ed. Ernest A. Moody. Cambridge, MA: The Mediaeval Academy of America, 1942.

Acutissimi philosophi reverendi Magistri Johannis Buridani subtilissime questiones super octo Phisicorum libros Aristotelis diligenter recognite et revise a Magistro Johanne Dullaert de Gandavo antea nusquam impresse. Paris, 1509. Facsimile, entitled *Johannes Buridanus, Kommentar zur Aristotelischen Physik.* Frankfurt: Minerva, 1964.

In Metaphysicen Aristotelis: Questiones argutissime Magistri loannis Buridani in ultima praelectione ab ipso recognitae et emissae ac ad archetypon diligenter repositae cum duplice indicio materiarum videlicet in fronte quaestionum in operis calce. Paris, 1518. Facsimile reprint titled *Johannes Buridanus Kommentar zur Aristotelischen Metaphysik.* Paris 1588 [should be 1518]. Frankfurt a. M.: Minerva G. M. B.H., 1964.

John Buridan: Sophisms on Meanings and Truth. Translated and with an introduction by Theodore Kermit Scott. New York: Appleton-Century-Crofts, 1966.

John Buridan on Self-Reference. Chapter Eight of Buridan's 'Sophismata,' translated with an Introduction, and a philosophical Commentary by G. E. Hughes. Cambridge: Cambridge University Press, 1982.

Burnett, Charles. "Scientific Speculations." In Peter Dronke, ed., *A History of Twelfth-Century Western Philosophy.* Cambridge: Cambridge University Press, 1988, 151–176.

Burtt, E. A. *The Metaphysical Foundations of Modern Physical Science.* Rev. ed. London: Routledge & Kegan Paul, 1932; reprinted 1950.

Cadden, Joan. "Science and Rhetoric in the Middle Ages: The Natural Philosophy of William of Conches." *Journal of the History of Ideas* 56 (Jan. 1995): 1–24.

Campanus of Novara. See Euclid.
Cantor, Norman. *The Meaning of the Middle Ages: A Sociological and Cultural History.* Boston: Allyn and Bacon, 1973.
Inventing the Middle Ages: The Lives, Works, and Ideas of the Great Medievalists of the Twentieth Century. New York: William Morrow and Company, 1991.
Chadwick, Henry. "Philo and the Beginnings of Christian Thought." In A. H. Armstrong, ed., *The Cambridge History of Later Greek and Early Medieval Philosophy.* Cambridge: Cambridge University Press, 1970, Part II, Chapters 8–11, 137–192.
Boethius: The Consolations of Music, Logic, Theology, and Philosophy. Oxford: Clarendon Press, 1981.
Chenu, M.-D. *Nature. Man, and Society in the Twelfth Century: Essays on New Theological Perspectives in the Latin West.* With a preface by Etienne Gilson; Selected, edited, and translated by Jerome Taylor and Lester K. Little. Chicago: The University of Chicago Press, 1968; originally published in French in 1957.
Chodorow, Stanley. "Law, Canon: After Gratian." In Joseph R. Strayer, ed., *Dictionary of the Middle Ages,* vol. 7. New York: Charles Scribner's Sons, 1986, 413–418.
Clagett, Marshall. *Greek Science in Antiquity.* London: Abelard-Schuman Ltd., 1957.
The Science of Mechanics in the Middle Ages. Madison, WI: University of Wisconsin Press, 1959.
"Some Novel Trends in the Science of the Fourteenth Century." In Charles S. Singleton, ed., *Art, Science, and History in the Renaissance.* Baltimore: The Johns Hopkins Press, 1967: 275–303.
"Adelard of Bath." In *Dictionary of Scientific Biography,* vol. 1 (1970), 61–64.
See also, Oresme, Nicole.
Clement of Alexandria. *Miscellanies.* Translated in *The Ante-Nicene Fathers: Translations of the Writings of the Fathers Down to A.D.* 325. Vol. 2: *Fathers of the Second Century: Hermas, Tatian, Athenagoras, Theophilus, and Clement of Alexandria (Entire).* American edition, chronologically arranged, with notes, prefaces and elucidations by A. Cleveland Coxe, D. D. Grand Rapids, MI: Wm. B. Eerdmans Publishing Co., 1983.
Cochrane, Louise. *Adelard of Bath: The First English Scientist.* London: British Museum Press, 1994.
Cohen, H. Floris. *The Scientific Revolution: A Historiographical Inquiry.* Chicago: University of Chicago Press, 1994.
Cohen, I. Bernard. *Introduction to Newton's 'Principia.'* Cambridge: Cambridge University Press, 1971.
Cohn, Norman. *Europe's Inner Demons: An Enquiry Inspired by the Great Witch-Hunt.* New York: New American Library, 1975.
Colish, Marcia. "Systematic Theology and Theological Renewal in the Twelfth Century." *Journal of Medieval and Renaissance Studies* 18, no. 2 (1988): 135–156.
Peter Lombard. 2 vols. Leiden: E. J. Brill, 1994.
Combes, André, and Francis Ruello. "Jean de Ripa I Sent. Dist. XXXVII: De modo inexistendi divine essentie in omnibus creaturis." *Traditio* 23 (1967): 191–267.
Condorcet, Antoine-Nicolas de. *Sketch for a Historical Picture of the Progress of the Human Mind.* Translated by June Barraclough, with an Introduction by Stuart Hampshire. London: Weidenfeld and Nicolson, 1955; first published 1795; first translated 1795.
Congar, Yves M.-J., O. P. *A History of Theology.* Translated and edited by Hunter Guthrie, S. J. Garden City, NY: Doubleday, 1968.

Copernicus, Nicholas. *Nicholas Copernicus On the Revolutions.* Ed. Jerzy Dobrzycki; trans. and commentary by Edward Rosen. London: Macmillan, 1978. First published 1543.

Copleston, Frederick, S. J. *A History of Philosophy.* Vol. 2: *Mediaeval Philosophy: Augustine to Scotus.* Westminster, MD: The Newman Press, 1957.

A History of Philosophy. Vol. 3: *Ockham to Suárez.* Westminster, MD: The Newman Press, 1953.

Courtenay, W. J. "Theology and Theologians from Ockham to Wyclif." In J. I. Catto and Ralph Evans, eds., *The History of the University of Oxford.* Vol. 2: *Late Medieval Oxford.* Oxford: Clarendon Press, 1992, 1–34.

"John of Mirecourt and Gregory of Rimini on Whether God can Undo the Past." In *Recherches de théologie ancienne et médiévale* 40 (Louvain, 1973), 147–174. Reprinted in William J. Courtenay, *Covenant and Causality in Medieval Thought: Studies in Philosophy, Theology and Economic Practice.* London: Variorum Reprints, 1984, VIIIb.

Cragg, Gerald R. *The Church and the Age of Reason 1648–1789.* New York: Atheneum, 1961. Vol. 4 in the *Pelican History of the Church:* first published in England, 1960, by Penguin Books, Ltd.

Crombie, A. C. "Science." In A. L. Poole, ed., *Medieval England.* 2 vols. Oxford: Clarendon Press, 1958, vol. 2, Chapter 18.

Medieval and Early Modern Science. Rev. 2nd ed. 2 vols. Garden City, NY: Doubleday Anchor Books, 1959.

Styles of Scientific Thinking in the European Tradition: The history of argument and explanation especially in the mathematical and biomedical sciences and arts. 3 vols. London: Gerald Duckworth, 1994.

Crosby, Alfred W. *The Measure of Reality: Quantification and Western Society, 1250–1600.* Cambridge: Cambridge University Press, 1997.

Crosby, H. Lamar. See Bradwardine, Thomas.

Cunningham, Andrew. "How the Principia Got its Name; or, Taking Natural Philosophy Seriously." *History of Science* 29 (1991): 377–392.

"The Identity of Natural Philosophy. A Response to Edward Grant." In *Early Science and Medicine,* 5, no. 3 (2000): 259–278.

"A Last Word." In *Early Science and Medicine,* 5, no. 3 (2000): 299–300.

Dales, Richard C., ed. *The Scientific Achievement of the Middle Ages.* Philadelphia: University of Pennsylvania Press, 1973.

Denifle, Heinrich, and Emil Chatelain. *Chartularium Universitatis Parisiensis.* 4 vols. Paris: Fratrum Delalain, 1889–1897.

Denomy, Alexander J. See Oresme, Nicole.

De Ripa, Jean. See Combes, André, and Francis Ruello.

Dicks, D. R. *Early Greek Astronomy to Aristotle.* Ithaca, NY: Cornell University Press, 1970.

Dijksterhuis, E. J. *The Mechanization of the World Picture.* Translated by C. Dikshoorn. Oxford: Clarendon Press, 1961; first published in Dutch in 1950.

Dod, Bernard G. "Aristoteles latinus." In Norman Kretzmann, Anthony Kenny, and Jan Pinborg, eds., *The Cambridge History of Later Medieval Philosophy from the Rediscovery of Aristotle to the Disintegration of Scholasticism 1100–1600.* Cambridge: Cambridge University Press, 1982, Chapter 2, 45–79.

Donahue, Charles, Jr. "Law, Civil – Corpus luris, Revival and Spread." In Joseph R. Strayer, ed., *Dictionary of the Middle Ages,* vol. 7. New York: Charles Scribner's Sons, 1986, 418–425.

Doucet, Victorinus. *Commentaire sur les Sentences: Supplément au Répertoire de M. F. Stegmüller.* Quaracchi: Collegii S. Bonaventurae ad Claras Aquas, 1954.

Drake, Stillman, trans. *Discoveries and Opinions of Galileo.* Garden City, NY: Doubleday Anchor Books, 1957.

Draper, John William. *History of the Conflict Between Religion and Science.* 8th ed. New York: Appleton, 1884; 1st ed., New York, 1874.

History of the Intellectual Development of Europe. Rev. ed. 2 vols. New York: Harper & Brothers, Publishers, 1876.

Duhem, Pierre. *Le Système du monde.* 10 vols. Paris: Hermann, 1913–1959.

Medieval Cosmology: Theories of Infinity, Place, Time, Void, and the Plurality of Worlds. Edited and translated by Roger Ariew. Chicago: University of Chicago Press, 1985.

Dunham, William Huse, Jr. "Parliament, English." In Joseph R. Strayer, ed., *Dictionary of the Middle Ages,* vol. 9. New York: Charles Scribner's Sons, 1987, 422–434.

Durandus de Sancto Porciano. *D. Durandi a Sancto Porciano, Ord. Praed. Et Meldensis Episcopi in Petri Lombardi Sententias Theologicas Commentariorum libri IIII. Nunc demum post omnes omnium editiones accuratissime recogniti et emendati. Auctoris vita, Indexque, Decisionum locupletissimus.* Venetiis [Venice], 1571, Ex Typographia Guerraea.

Erasmus of Rotterdam. *Praise of Folly.* Trans. Betty Radice; revised with a new introduction and notes by A. H. T. Levi. London: Penguin Books, 1993; first published 1971.

Euclid. *Euclidis Megarensis mathematici clarissimi Elementorum geometricorum libri XV. Cum expositione Theonis in priores XIII a Bartholomaeo Veneto Latinitate donata. Campani in omnes, et Hypsicles Alexandrini in duos postremos.* Basel, 1546.

The Thirteen Books of Euclid's "Elements." Translated with introduction and commentary by Sir Thomas L. Heath. 3 vols. 2nd ed., revised with additions. New York: Dover Publications, 1956.

Evans, G. R. *Anselm and a New Generation.* Oxford: Clarendon Press, 1980.

Old Arts and New Theology. Oxford: Clarendon Press, 1980.

Alan of Lille: The Frontiers of Theology in the Later Twelfth Century. Cambridge: Cambridge University Press, 1983.

Ferguson, Wallace K. *The Renaissance in Historical Thought: Five Centuries of Interpretation.* Cambridge, MA: The Riverside Press, 1948.

Finocchiaro, Maurice A., ed. and trans. *The Galileo Affair: A Documentary History.* Berkeley/Los Angeles: University of California Press, 1989.

Fisher, N. W., and Sabetai Unguru. "Experimental Science and Mathematics in Roger Bacon's Thought." *Traditio* 27 (1971): 353–378.

Fix, Andrew C. *Prophecy and Reason: The Dutch Collegiants in the Early Enlightenment.* Princeton: Princeton University Press, 1991.

Fraser, Russell. *The Dark Ages & the Age of Gold.* Princeton: Princeton University Press, 1973.

Frede, Michael. "Aristotle's Rationalism." In Michael Frede and Gisela Striker, eds. *Rationality in Greek Thought.* Oxford: Clarendon Press, 1996, 157–173.

French, Roger. "Astrology in Medical Practice." In Luis García-Ballester, Roger French, Jon Arrizabalaga, and Andrew Cunningham, eds., *Practical Medicine from Salerno to the Black Death.* Cambridge: Cambridge University Press, 1994, 30–59.

French, Roger, and Andrew Cunningham. *Before Science: The Invention of the Friars' Natural Philosophy.* Aldershot Hants, Eng.: Scolar Press, 1996.

Galilei, Galileo. *Dialogues Concerning Two New Sciences by Galileo Galilei.* Translated by Henry Crew and Alfonso de Salvio, with an Introduction by Antonio Favaro. New York: Dover Publications, n.d.; copyright by the Macmillan Company, 1914.

Galileo Galilei "On Motion" and "On Mechanics" Comprising "De Motu" (ca. 1590), translated with introduction and notes by I. E. Drabkin, and *"Le Meccaniche" (ca. 1600),* translated with introduction and notes by Stillman Drake. Madison, WI: University of Wisconsin Press, 1960.

Dialogue Concerning the Two Chief World Systems – Ptolemaic & Copernican. Translated by Stillman Drake, foreword by Albert Einstein. Berkeley/Los Angeles: University of California Press, 1962.

Two New Sciences, Including Centers of Gravity & Force of Percussion. Translated, with Introduction and Notes, by Stillman Drake. Madison: The University of Wisconsin Press, 1974.

García-Ballester, Luis. "Introduction." In Luis García-Ballester, Roger French, Jon Arrizabalaga, and Andrew Cunningham, eds., *Practical Medicine from Salerno to the Black Death.* Cambridge: Cambridge University Press, 1994, 1–29.

García y Garcia, Antonio. "The Faculties of Law." In Hilde de Ridder-Symoens, ed. *A History of the University in Europe.* Vol. 1: *Universities in the Middle Ages.* Cambridge: Cambridge University Press, 1992, Chapter 12, 388–408.

Gassendi, Pierre. *Pierre Gassend, Dissertations en Forme de Paradoxes contre les Aristotéliciens (Exercitationes Paradoxicae Adversus Aristoteleos) Livres I et II.* Texte établi, traduit et annoté par Bernard Rochot. Paris: Librairie Philosophique J. Vrin, 1959.

Gaybaa, B. P. *Aspects of the Mediaeval History of Theology 12th to 14th Centuries.* Pretoria: University of South Africa, 1988.

Gelber, Hester Goodenough. See Holkot, Robert.

Gibson, Margaret. "The *Opuscula Sacra* in the Middle Ages." In Margaret Gibson, ed., *Boethius: His Life, Thought and Influence.* Oxford: Basil Blackwell, 1981, 214–234.

"Latin Commentaries on Logic before 1200." *Bulletin de philosophie médiévale* 24 (1982): 54–64.

Gillispie, Charles C., ed. *Dictionary of Scientific Biography.* 16 vols. New York: Charles Scribner's Sons, 1970–1980.

Gilson, Etienne. *Reason and Revelation in the Middle Ages.* New York: Charles Scribner's Sons, 1938.

History of Christian Philosophy in the Middle Ages. London: Sheed and Ward, 1955.

Gollancz, Hermann, ed. and trans. *Dodi Ve-Nechdi ("Uncle and Nephew"). The Work of Berachya Hanakdan, now edited from MSS. at Munich and Oxford, with an English Translation, Introduction etc. to which is added the first English Translation from the Latin of Adelard of Bath's "Quaestiones Naturales."* London: Humphrey Milford Oxford University Press, 1920.

Grant, Edward. Essay Review of Marshall Clagett (ed. and trans.), *Nicole Oresme and the Medieval Geometry of Qualities and Motions. A Treatise on the Uniformity and Difformity of Intensities known as "Tractatus de configurationibus qualitatum et motuum"* (Madison/Milwaukee: University of Wisconsin Press, 1968; London, 1969). Reviewed in *Studies in History and Philosophy of Science* 3 (1972): 167–182.

A Source Book in Medieval Science. Cambridge, MA: Harvard University Press, 1974.

"The Concept of *Ubi* in Medieval and Renaissance Discussions of Place." In Nancy G. Siraisi and Luke Demaitre, eds, *Science, Medicine, and the University 1200–1500, Essays in Honor of Pearl Kibre,* Pt. 1. *Manuscripta* 20 (1976): 71–80.

Much Ado About Nothing: Theories of Space and Vacuum from the Middle Ages to the Scientific Revolution. Cambridge: Cambridge University Press, 1981.

"Medieval Departures from Aristotelian Natural Philosophy." In Stefano Caroti, ed., *Studies in Medieval Natural Philosophy.* [Florence:] Olschki, 1989, 237–256.

"Jean Buridan and Nicole Oresme on Natural Knoweldge." In *Vivarium* 31 (1993): 84–105.

Planets, Stars, & Orbs: The Medieval Cosmos, 1200–1687. Cambridge: Cambridge University Press, 1994.

The Foundations of Modern Science in the Middle Ages: Their Religious, Institutional, and Intellectual Contexts. Cambridge: Cambridge University Press, 1996.

"Nicole Oresme, Aristotle's *On the Heavens,* and the Court of Charles V." In *Texts and Contexts in Ancient and Medieval Science: Studies on the Occasion of John E. Murdoch's Seventieth Birthday.* Edited by Edith Sylla and Michael McVaugh. Leiden: Brill, 1997, 187–207.

"God, Science, and Natural Philosophy in the Late Middle Ages." In Lodi Nauta and Arjo Vanderjagt, eds., *Between Demonstration and Imagination, Essays in the History of Science and Philosophy Presented to John D. North.* Leiden: Brill, 1999, 243–267.

"God and Natural Philosophy: The Late Middle Ages and Sir Isaac Newton." In *Early Science and Medicine,* 5, no. 3 (2000): 279–298.

See also, Oresme, Nicole.

Gregory of Rimini. *Gregorii Arimensis OESA, Lectura super primum et secundum Sententiarum.* 7 vols. Berlin: Walter de Gruyter, 1979–1987.

Guerlac, Rita. *Juan Luis Vives Against the Pseudodialecticians: A Humanist Attack on Medieval Logic. The Attack on the Pseudodialecticians* and *On Dialectic,* Book III, v, vi, vii from *The Causes of the Corruption of the Arts with an Appendix of Related Passages by Thomas More.* The Texts, with translation, introduction, and notes by Rita Guerlac. Dordrecht, Holland: D. Reidel Publishing Co., 1978.

Hahn, Hans. "Infinity." In James R. Newman, ed., *The World of Mathematics.* 4 vols. New York: Simon and Schuster, 1956, vol. 3, 1593–1611.

Hampshire, Stuart. *The Age of Reason: The 17th Century Philosophers.* Selected, with Introduction and Commentary. New York: Mentor Books, 1956.

Hansen, Bert. See Oresme, Nicole.

Haskins, Charles Homer. *The Renaissance of the Twelfth Century.* Cleveland/New York: Meridian Books, World Publishing Co., 1964; first published 1927.

Hissette, Roland. *Enquête sur les 219 Articles Condamnés à Paris le 7 Mars 1277.* Louvain: Publications Universitaires; Paris: Vander-Oyez, S.A., 1977.

Hobbes, Thomas. *The English Works of Thomas Hobbes of Malmesbury.* Now First collected by Sir William Molesworth. 11 vols. London: John Bohn, 1839–1845; reprint, Darmstadt: Scientia Verlag Aalen, 1966.

Vol. 3: *Leviathan, or The Matter, Form, and Power of a Commonwealth Ecclesiastical and Civil.*

Höffding, Harald. *A History of Modern Philosophy: A Sketch of the History of Philosophy from the Close of the Renaissance to Our Own Day.* Vol. 1. London: Macmillan and Co., Limited, 1935; first edition 1900.

Holkot (or Holcot), Robert. *In quattuor libros Sententiarum quaestiones.* Lyon, 1518. Facsimile reprint, Frankfurt: Minerva, 1967.

Exploring the Boundaries of Reason: Three Questions on the Nature of God by Robert Holcot, OP. Ed. Hester Goodenough Gelber. *Studies and Texts* 62. Toronto: Pontifical Institute of Mediaeval Studies, 1983.

Hollister, C. Warren. *Medieval Europe: A Short History,* 7th ed. New York: McGraw-Hill, 1994.

Holmes, George, ed. *The Oxford History of Medieval Europe.* Oxford: Oxford University Press, 1992.

Holopainen, Toivo, J. *Dialectic and Theology in the Eleventh Century.* Leiden: E. J. Brill, 1996.

Huff, Toby. *The Rise of Early Modern Science: Islam, China, and the West.* Cambridge: Cambridge University Press, 1993.

Hugh of Saint Victor. *The "Didascalicon" of Hugh of St. Victor: A Medieval Guide to the Arts.* Translated from the Latin with an Introduction and Notes by Jerome Taylor. New York: Columbia University Press, 1961.

Hugolin of Orvieto. *Hugolini de Urbe Veteri OESA Commentarius in Quattuor Libros Sententiarum.* Edited by Willigis Eckermann O. S. A. 4 vols. Würzburg: Augustinus Verlag, 1980–1988.

Hugonnard-Roche, Henri. "L'hypothétique et la nature dans la physique parisienne du xive siècle." In Stefano Caroti and Pierre Souffrin, eds., *La nouvelle physique du xive siècle.* Florence: Leo S. Olschki, 1997, 161–177.

Iorio, Dominick A. *The Aristotelians of Renaissance Italy: A Philosophical Exposition.* Lewiston, NY: The Edwin Mellen Press, 1991.

Irving, Washington. *The Life and Voyages of Christopher Columbus.* Ed. John Harmon McElroy. Boston: Twayne Publishers, 1981.

James, Edward. "The Northern World in the Dark Ages 400–900." In George Holmes, ed., *The Oxford History of Medieval Europe.* Oxford: Oxford University Press, 1992, 59–108.

Jardine, Lisa. "Humanism and the Teaching of Logic." In Kretzmann et al., eds., *The Cambridge History of Later Medieval Philosophy from the Rediscovery of Aristotle to the Disintegration of Scholasticism 1100–1600.* Cambridge: Cambridge University Press, 1982, 797–807.

John of Damascus, Saint. *The Fount of Knowledge,* translated by Frederic H. Chase, Jr., in *Saint John of Damascus, Writings.* In *The Fathers of the Church, A New Translation.* Vol. 37. New York: Fathers of the Church, Inc., 1958.

John of Salisbury. *The "Metalogicon" of John of Salisbury: A Twelfth-Century Defense of the Verbal and Logical Arts of the Trivium.* Translated with an Introduction & Notes by Daniel D. McGarry. Berkeley/Los Angeles: University of California Press, 1955.

Johonnot, James, compiler and arranger. *Ten Great Events in History.* New York, 1887.

Kant, Immanuel. *Immanuel Kant's Critique of Pure Reason.* Translated by Norman Kemp Smith; unabridged ed. New York: St. Martin's Press, 1965; first published 1929.

Kantorowicz, Ernst H. "Kingship Under the Impact of Scientific Jurisprudence." In Marshall Clagett, Gaines Post, and Robert Reynolds, eds., *Twelfth-Century Europe and the Foundations of Modern Society.* Madison: The University of Wisconsin Press, 1966, 89–111.

Kibre, Pearl. "Logic and Medicine in Fourteenth Century Paris." In A. Maierù and A. Paravicini Bagliani, eds., *Studi sul XIV secolo in memoria di Anneliese Maier.* Rome: Edizioni di storia e letteratura, 1981, 415–420.

Kieckhefer, Richard. *Magic in the Middle Ages.* Cambridge: Cambridge University Press, 1990; first published 1989.

Kilvington, Richard. *The Sophismata of Richard Kilvington.* Introduction, Translation, and Commentary by Norman Kretzmann and Barbara Ensign Kretzmann. Cambridge: Cambridge University Press, 1990.

Kilwardby, Richard, O. P. *De ortu scientiarum.* Ed. Albert G. Judy, O. P. London: The British Academy, 1976.

King, Peter. "Mediaeval Thought-Experiments: The Metamethodology of Mediaeval Science." In Tamara Horowitz and Gerald J. Massey, eds., *Thought Experiments in Science and Philosophy.* Savage, MD: Rowman & Littlefield, 1991, 43–64.

Kneale, William, and Martha Kneale. *The Development of Logic.* Oxford: Clarendon Press, 1962.

Knowles, David. *The Evolution of Medieval Thought.* Baltimore: Helicon Press, 1962.

Knuutila, Simo (S. K.). "future contingents." In Robert Audi, ed., *The Cambridge Dictionary of Philosophy.* Cambridge: Cambridge University Press, 1995, 290.

Kors, Alan C., and Edward Peters, eds. *Witchcraft in Europe 1100–1700: A Documentary History.* Philadelphia: University of Pennsylvania Press, 1972.

Krafft, Fritz. "Guericke (Gericke), Otto Von." In *Dictionary of Scientific Biography,* vol. 5 (1972), 574–576.

Kramer, Heinrich, and James Sprenger *The Malleus Maleficarum of Heinrich Kramer and James Sprenger.* Translated with an Introduction, Bibliography & Notes by the Reverend Montague Summers. New York: Dover Publications, 1971; first published in 1928 by John Rodker, London.

Kren, Claudia. See Oresme, Nicole.

Kretzmann, Norman, Anthony Kenny, and Jan Pinborg, eds. *The Cambridge History of Later Medieval Philosophy from the Rediscovery of Aristotle to the Disintegration of Scholasticism 1100–1600.* Cambridge: Cambridge University Press, 1982.

The Metaphysics of Creation: Aquinas's Natural Theology in Summa contra gentiles II. Oxford: Oxford University Press, 1999.

LaMonte, John L. *The World of the Middle Ages: A Reorientation of Medieval History.* New York: Appleton-Century-Crofts, Inc., 1949.

Lawn, Brain. *The Rise and Decline of the Scholastic 'Quaestio Disputata' With Special Emphasis on its Use in the Teaching of Medicine and Science.* Leiden: E. J. Brill, 1993.

Leff, Gordon. "The *Trivium* and the Three Philosophies." In H. de Ridder-Symoens, ed., *A History of the University in Europe.* Vol. 1: *Universities in the Middle Ages.* Cambridge: Cambridge University Press, 1992, Chapter 10.1, 307–336.

Leijenhorst, Cornelis Hendrik. *Hobbes and the Aristotelians: The Aristotelian Setting of Thomas Hobbes's Natural Philosophy.* Ph.D. diss., University of Utrecht, 1998.

Lindberg, David C. *The Beginnings of Western Science: The European Scientific Tradition in Philosophical, Religious, and Institutional Context, 600 B.C. to A.D. 1450.* Chicago: University of Chicago Press, 1992.

Little A. G. "The Franciscan School at Oxford in the Thirteenth Century." In *Archivum Franciscanum Historicum* 19 (1926): 803–874.

Lloyd, G. E. R. *Aristotle: The Growth and Structure of His Thought.* Cambridge: Cambridge University Press, 1968.

Early Greek Science: Thales to Aristotle. New York: W. W. Norton & Co., 1970.

Locke, John. *John Locke An Essay Concerning Human Understanding.* Edited with an Introduction by Peter H. Nidditch. Oxford: Clarendon Press, 1975.

Logan, F. Donald. "Vikings." In Joseph R. Strayer, ed., *Dictionary of the Middle Ages,* vol. 12. New York: Charles Scribner's Sons, 1989, 422–437.

Lohr, Charles H. "Medieval Latin Aristotle Commentaries," in *Traditio,* vols. 23 (1967): 33–413 (Authors A-F); 24 (1968): 149–245 (Authors G-I); 26 (1970): 135–216 (Authors:

Jacobus-Johannes Juff); 27 (1971): 251–351 (Authors: Johannes de Kanthi-Myngodus); 28 (1972): 281–396 (Authors: Narcissus-Richardus); 29 (1973): 93–197 (Authors: Robertus-Wilgemus); 'Supplementary Authors,' 30 (1974): 119–144.

Latin Aristotle Commentaries, II: Renaissance Authors. Florence: Leo S. Olschki Editore, 1988.

Long, R. James. "The Science of Theology According to Richard Fishacre: Edition of the Prologue to His *Commentary on the Sentences.*" In *Mediaeval Studies* 34 (1972): 71–98.

Luscombe, D. E. "Peter Abelard." In Peter Dronke, ed., *A History of Twelfth-Century Western Philosophy.* Cambridge: Cambridge University Press, 1988, Chapter 10, 279–307.

Luther, Martin. *Luther's Works,* Vol. 31: *Career of the Reformer: I.* Ed. Harold J. Grimm. Philadelphia: Muhlenberg Press, 1957.

Mahoney, Michael S. "Ramus, Peter." In *Dictionary of Scientific Biography,* vol. 11 (1975), 286–290.

Maier, Anneliese. *Die Vorläufer Galileis im 14. Jahrhundert. Studien zur Naturphilosophie der Spätscholastik.* Rome: Edizioni di Storia e Letteratura, 1949.

Ausgehendes Mittelalter: Gesammelte Aufsätze zur Geistesgeschichte des 14. Jahrhunderts. Rome: Edizioni di Storia e Letteratura, 1964.

Maierù, A. "Logique et théologie trinitaire dans le moyen-âge tardif: Deux solutions en présence." In Monika Asztalos, *The Editing of Theological and Philosophical Texts from the Middle Ages.* Stockholm: Almqvist & Wiksell International, 1986: 185–212.

Manchester, William. *A World Lit Only by Fire: The Medieval Mind and the Renaissance: Portrait of an Age.* Boston: Little, Brown and Company, 1992.

Marenbon, John. *From the Circle of Alcuin to the School of Auxerre: Logic, Theology and Philosophy in the Early Middle Ages.* Cambridge: Cambridge University Press, 1981.

The Philosophy of Peter Abelard. Cambridge: Cambridge University Press, 1997.

Marsden, George M. *Religion and American Culture.* San Diego: Harcourt, Brace, Jovanovich, 1990.

Marshall, Peter. See Oresme, Nicole.

Marsilius of Inghen. *Questiones subtilissime Johannis Marcilii Inguen super octo libros Physicorum secundum nominalium viam.* Cum tabula in the fine libri posita. Lyon, 1518. Facsimile reprint titled *Johannes Marsilius von Inghen, Kommentar zur Aristotelischen Physik.* Lugduni, 1518. Frankfurt a. M.: Minerva G. M. B. H., 1964.

Matthews, John. "Anicius Manlius Severinus Boethius." In Margaret Gibson, ed., *Boethius: His Life, Thought and Influence.* Oxford: Basil Blackwell, 1981, 15–43.

Maurer, Armand. *Medieval Philosophy.* New York: Random House, 1962; reprinted 1969.

McCluskey, Stephen C. *Astronomies and Cultures in Early Medieval Europe.* Cambridge: Cambridge University Press, 1998.

McKeon, Charles King. *A Study of the "Summa philosophiae" of the Pseudo-Grosseteste.* New York: Columbia University Press, 1948.

McKeon, Richard, ed. and trans. *Selections from Medieval Philosophers.* Vol. 1: *Augustine to Albert the Great.* With Introductory Notes by Richard McKeon. New York: Charles Scribner's Sons, 1929.

Mele, Alfred R. (A.R.M.). "theoretical reason." In Robert Audi, ed., *The Cambridge Dictionary of Philosophy.* Cambridge: Cambridge University Press, 1995, 796.

Menut, Albert D. See Oresme, Nicole.

Mercer, Christia. "The Vitality and Importance of Early Modern Aristotelianism." In Tom Sorrell, ed., *The Rise of Modern Philosophy: The Tension between the New and Traditional Philosophies from Machiavelli to Leibniz.* Oxford: Clarendon Press, 1993, 33–67.

Michelet, Jules. *Satanism and Witchcraft: A Study in Medieval Superstition.* Translated by A. R. Allinson. London: Tandem Books, 1965.

History of the French Revolution. Translated by Charles Cocks; edited with an Introduction by Gordon Wright. Chicago: The University of Chicago Press, 1967.

Minio-Paluello, Lorenzo. "Boethius." In *Dictionary of Scientific Biography,* vol. 2 (1970), 228–236.

Mommsen, T. E. "Petrarch's Conception of the 'Dark Ages.'" *Speculum* 17 (1942): 226–242.

Montgomery, Scott L. *Science in Translation: Movements of Knowledge Through Cultures and Time.* Chicago: The University of Chicago Press, 2000.

Mossner, Ernest Campbell. *Bishop Butler and the Age of Reason: A Study in the History of Thought.* New York: The Macmillan Company, 1936.

Muckle, J. T. See Abelard, Peter.

Murdoch, John E. "*Mathesis in Philosophiam Scholasticam Introducta:* The Rise and Development of the Application of Mathematics in Fourteenth Century Philosophy and Theology." In *Arts Libéraux et Philosophie au Moyen Age.* Montreal: Institut d'Études Médiévales, 1969, 215–249.

"Bradwardine, Thomas." In *Dictionary of Scientific Biography,* vol. 2 (1970), 390–397.

"Euclid: Transmission of the Elements." In *Dictionary of Scientific Biography,* vol. 4 (1971), 437–459.

"Logic." In Edward Grant, *A Source Book in Medieval Science.* Cambridge, MA: Harvard University Press, 1974, 77–89.

"Atomism." In Edward Grant, *A Source Book in Medieval Science.* Cambridge, MA: Harvard University Press, 1974, 312–324.

"From Social into Intellectual Factors: An Aspect of the Unitary Character of Late Medieval Learning." In *The Cultural Context of Medieval Learning,* edited with an Introduction by John E. Murdoch and Edith Dudley Sylla. Dordrecht, Holland: D. Reidel, 1975, 271–339.

"The Analytic Character of Late Medieval Learning: Natural Philosophy Without Nature." In Lawrence D. Roberts, ed., *Approaches to Nature in the Middle Ages.* Binghamton, NY: Center for Medieval & Early Renaissance Studies, 1982, 171–213.

"Infinity and Continuity." In Norman Kretzmann, Anthony Kenny, and Jan Pinborg, eds., *The Cambridge History of Later Medieval Philosophy, from the Rediscovery of Aristotle to the Disintegration of Scholasticism 1100–1600.* Cambridge: Cambridge University Press, 1982, Chapter 28, 564–591.

"The Involvement of Logic in Late Medieval Natural Philosophy." In Stefano Caroti, ed.; Introduction of John E. Murdoch, *Studies in Medieval Natural Philosophy.* [Florence]: Leo S. Olschki, 1989, 3–28.

Murray, Alexander. *Reason and Society in the Middle Ages.* Oxford: Oxford University Press, 1978.

Nachod, Hans. "Francesco Petrarca." Introduction and translations of "On His Own Ignorance and That of Many Others" and "A Disapproval of an Unreasonable Use of the Discipline of Dialectic." In Ernst Cassirer, Paul Oskar Kristeller, and John Herman Randall, Jr., eds., *The Renaissance Philosophy of Man.* Chicago: University of Chicago Press, 1948, 23–143.

Nathanson, Stephen. *The Ideal of Rationality.* Atlantic Highlands, NJ: Humanities Press International, 1985.

Newell, John. "Rationalism in the School of Chartres." *Vivarium* 21 (1983): 108–126.

Newton, Isaac. *Sir Isaac Newton's Mathematical Principles of Natural Philosophy and His System of the World.* Translated into English by Andrew Motte in 1729. The translations revised, and supplied with an historical and explanatory appendix, by Florian Cajori. Berkeley: University of California Press, 1947.

Nicholas, David. *The Evolution of the Medieval World.* New York: Longman Publishing, 1992.

Niewöhner, Friedrich, and Olaf Pluta, eds. *Atheismus im Mittelalter und in der Renaissance.* Wiesbaden: Harrassowitz, 1999.

Nisbet, Robert. *History of the Idea of Progress.* New York: Basic Books, Inc., 1980.

Normore, Calvin. "Future Contingents." In Norman Kretzmann, Anthony Kenny, and Jan Pinborg, eds., *The Cambridge History of Later Medieval Philosophy from the Rediscovery of Aristotle to the Disintegration of Scholasticism 1100–1600.* Cambridge: Cambridge University Press, 1982, Chapter 18, 358–381.

North, John. "The *Quadrivium.*" In H. de Ridder-Symoens, ed., *A History of the University in Europe.* Vol. 1: *Universities in the Middle Ages.* Cambridge: Cambridge University Press, 1992, Chapter 10.2, 337–359.

Norton, Arthur O. *Readings in the History of Education: Mediaeval Universities.* Cambridge, MA: Harvard University Press, 1909.

Nozick, Robert. *The Nature of Rationality.* Princeton: Princeton University Press, 1993.

Ockham, William. *Venerabilis Inceptoris Guillelmi de Ockham, Quaestiones in Librum Secundum Sententiarum (Reportatio).* Ed. Gedeon Gal, O. F. M., and Rega Wood. St. Bonaventure, NY: St. Bonaventure University, 1981. This volume forms part of the series *Guillelmi de Ockham Opera Philosophica et Theologica ad fidem codicum manuscriptorum edita.* Cura Instituti Franciscani, Universitatis S. Bonaventurae.

Olivi, Peter John. *Fr. Petrus Iohannis Olivi, O.F.M. Quaestiones in Secundum Librum Sententiarum.* Edited by Bernard Jansen, S. I. 3 vols. Ad Claras Aquas [Quaracchi]: Ex Typographia Collegii S. Bonaventurae, 1922–1926.

Ong, Walter. *Ramus, Method, and the Decay of Dialogue: From the Art of Discourse to the Art of Reason.* Cambridge, MA: Harvard University Press, 1958.

Oresme, Nicole. *The "Questiones super De celo" of Nicole Oresme* by Claudia Kren. Ph.D. diss., University of Wisconsin, 1965.

Nicole Oresme "De proportionibus proportionum" and "Ad pauca respicientes." Edited with Introductions, English Translations, and Critical Notes by Edward Grant. Madison/Milwaukee: The University of Wisconsin Press, 1966.

Nicole Oresme and the Medieval Geometry of Qualities and Motions: A Treatise on the Uniformity and Difformity of Intensities Known as "Tractatus de configurationibus qualitatum et motuum" Edited with an Introduction, English Translation, and Commentary by Marshall Clagett. Madison: University of Wisconsin Press, 1968.

Nicole Oresme: Le Livre du ciel et du monde. Edited by Albert D. Menut and Alexander J. Denomy. Translated with an Introduction by Albert D. Menut. Madison: The University of Wisconsin Press, 1968.

Nicholas Oresme's 'Questiones super libros Aristotelis De anima': A Critical Edition with Introduction and Commentary by Peter Marshall. Ph.D. diss., Cornell University, 1980.

Nicole Oresme and the Marvels of Nature: A Study of his "De causis mirabilium." With Critical Edition, Translation and Commentary by Bert Hansen. Toronto: Pontifical Institute of Mediaeval Studies, 1985.

Nicole Oresme Quaestiones super De generatione et corruptione. Edited by Stefano Caroti. Munich: Verlag der Bayerischen Akademie der Wissenschaften, 1996.

Origen. "Letter of Origen to Gregory." In Allan Menzies, D. D., ed., *The Ante-Nicene Fathers: Translations of the Writings of the Fathers Down to A.D. 325*. Original supplement to the American edition, 5th ed. Grand Rapids, MI: Wm. B. Eerdmans Publishing Co., 1980, vol. 10.

Palmer, R. R. *A History of the Modern World.* 2nd ed., revised with the collaboration of Joel Colton. New York: Alfred A. Knopf, 1957; first published 1950.

Pei, Mario. *The Story of Language.* London: George Allen & Unwin Ltd., 1952.

Peters, Edward. *Europe and the Middle Ages.* 2nd ed. Englewood Cliffs, NJ: Prentice Hall, 1989.

Pieper, Josef. *Scholasticism: Personalities and Problems of Medieval Philosophy.* New York: McGraw-Hill Book Co., 1964; first published in German under the title *Scholastik.*

Piltz, Anders. *The World of Medieval Learning.* Translated into English by David Jones. Totowa, NJ: Barnes & Noble Books, 1981; first published in Swedish, 1978.

Plantinga, Alvin, ed. *The Ontological Argument: From St. Anselm to Contemporary Philosophers,* with an Introduction by Richard Taylor. Garden City, NY: Doubleday & Co., 1965.

Plato *Plato's Cosmology: The "Timaeus" of Plato translated with a running commentary.* By Francis M.Cornford. New York: The Liberal Arts Press, 1957.

Post, Gaines. *Studies in Medieval Legal Thought: Public Law and the State, 1100–1322.* Princeton: Princeton University Press, 1964.

Raedts, Peter. *Richard Rufus of Cornwall and the Tradition of Oxford Theology.* Oxford: Clarendon Press, 1987.

Randall, John Herman, Jr. *The Making of the Modern Mind: A Survey of the Intellectual Background of the Present Age.* Rev. ed. Boston: Houghton Mifflin Co., 1940.

Randles, W. G. L. *The Unmaking of the Medieval Christian Cosmos, 1500–1760: From Solid Heavens to Boundless Aether.* Aldershot Hants, Eng.: Ashgate, 1999.

Reynolds, Roger E. "Law, Canon: To Gratian." In Joseph R. Strayer, ed., *Dictionary of the Middle Ages,* vol. 7. New York: Charles Scribner's Sons, 1986, 395–413.

Richard of Middleton. *Clarissimi theologie magistri Ricardi de Media Villa ... Super quatuor libros Sententiarum Petri Lombardi questiones subtilissimae.* 4 vols. Brixia [Brescia], 1591. Facsimile, Frankfurt: Minerva, 1963.

Ridder-Symoens, Hilde de, ed. *A History of the University in Europe.* Vol. 1: *Universities in the Middle Ages.* Cambridge: Cambridge University Press, 1992.

Robinson, James Harvey, ed. and trans. *Francesco Petrarca: The First Modern Scholar and Man of Letters.* New York: G. P. Putnam, 1898.

Rogers, Elizabeth Frances. *St. Thomas More: Selected Letters.* New Haven: Yale University Press, 1961.

Rummel, Erika. *The Humanist-Scholastic Debate in the Renaissance & Reformation.* Cambridge, MA: Harvard University Press, 1995.

Russell, Jeffrey Burton. *Witchcraft in the Middle Ages.* Ithaca, NY: Cornell University Press, 1972.

Inventing the Flat Earth: Columbus and Modern Historians. New York: Praeger, 1991.

Sambursky, Samuel. *The Physics of the Stoics.* London: Routledge and Paul, 1959.

Schmitt, Charles. *Aristotle and the Renaissance.* Cambridge, MA: Published for Oberlin College by Harvard University Press, 1983.

Schrag, Calvin O. *The Resources of Rationality: A Response to the Postmodern Challenge.* Bloomington: Indiana University Press, 1992.

Serene, Eileen. "Demonstrative Science." In Norman Kretzmann, Anthony Kenny, and Jan Pinborg, eds., *The Cambridge History of Later Medieval Philosophy from the Rediscovery of Aristotle to the Disintegration of Scholasticism 1100–1600*. Cambridge: Cambridge University Press, 1982, 496–517.

Singer, Charles. *A Short History of Science to the Nineteenth Century.* Oxford: Clarendon Press, 1941.

Siraisi, Nancy. "The Faculty of Medicine." In H. de Ridder-Symoens, ed., *A History of the University in Europe.* Vol. 1: *Universities in the Middle Ages.* Cambridge: Cambridge University Press, 1992, 360–387.

Smith, T. V., and Marjorie Grene, eds. *From Descartes to Kant: Readings in the Philosophy of the Renaissance and Enlightenment.* Chicago: University of Chicago Press, 1940.

Sorabji, Richard. "Infinity and Creation." In *Philoponus and the Rejection of Aristotelian Science.* Ed. Richard Sorabji. Ithaca, NY: Cornell University Press, 1987, 164–178.

Southern, R. W. *The Making of the Middle Ages.* New Haven: Yale University Press, 1953.

Spade, Paul Vincent. "Insolubilia." In Norman Kretzmann, Anthony Kenny, and Jan Pinborg, eds., *The Cambridge History of Later Medieval Philosophy from the Rediscovery of Aristotle to the Disintegration of Scholasticism 1100–1600.* Cambridge: Cambridge University Press, 1982, Chapter 12, 246–253.

Article "Syncategoremata." In Robert Audi, ed., *The Cambridge Dictionary of Philosophy.* Cambridge: Cambridge University Press, 1995, 783.

"The History and Kinds of Logic." *Encyclopedia Britannica* (1995). CD-Rom disk: the section on "The 'properties of terms' and discussions of fallacies"; also section on "The theory of supposition."

Thoughts, Words, and Things: An Introduction to Late Mediaeval Logic and Semantic Theory. Version 1.1 on Spade's Web site (copyright, 1998), Chapter 2: "Thumbnail Sketch of the History of Logic to the End of the Middle Ages."

Spencer, Lloyd, and Andrzej Krauze. *Introducing the Enlightenment.* New York: Totem Books, 1997.

Steenberghen, Fernand van. *The Philosophical Movement in the Thirteenth Century.* Edinburgh: Nelson, 1955.

Stegmüller, Friedrich. *Reportorium Commentatorium in Sententias Petri Lombardi.* 2 vols. Würzburg: Schöningh, 1947.

Stiefel, Tina. *The Intellectual Revolution in Twelfth-Century Europe.* New York: St. Martin's Press, 1985.

Strayer, Joseph R., ed. *Dictionary of the Middle Ages.* 13 vols. New York: Charles Scribner's Sons, 1982–1989.

Strayer, Joseph R., and Dana Carleton Munro. *The Middle Ages 395–1500.* New York: Appleton-Century-Crofts, 1942.

Stump, Eleonore. "Obligations: From the Beginning to the Early Fourteenth Century." In Norman Kretzmann, Anthony Kenny, and Jan Pinborg, eds., *The Cambridge History of Later Medieval Philosophy from the Rediscovery of Aristotle to the Disintegration of Scholasticism 1100–1600.* Cambridge: Cambridge University Press, 1982, Chapter 16, 315–334.

Sylla, Edith Dudley. "Autonomous and Handmaiden Science: St. Thomas Aquinas and William of Ockham on the Physics of the Eucharist." In *The Cultural Context of Medieval Learning,* edited with an Introduction by John E. Murdoch and Edith Dudley Sylla. Dordrecht, Holland: D. Reidel, 1975, 349–396.

"The Oxford Calculators." In Norman Kretzmann, Anthony Kenny, and Jan Pinborg, eds., *The Cambridge History of Later Medieval Philosophy from the Rediscovery of Aristotle to the Disintegration of Scholasticism 1100–1600.* Cambridge: Cambridge University Press, 1982, Chapter 27, 540–563.

"Galileo and Probable Arguments." In Daniel O. Dahlstrom, ed., *Nature and Scientific Method. Studies in Philosophy and the History of Philosophy,* vol. 22. Washington, DC: Catholic University of America Press, 1991, 211–234.

Symonds, John Addington. *Renaissance in Italy: The Age of the Despots.* New ed. London: Smith, Elder, & Co., 1901.

Synan, Edward A. "Introduction: Albertus Magnus and the Sciences." In James A. Weisheipl, O. P., ed., *Albertus Magnus and the Sciences: Commemorative Essays 1980.* Toronto: The Pontifical Institute of Mediaeval Studies, 1980, 1–12.

Taylor, Henry Osborn. *The Medieval Mind.* 2 vols. 4th ed. Cambridge, MA: Harvard University Press, 1951.

Tertullian. *On Prescription Against Heretics.* Translated by Peter Holmes in Alexander Roberts and James Donaldson, eds., *The Ante-Nicene Fathers.* 10 vols. New York: Charles Scribner's Sons, 1896–1903, vol. 3.

Thijssen, J. M. M. H. *Censure and Heresy at the University of Paris 1200–1400.* Philadelphia: University of Pennsylvania Press, 1998.

Thomas Aquinas. *Scriptum super libros Sententiarum Magistri Petri Lombardi Episcopi Parisiensis.* New ed. 4 vols. Paris: P. Lethielleux, 1929–1947.

Bks. 1–2: ed. R. P. Mandonnet, vols. 1–2 (1929).

Bks. 3–4, dist. 22: ed. R. P. Maria Fabianus Moos, vols. 3–4 (1933, 1947).

Introduction to Saint Thomas Aquinas. Edited with an Introduction by Anton C. Pegis. New York: The Modern Library, 1948.

Opera omnia secundum impressionem Petri Fiaccadori Parmae 1852–1873, photolithographice reimpressa cum nova introductione generali anglice scripta a Vernon J. Bourke. New York: Musurgia Publishers, 1948–1950.

S. Thomae Aquinatis In Aristotelis libros De caelo et mundo: De generatione et corruptione: Meteorologicorum Expositio cum textus ex recensione leonina. Ed. Raymundus M. Spiazzi, O. P. Turin: Marietti, 1952.

S. Thomae Aquinatis In octo libros De physico auditu sive Physicorum Aristotelis commentaria. Ed. P. Fr. Angeli-M. Pirotta, O. P. Naples: M. D'Auria Pontificius Editor, 1953.

Commentary on Aristotle's "Physics" by St. Thomas Aquinas. Translated by Richard J. Blackwell, Richard J. Spath, and W. Edmund Thirlkel; Introduction by Vernon J. Bourke. New Haven: Yale University Press, 1963.

Summa Theologiae. 60 vols. Latin text with English translation. Ed. and trans. T. Gilby et al. Blackfriars/McGraw-Hill, 1964–1976.

Vol. 2: *Existence and Nature of God* (1a.2–11). Latin text, English translation, Introduction, Notes, Appendices & Glossary by Timothy McDermott, O. P.; Additional Appendices by Thomas Gilby, O. P. Blackfriars, 1964.

Vol. 8: *Creation, Variety and Evil* (1a. 44–49). Latin text, English translation, Introduction, Notes, Appendices & Glossary by Thomas Gilby, O. P. Blackfriars, 1967.

Vol. 9: *Angels* (prima pars, questions 50–64), Latin text, English translation, Introduction, Notes, Appendices & Glossary by Kenelm Foster, O. P. Blackfriars, 1968.

Vol. 10: *Cosmogony* (1a. 65–74). Latin text, English translation, Introduction, Notes, Appendices & Glossary by William A. Wallace, O. P. Blackfriars, 1967.

See also, Vollert, Cyril.

Thomas of Strasbourg. *Thomae ab Argentina, Eremitarum divi Augustini prioris generalis qui floruit anno Christi 1345. Commentaria in IIII libros Sententiarum.* Venice: Ex officina Stellae, lordani Ziletti, 1564. Facsimile, Ridgewood, NJ: Gregg, 1965.

Thompson, James Westfall, and Edgar Nathaniel Johnson. *An Introduction to Medieval Europe 300–1500.* New York: W. W. Norton, 1937.

Thorndike, Lynn. *A History of Magic and Experimental Science.* 8 vols. New York: Columbia University Press, 1923–1958.

Science and Thought in the Fifteenth Century: Studies in the History of Medicine and Surgery, Natural and Mathematical Science, Philosophy and Politics. New York: Columbia University Press, 1929.

University Records and Life in the Middle Ages. New York: Columbia University Press, 1944.

Toomer, G. J. "Campanus of Novara." In *Dictionary of Scientific Biography,* vol. 3 (1971), 23–29.

Trevor-Roper, H. R. The European Witch-Craze of the 16th and 17th Centuries. Harmondsworth, Middlesex, Eng.: Pelican Books, 1969.

Trinkaus, Charles. "Lorenzo Valla's Anti-Aristotelian Natural Philosophy." In *I Tatti Studies* (1993): 279–325.

Turner, Ralph V. "Magna Carta." In Joseph R. Strayer, ed., *Dictionary of the Middle Ages,* vol. 8. New York: Charles Scribner's Sons, 1987, 41–42.

Verbeke, Gerard. "Philosophy and Heresy: Some Conflicts Between Reason and Faith." In W. Lourdaux and D. Verhelst, *The Concept of Heresy in the Middle Ages (11th–13th C.).* Proceedings of the International Conference, Louvain, May 13–16, 1973. Leuven: Leuven University Press; The Hague: Martinus Nijhoff, 1976, 172–197.

Verger, Jacques. "Patterns." In Hilde de Ridder-Symoens, ed. *A History of the University in Europe,* Vol. 1: *Universities in the Middle Ages.* Cambridge: Cambridge University Press, 1992, Chapter 2: 35–74.

Vives, Juan Luis. See Guerlac, Rita.

Vollert, Cyril, S. J., Lottie H. Kendzierski, and Paul M. Byrne, trans. *St. Thomas Aquinas, Siger of Brabant, St. Bonaventure On the Eternity of the World ("De Aeternitate Mundi").* Translated from the Latin with an Introduction. Milwaukee, WI: Marquette University Press, 1964.

St. Thomas Aquinas: trans. Cyril Vollert, 1–73.

Siger of Brabant: trans. Lottie H. Kendzierski, 75–98.

St. Bonaventure: trans. Paul M. Byrne, 99–117.

Voltaire. *Oeuvres complètes de Voltaire, nouvelle édition avec notices, préfaces, variantes, table analytique, les notes de tous les commentateurs et des notes nouvelles, conforme pour le texte à l'édition de Beuchot, enrichie des découvertes les plus récentes et mise au courant des travaux qui ont paru jusqu'à ce jour; précédée de la vie de Voltaire par Condorcet, et autres études biographiques.* 52 vols. Paris: Garnier Frères, 1877–1885. Vols. 11–12 (1878): *Essai sur les Moeurs, I, II.*

The Complete Works of Voltaire, W. H. Barber, general ed. Oxford: The Voltaire Foundation, Taylor Institution Oxford. Vol. 33 (1987), Jeroom Vercruysse, ed.: *Oeuvres alphabétiques* I, Articles pour l'Encyclopédie.

Walker, Williston. *A History of the Christian Church.* New York: Scribner's, 1949.

Wallace, William A. *Causality and Scientific Explanation.* 2 vols. Ann Arbor: The University of Michigan Press, vol. 1, 1972; vol. 2, 1974.

Weinberg, Julius. *A Short History of Medieval Philosophy.* Princeton: Princeton University Press, 1964.

Weisheipl, James A., O. P. "The Nature, Scope, and Classification of the Sciences." In David C. Lindberg, ed., *Science in the Middle Ages.* Chicago: University of Chicago Press, 1978, 461–482.

Westfall, Richard S. *Science and Religion in Seventeenth-Century England.* N.p.: Archon Books, 1970; first published by Yale University Press, 1958.

Never at Rest: A Biography of Isaac Newton. Cambridge: Cambridge University Press, 1980.

Whewell, William. *The Historical and Philosophical Works of William Whewell.* 10 vols. Collected and edited by G. Buchdahl and L. L. Laudan. *History of the Inductive Sciences,* vols. 2–4; first published 1837; 3d ed. 1857. London: Frank Cass & Co., 1967.

White, Lynn, Jr. *Medieval Technology and Social Change.* Oxford: Oxford University Press, 1962.

Medieval Religion and Technology: Collected Essays. Berkeley: University of California Press, 1978.

"Agriculture and Nutrition: Northern Europe." In Joseph R. Strayer, ed., *Dictionary of the Middle Ages,* vol. 1. New York: Charles Scribner's Sons, 1982, 89–96.

William of Sherwood. *William of Sherwood's "Introduction to Logic."* Translated with an Introduction and Notes by Norman Kretzmann. Minneapolis: University of Minnesota Press, 1966.

Wilson, Curtis. *William Heytesbury: Medieval Logic and the Rise of Mathematical Physics.* Madison, WI.: The University of Wisconsin Press, 1956.

INDEX

Page numbers cited directly after a semicolon following a final text subentry refer to relatively minor mentions in the text of the main entry.